Anonymous

Mountaineering

Anonymous

Mountaineering

ISBN/EAN: 9783337317027

Printed in Europe, USA, Canada, Australia, Japan

Cover: Foto ©berggeist007 / pixelio.de

More available books at **www.hansebooks.com**

# The Badminton Library

OF

SPORTS AND PASTIMES

EDITED BY

HIS GRACE THE DUKE OF BEAUFORT, K.G.

ASSISTED BY ALFRED E. T. WATSON

MOUNTAINEERING

PRINTED BY
SPOTTISWOODE AND CO., NEW-STREET SQUARE
LONDON

Mont Blanc and the Aiguille du Géant from Mont Mallet

MOUNTAINEERING

BY

C. T. DENT

WITH CONTRIBUTIONS BY W. M. CONWAY, D. W. FRESHFIELD,
C. E. MATHEWS, C. PILKINGTON, SIR F. POLLOCK,
H. G. WILLINK, AND AN INTRODUCTION
BY MR JUSTICE WILLS

MARIE TOURNIER (*dit* L'OISEAU) THE LAST OF DE SAUSSURE'S GUIDES
From a photograph in the possession of the Alpine Club

ILLUSTRATIONS BY H. G. WILLINK AND OTHERS

Second Edition

LONDON
LONGMANS, GREEN, AND CO.
1892

All rights reserved

DEDICATION

TO

H.R.H. THE PRINCE OF WALES

BADMINTON : *May* 1885.

HAVING received permission to dedicate these volumes, the BADMINTON LIBRARY of SPORTS and PASTIMES, to HIS ROYAL HIGHNESS THE PRINCE OF WALES, I do so feeling that I am dedicating them to one of the best and keenest sportsmen of our time. I can say, from personal observation, that there is no man who can extricate himself from a bustling and pushing crowd of horsemen, when a fox breaks covert, more dexterously and quickly than His Royal Highness ; and that when hounds run hard over a big country, no man can take a line of his own and live with them better. Also, when the wind has been blowing hard, often have I seen His Royal Highness knocking over driven grouse and partridges and high-rocketing pheasants in first-rate workmanlike style. He is held to be a good yachtsman, and as Commodore of the Royal Yacht Squadron is looked up to by those who love that pleasant and exhilarating pastime. His encouragement of racing is well known, and his attendance at the University, Public School, and other important Matches testifies to his being, like most English gentlemen, fond of all manly sports. I consider it a great privilege to be allowed to dedicate these volumes to so eminent a sportsman as His Royal Highness the Prince of Wales, and I do so with sincere feelings of respect and esteem and loyal devotion.

BEAUFORT.

BADMINTON

PREFACE

A FEW LINES only are necessary to explain the object with which these volumes are put forth. There is no modern encyclopædia to which the inexperienced man, who seeks guidance in the practice of the various British Sports and Pastimes, can turn for information. Some books there are on Hunting, some on Racing, some on Lawn Tennis, some on Fishing, and so on ; but one Library, or succession of volumes, which treats of the Sports and Pastimes indulged in by Englishmen—and women—is wanting. The Badminton Library is offered to supply the want. Of the imperfections which must be found in the execution of such a design we are conscious. Experts often differ. But this we may say, that those who are seeking for knowledge on any of the subjects dealt with will find the results of many years' experience written by men who are in every case adepts at the Sport or Pastime of which they write. It is to point the way to success to those who are ignorant of the sciences they aspire to master, and who have no friend to help or coach them, that these volumes are written.

To those who have worked hard to place simply and clearly before the reader that which he will find within, the best thanks of the Editor are due. That it has been no slight labour to supervise all that has been written, he must acknowledge; but it has been a labour of love, and very much lightened by the courtesy of the Publisher, by the unflinching, indefatigable assistance of the Sub-Editor, and by the intelligent and able arrangement of each subject by the various writers, who are so thoroughly masters of the subjects of which they treat. The reward we all hope to reap is that our work may prove useful to this and future generations.

THE EDITOR.

NOTE

THE assistance rendered by the contributors in the preparation of this volume is by no means limited to the chapters they have severally written. In order to increase the authority of the work, and, at the same time, to diminish the appearance of patchwork that often results from collaboration, I have throughout consulted my colleagues freely, and have to acknowledge valuable help received from them. In addition, my best thanks are due, for many suggestions and much friendly aid, to Captain W. DE W. ABNEY, MELCHIOR ANDEREGG, Professor T. G. BONNEY, Mr. E. WHYMPER, and Mr. J. H. WICKS.

C. T. D.

CONTENTS

CHAPTER		PAGE
I.	THE EARLY HISTORY OF MOUNTAINEERING *By Sir Frederick Pollock*	1
II.	EQUIPMENT AND OUTFIT *By C. T. Dent*	39
III.	MOUNTAINEERING AND HEALTH *By C. T. Dent*	75
IV.	THE PRINCIPLES OF MOUNTAINEERING *By C. T. Dent*	90
V.	RECONNOITRING *By C. T. Dent*	129
VI.	SNOWCRAFT *By C. T. Dent*	149
VII.	ROCK CLIMBING *By C. T. Dent*	215
VIII.	MAPS AND GUIDE-BOOKS *By W. M. Conway*	263
IX.	MOUNTAINEERING BEYOND THE ALPS *By Douglas W. Freshfield*	280
X.	CLIMBING WITHOUT GUIDES *By Charles Pilkington*	307
XI.	HILL CLIMBING IN THE BRITISH ISLES *By Charles Pilkington*	325

CHAPTER		PAGE
XII.	THE RECOLLECTIONS OF A MOUNTAINEER	348
	By C. E. Mathews	
XIII.	SKETCHING FOR CLIMBERS	380
	By H. G. Willink	
XIV.	CAMPING	392
	By C. T. Dent	
XV.	PHOTOGRAPHY	402
	By C. T. Dent	
	GLOSSARY	414
	INDEX	419

ILLUSTRATIONS

(Reproduced by Walker & Boutall)

PLATES

	ARTIST	TO FACE P.
Mont Blanc and the Aiguille du Géant from Mont Mallet .	Photographed by W. F. Donkin, by permission of Messrs. Spooner	Frontispiece
The Lower Grindelwald Glacier	From Merian's 'Topographia Helvetiæ,' 1654	20
Monsieur Desaussure, son fils, et ses guides arrivant au Glacier du Tacul au Grand Géant, où ils ont habité 17 jours, sous des tentes, en Juillet, 1788	From an old print in the possession of the Alpine Club	28
An Ice Slope .	H. G. Willink	52
Serve him right	,, ,,	104
The Beispielspitz .	,, ,,	136
Backing up .	,, ,,	176
'Up you come'	,, ,,	190
On the Messer Grat	,, ,,	204
Crack Climbers	,, ,,	232
The Pass in Sight .	,, ,,	308
British Hill Weather .	,, ,,	336
'Rest after Toyle' .	,, ,,	394

ILLUSTRATIONS IN TEXT

	ARTIST	PAGE
Marie Tournier	H. G. Willink	Title-page
Alpenrose	,, ,,	xvii
Edelweiss	,, ,,	xx

	ARTIST	PAGE
A Friend of Scheuchzer's	H. G. Willink	1
Wilderwurm Gletscher	,, ,,	19
Training	,, ,,	38
Hamlet iii. 1, 81	,, ,,	39
A regular Nailer	,, ,,	44
The Apparel oft proclaims the Man	,, ,,	46
The 'Mummery' Tent	Ellis Carr	63
The Nightly Task	H. G. Willink	74
A Chamber Scene	,, ,,	75
An Awkward Drop	,, ,,	89
Les Tricoteuses	,, ,,	90
Ein Junger	,, ,,	92
All very well at the Curragh	,, ,,	93
Footmarks of a cantering Man	,, ,,	98
Rope Knots	Ellis Carr	100
Ditto	,,	101
Player in hand	H. G. Willink	107
On the Tit-für-tatt Berg	,, ,,	109
The Axe Fiend	,, ,,	111
Vingt-et-un	,, ,,	114
Pickel'd Herren	,, ,,	128
Übermorgen's Arbeit	,, ,,	129
A Chamoniard	,, ,,	148
Incidence and Reflection	,, ,,	149
Map-diagram of a Glacier	,, ,,	154
Diagram of Crevasses	,, ,,	155
A Plunger	,, ,,	161
Diagram of a Snow Cornice	,, ,,	162
Step-cutting, A.	,, ,,	167
Ditto B.	,, ,,	169
Ditto C.	,, ,,	170
A Dry Glacier	,, ,,	171
Step-Cutting, D.	,, ,,	172
Ditto E.	,, ,,	174
Ditto F.	,, ,,	175
'The Right Angle'	,, ,,	181
Bowline on a Bight	Ellis Carr	184

ILLUSTRATIONS

	ARTIST	PAGE
(TO BE CONTINUED)	H. G. *Willink*	186
A PROFESSOR ON A T-TABLE . . .	,, ,,	186
DIAGRAMS OF BERGSCHRUNDS . . .	,, ,,	189
A MOOT POINT	,, ,,	194
GLÜCK	,, ,,	196
A FOUR-POSTER	,, ,,	205
A MOUNTAIN ARAB	,, ,,	206
A 'SCHNEE-ZUG'	,, ,,	214
OVER IT GOES !	,, ,,	215
TYPES OF MOUNTAIN FORMATION :—		
LIMESTONE	,, ,,	218
'WRITING-DESK'	,, ,,	219
DOLOMITE	,, ,,	220
AIGUILLES (MASS)	,, ,,	221
AIGUILLES (DETAIL)	,, ,,	222
GRANITE	,, ,,	222
A VERY WORM	,, ,,	225
'KOMMEN SIE NUR !'	,, ,,	227
GIVING A HAND	,, ,,	230
TAKING THE OATH AND HIS SEAT . .	,, ,,	233
HOOKEY WALKER	,, ,,	237
A WILL AND A WAY	,, ,,	241
UNSTABLE EQUILIBRIUM . .	,, ,,	247
EXPERIENCE DOES IT	,, ,,	254
STRATIFICATION AND SURFACE (DIAGRAM)	,, ,,	258
'THEY'RE GOIN' VERY SLOWLY !' . .	,, ,,	262
TOPOGRAPHERS	,, ,,	263
EIN RÜCKEIMER	,, ,,	279
CARDUUS RHODANOGLETSCHERENSIS . .	,, ,,	280
'SIC ITUR AD ASTRA'	,, ,,	306
A ROTIFER	,, ,,	307
'WILL IT HOLD ?'	,, ,,	313
LETTING HIM DOWN GENTLY . . .	,, ,,	323
CALLER HERREN	,, ,,	324
πᾶ βῶ; πᾶ στῶ;	,, ,,	325
A STEEP SLOPE	,, ,,	330
A GENERAL DEPRESSION	,, ,,	347

	ARTIST	PAGE
AN I GLASS	*H. G. Willink*	348
J. D. FORBES	,, ,,	351
JOHN BALL	,, ,,	353
THE RAW MATERIAL	,, ,,	364
ONE OF THE OLD GUARD	,, ,,	366
'RUDE DONATUS'	,, ,,	371
ZERMATTERINNE	,, ,,	379
STONECROP	,, ,,	380
DIAGRAMS SHOWING PERSPECTIVE ILLUSION	,, ,,	386
ROUGH SKETCH OF THE BEISPIELSPITZ	,, ,,	387
ON THE ZERMATT WALL	,, ,,	391
TRAVELLERS' RESTING-PLACES	,, ,,	392
'SUB SCOPULO'	,, ,,	401
A GLASS I	,, ,,	402
CAMARADERIE	,, ,,	413
A LENGTHY OBSERVATION	,, ,,	416
'AUF WIEDERSEHEN'	,, ,,	417
'FINIS CORONAT OPUS'	,, ,,	439

INTRODUCTION

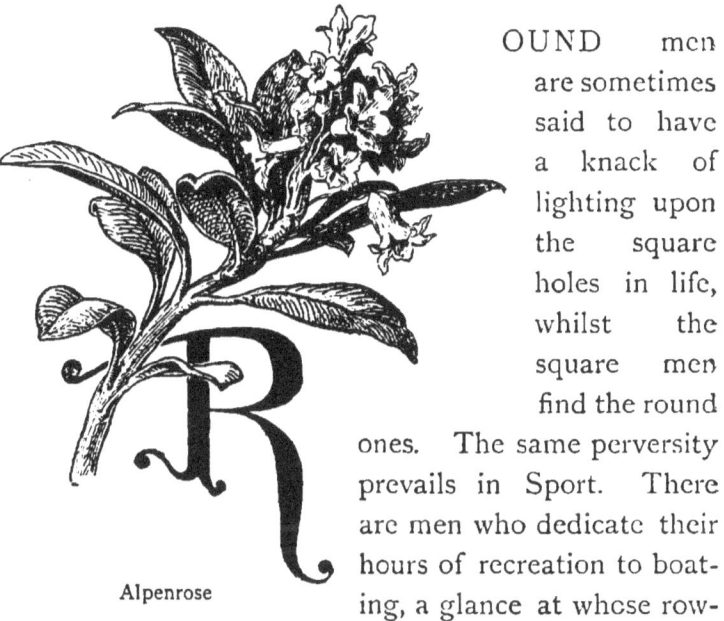

Alpenrose

ROUND men are sometimes said to have a knack of lighting upon the square holes in life, whilst the square men find the round ones. The same perversity prevails in Sport. There are men who dedicate their hours of recreation to boating, a glance at whose rowing form shows that nature meant them for the mountain-side; others, bent on distinction as mountaineers, betray their natural bent by a passion for sliding seats. A book on Mountaineering may help to redress this want of symmetry. A natural magnetism will draw in the right direction a reader whose inborn, though undeveloped, instincts are those of the mountaineer, and a like influence may perhaps turn aside from an uncongenial pursuit one in whose nature the spirit of mountain adventure finds only a repellent pole.

To precipitate the subtle essence of thoughts, actions and habits which have become instinctive, and crystallise it into language is no easy task. How far it has been successfully attempted the reader must judge for himself. It will occur to some, no doubt, that it takes many words to describe a simple movement; and that no description will teach a man to climb—still less to be a mountaineer—any more than to dance. None the less may it be of use to enunciate some axioms of the craft, just as it may serve a purpose to tell the would-be sportsman how to hold a gun, or the embryo Grace how to wield a bat or deliver a ball. Even experts may correct faults ; and he must have great good fortune or little modesty—or both—who can learn nothing from the experience of others and from the precision which the very effort at system imposes upon a writer.

Nothing in the history of modern sport has been more striking than the rapid growth of interest in Mountaineering. No wonder: few sports, perhaps few pursuits, afford keener or more lasting enjoyment, or contribute more to the acquisition of self-reliance, patience, and self-restraint. The Introduction, however, must not depart from the tone of the book. The authors of the following chapters have felt that the spirit which rejoices in mountain craft is either in a man, or is absent. In the one case, good wine needs no bush ; in the other the cup of mountain nectar has no taste, and the book of mountain wisdom will be written in an unknown tongue. There has been no desire to whip up recruits for the Alpine Club ; still less to incite to rash or precipitate enterprise those who would not spontaneously seek the snow-field or the aiguille. There is, indeed, throughout these pages, a key-note of warning rather than of encouragement. The Alps have been styled the playground of Europe; but the playfield may be the field of death if incautiously approached or ignorantly dallied with. Nature trusts her secrets to those only who hold her in awe and reverently own her mystery and reserve. To such as woo her with respectful admiration and with patient suit she tells her story freely and with ever-growing kindness. Woe betide those who think they can know her at first sight, or win her confidence without sympathy or effort.

Those who sound the warning know what they are talking about. I myself belong to a generation of mountain lovers already past, and have been rather a pioneer than one of the good soldiers by whom the triumphs of mountaineering have been won. It is very satisfactory to me that men younger, bolder, better trained in mountain craft than I was, inculcate the spirit of caution which I have long advocated, and teach the immorality as well as the folly of running into useless danger and risking valuable lives.

There are three things specially to be dreaded in the mountains, as beyond human control and occasionally beyond human foresight: bad weather, falling stones, steep grass slopes, with herbage either short and dry, or long and wet or frozen. I do not think it possible for anyone who has not felt it to have any idea what very bad weather means in high places, even in places by no means of the highest; or to imagine the rapidity with which, under unsettled atmospherical conditions, the destructive forces of nature can be raised, and the worst assaults of the elements delivered. Falling stones may come from the most unexpected places, and I have seen from my own Alpine home a whole flake of mountain-side peel off without warning, and sweep with a cannonade of thirty hours' duration a gully that I and mine have used for years as a highway to the upper world. Slopes of grass look so easy, and are so treacherous that it is scarcely possible to secure for them the respect which they have a deadly fashion of enforcing.

There are few other dangers which care and knowledge will not eliminate. To narrow to the utmost the area of unavoidable danger, to press the necessary care, to show how the necessary knowledge can be gained, and to make it easier of acquisition, has been one aim of this compilation. It is enriched with matter historical, scientific, and practical, which ought to make it not the least attractive of the series.

ALFRED WILLS.

Edelweiss

MOUNTAINEERING

CHAPTER I

THE EARLY HISTORY OF MOUNTAINEERING

BY SIR FREDERICK POLLOCK

A friend of Scheuchzer's

OUNTAINEERING, as the term has been used and understood for about a generation, is not the same thing as mountain travelling. When we speak of mountaineering, we imply that the traveller goes to the mountains for the sake of something that is to be found there. His motive may be the scientific curiosity of a geographer, a geologist, a botanist, or a natural philosopher in other branches. It may be the love of mountain scenery, or of forms of adventure and exercise remote from the common round of his every-day life. Oftentimes the mountaineer's pleasure is happily compounded of all or several of these elements. In every case there is something that turns his free choice to the mountain heights. A traveller who goes among mountains merely because that is the only way for him to attain some further object, or the least inconvenient way, may possibly become a mountaineer in the course of his journey, but he is not one when he begins it. He is a true apprentice of the craft only when he can say from a full heart *Labor ipse voluptas*. Hannibal was not a mountaineer, nor Bonaparte, nor the kings and emperors who, through the middle ages, fared to Rome by way of the Mont Cenis or the Brenner;[1] still less that nameless English clerk of the tenth century who was so impressed by the horrors of the Alps that in the charters framed by him he threatened any profane violator of the gift, not with the fires of Erebus, but with the icy breath of Pennine fiends.[2] The history of mountaineering proper has no direct concern with the journeys undertaken during the last twenty centuries or more by the people, famous or obscure, warriors, pilgrims, or travellers, bent on business or pleasure in Italy, who crossed the Alps because they stood in the way, and would have been better pleased had there not been Alps to cross. Indeed the necessity of traversing mountain regions does not seem of itself likely to produce any desire to explore them beyond that of a practical road-maker. And, in fact, the decisive impulses were quite unconnected with the well-worn Alpine routes of the middle ages, and took a quite different direction. If Alpine exploration had been determined by the known lines of travel, Mont Blanc would have been gradually approached from the valley of Aosta, and

[1] For particulars of such journeys see E. Oehlmann, *Die Alpenpässe im Mittelalter*, 1878-9. This book does not deal with the legends of high glacier passes having been used in the Middle Ages, as to which see E. Richter, *Geschichte der Schwankungen der Alpengletscher* (from vol. xxii. of *Ztschr. des D. u. Oe. A. V.*). Wien, 1891.

[2] *Alpine Journal*, xi. 294; Coolidge, *Swiss Travel and Swiss Guide-books*, 160; Kemble, *Cod. Dipl.* ii. pp. 208-262, in nine charters ranging in date from 938 to 946. The form is 'perpessus sit gelidis glacierum flatibus et pennino exercitu malignorum spirituum.'

THE EARLY HISTORY OF MOUNTAINEERING

Courmayeur, not Chamonix, would have been the first headquarters of mountaineering enterprise. But Chamonix and its glaciers were an object of undefined curiosity to Windham and Pococke, of noble scientific ambition to De Saussure, and of artistic enthusiasm to Bourrit, because they lay so much off the main lines as to be to all practical intent an undiscovered country for the good people of Geneva, much more for the European world of letters.[1] The way to the heights was pointed out by men to whom travel, as travel, was of little importance, and who were inspired by sentiments of art, science, and adventure in various proportions.

After such men had set the example, but only then, the mountains came to be regarded as a normal resort of persons travelling for pleasure, and energy which in former generations would have been divided between field sports at home and the performance of the Grand Tour abroad was directed to Alpine excursions. It is needless to set forth that the taste for travelling and the taste for mountain climbing are in themselves distinct things; they are perhaps now more likely to be found together than they have been at any former time; still we have many good mountaineers who are no travellers, travelling only so far as needful to find their climbing, and travellers who are no mountaineers, climbing (if at all) only when they cannot accomplish their travels otherwise.

What concerns us here, therefore, is the history not of travel routes or topography, but of a taste and a pursuit. The

---

[1] Chamonix has a continuous and particularly well authenticated medieval history going back to the eleventh century (A. Perrin, *Histoire de la Vallée et du prieuré de Chamonix*, Chambéry, 1887); nor did Windham and Pococke, although they were not acquainted with the fact, ever assert the contrary. M. Durier has well pointed out that they are not answerable for the fictions and exaggerations of later writers who talked of Chamonix as if it had been in Central Africa (*Le Mont-Blanc*, chap. iv.). As for Windham and Pococke's party going 'well arm'd,' it was the practice of travellers on every high road in Europe in 1741 and long afterwards. They would have had to go at least as well armed in some parts of Great Britain. Besides, their arms were carried partly with a view to sport. See, however, De Saussure, *Voyages dans les Alpes*, § 732, which seems to be the origin of the statements in later books.

modern legends of disused medieval glacier passes in the Alps would not be relevant to our topic even if they were as true as they are, almost without exception, demonstrably false. And the fathers of mountaineering were not those who first attained any given height above the sea-level (in which case Europe would have to give place to the traders of Thibet), but those who first perceived in mountain climbing something else than mere labour, and were bold enough to proclaim its delights. We may read in Livy how Philip V. of Macedon ascended some ridge of the Balkans in search of a view comprising the Alps, the Adriatic, the Danube, and the Black Sea; how his chief difficulty was making way through forest (still a very possible experience in roadless countries, the Selkirks for example); and how he saw nothing but mist, and so, being unable to check the popular estimate of the view, wisely refrained from contradicting it. And we may read in Sallust how Marius, with the aid of a local soldier who had found the way by accident, and acted as guide, got his men up a rock supposed inaccessible to troops, and took a Numidian hill-fort in rear with complete success. But these things are mere detached curiosities. As such they were recorded, and such they remained. From the point of view now taken, the continuous history of mountaineering can be carried back only a few generations. The recognition of climbing among the forms of active recreation open to civilised men may be dated from the time when persons who ascended Mont Blanc were no longer deemed to have performed an astonishing feat, or expected to write a little book about it. Albert Smith's ascent in 1851 was the last, or nearly the last, of the old school, as his lecture on Mont Blanc, afterwards preserved in book form, was the last, or nearly the last, of the old-fashioned narratives. Albert Smith himself contributed in no small measure to make it so. He could not be called a mountaineer, but his admiration of the mountains was perfectly genuine, and his work doubtless kindled Alpine enthusiasm in many young Englishmen of the generation who, from ten to fifteen years later, achieved the

THE EARLY HISTORY OF MOUNTAINEERING 5

conquest of the great peaks and passes.[1] Also the year 1849 was the last in which no ascent of Mont Blanc was made, and from 1853 the frequency of ascents increased rapidly. Thus the modern period of mountaineering may be fairly said to have opened with the second half of the nineteenth century. But if we are to do justice to the precursors of the art, we must go much farther back.

The earliest deliberate attack on an Alpine peak of which we have any record was the unsuccessful attempt on the Roche Melon near Susa (over 11,600 feet), shortly described by the anonymous eleventh-century chronicler of Novalesa. Down to the last century this mountain was reputed the highest in Savoy. In the early middle ages it was called *Mons Romuleus*, from Romulus, a prince of monstrous ambition and avarice ('rex elefantiosissimus' may thus be rendered with the help of Ducange), who was said to have amassed a huge treasure somewhere in its flanks. An old man who himself told the story to the chronicler tried the ascent, with a certain Count Clement, on a day of brilliant promise; but after making considerable progress they were baffled by mist, and, it seems, also by falling stones, if we may so far trust the story in details ('visum erat ... illis ut desuper lapides mitterentur'). A procession of clerks who repeated the adventure, armed with all spiritual power, singing litanies and 'Vexilla regis,' had no better fortune. In 1358, however, Rotario d'Asti not only achieved the ascent, but built a chapel close to the top, in performance of some vow, perhaps on behalf of his city. For many generations there has been a regular pilgrimage to this chapel in August; the present king of Italy and several of his ancestors, beginning with Charles Emmanuel II. in 1659, have made the ascent.[2]

[1] Durier, *Le Mont-Blanc*, ch. v. note ii. Mr. L. Stephen has favoured me with private information to the same effect. Albert Smith's book was first printed in 1852, and published in 1853.

[2] *Chron. Noval.* ii. 5; Martelli e Vaccarone, *Guida delle Alpi occidentali*, i. 350. Cf. letter on 'Mountain Sanctuaries,' signed 'W. A. B. C.,' in *Guardian*, June 17, 1891, p. 956. Apparently the tradition connecting

In the last quarter of the thirteenth century Peter III. of Aragon, whose praise is in Dante's 'Purgatory,' ascended Canigou (2,787 m.) in the Pyrenees. As Fra Salimbene tells the tale in his chronicle,[1] this king, being of a high courage and daring temper, had a mind to see what might be on the summit of this hill where no son of man was known to have trod. So he took to himself two knights of his companions whom he loved, and, having equipped themselves with provisions and fitting instruments (*congruentibus armis*, possibly alpenstocks), they set forth. They met with a thunderstorm at some considerable height, but persevered. Afterwards, however, the two knights 'began to fail in such wise that for exceeding weariness and for fear of the thunder they could scarce breathe.' Peter therefore, like Mr. Bryce on Ararat, completed the ascent alone. When he came down he told his companions how he had found a lake on the summit and cast a stone into it, whereupon there came out a great and terrible dragon, and flew away, breathing out a vapour which darkened the air. Moreover he gave them leave to repeat the story as much as they pleased. We may suppose that he had accurately measured their credulity. It is by no means clear that this royal jest was taken seriously by Fra Salimbene, whose name is singularly apt for an early recorder of climbing, and who was otherwise a man of some mark. There is no reason, however, to doubt that King Peter did reach the summit. According to modern accounts there is no difficulty about the ascent, though the last piece is a steep rock gully where hands as well as feet must be used.

Rotario's vow with a captivity among the infidels has not any real foundation. The date of the ascent itself is attested by Rotario's own inscription on a statue of the Virgin which he placed in the chapel, and which is now at Susa.

[1] *Bollett. del C.A.I.* xxiii. 155, in appendix to Signor Uzielli's article 'Leonardo da Vinci e le Alpi.' The article itself is a storehouse of early Alpine and mountaineering references. Fra Salimbene's rather ambiguous sentence of commentary runs thus : ' Videtur mihi quod opus Petri Aragonum possit connumerari cum operibus Alexandri, qui in multis terribilibus negotiis et operibus voluit exsperiri ut laudem in posterum mereretur.'

We can hardly count Petrarch as a spiritual father of those who climb on the strength of his ascent of Mont Ventoux near Vaucluse, accomplished in 1339, which is described with some minuteness at the opening of the fourth book of his collected letters. But, being Petrarch, he must not be forgotten in this place. One element of mountaineering was indeed present ; he made the expedition from pure curiosity to see what the top of a high hill was like, 'sola videndi insignem loci altitudinem cupiditate ductus.' And when he reached the top, after losing much time by casting about at lower levels in search of an easier way up which was not to be found, and rejoining his companion with much trouble, he took note of the view. But there is nothing to show that he enjoyed the walk, or that the view was more to him than the strange fact of seeing so much of the world at once. At any rate he did not follow up his performance, nor did it become an example to others. Dante, though no one remarkable ascent can be ascribed to him, had much more of a climber's feelings and observation, as both an English and an Italian mountaineer have shown.[1] There is nothing in his descriptions, however, to suggest any very high or remote excursion.

Again the Mont Aiguille, in the parts of Grenoble, has a curious isolated history. It is a very steep and difficult rock-mass, called one of the Seven Wonders of Dauphiné, with trees and grass on its flat top ; and it was reputed inaccessible until in 1492 Charles VIII. of France commanded his chamberlain, Julien de Beaupré, to ascend it, which he did by means of ropes and ladders. No other ascent is recorded until 1834, so that the effect of Julien de Beaupré's feat on the custom and art of mountaineering in general may safely be set down as stark naught. In our own time chains have been

[1] *The Mountains of Dante*, A.J. x. 400, by D. W. Freshfield ; *Dante Alpinista*, Bollett. del C.A.I. xx. 12, by O. Brentari. Signor Brentari perhaps proves too much when he undertakes to show not only that Dante was a mountaineer, but that Virgil has all the qualities of an excellent guide. I am not aware that Shakespeare has yet been proved a mountaineer. It would be about the only thing left to prove him.

provided by the Section de l'Isère of the French Alpine Club. Mr. Coolidge, whose notes of an ascent made in 1881 were published by Mr. F. Gardiner in the 'Alpine Journal' in 1889, is, I believe, the only English-speaking mountaineer who has yet ascended the Mont Aiguille. Few heights in Europe of historic fame have been so much left alone by the Alpine Club. On the whole, it must be admitted that the Middle Ages give us nothing but stray curiosities.

In the sixteenth century we may claim, as in some sort a precursor of mountaineering, one of the most extraordinary men of any time, Leonardo da Vinci. His drawings show an admirable sense of mountain form, and there is ample evidence of his acquaintance with the hill country of Lombardy. But more than this, Leonardo was at some time on a height of the main chain of the Alps 'dividing France from Italy,' then called Monboso, so high, he says in a note still extant, that little snow falls there, but rather hail, as he observed about the middle of July ; and he noticed that the sky overhead was dark, but the sun more brilliant than in the plains. The remark about hail is either a hasty generalisation or, as Mr. Freshfield suggests, an attempt to account for the granular structure of glacier ice. He expressly says that 'no mountain hath his basis at the like height.' These observations point to Leonardo having made a fairly high expedition, and Signor Uzielli, in an elaborate paper published in the 'Journal of the Italian Alpine Club,' for 1889, has given strong reasons for holding that Monboso signifies some part of the Monte Rosa group.[1] Indeed, that is still a local name of a certain spur of Monte Rosa (some say of the whole mass) in the parts of the Val Sesia, the direction in which Leonardo would probably have come if he came that way at all. The gravest objection to Signor Uzielli's theory is that no part of the Monte Rosa group could be said, in any sense to which we

---

[1] See Mr. Freshfield's note on Signor Uzielli's paper, *Proc. R.G.S.*, May 1892.

THE EARLY HISTORY OF MOUNTAINEERING 9

are accustomed, to divide France from Italy. Still it is possible that Leonardo regarded the Valais as a mere incident on the road from Italy into France ; and it would be the natural road for the centre and north of France for a traveller starting from Milan. Moreover it is known that maps of the sixteenth century did use geographical expressions in a large way without regard to the details of political boundaries, and treated everything north of the Alps and west of the Rhine as being in some sense French. And perhaps it is not insignificant that what modern diplomacy would call the sphere of French influence extended at that time as far up the Valais as Sion. On the whole, it may be accepted as a probable opinion, though not beyond reasonable doubt, that Leonardo did attain a considerable height somewhere in the Alagna region of the Monte Rosa mass. Mr. Freshfield thinks he 'may have got as far as the rocks above the Col d'Ollen, where, at a height of 10,000 feet, the inscription "A.T.M. 1615" has been found cut in the crags.'

Proceeding to certainties, we find that as early as the middle of the sixteenth century the humanists of Switzerland studied and visited the mountains in the truest mountaineering spirit. There is reason to think that their work was arrested only by the troublous time of religious controversies and wars which checked all civilisation throughout Central Europe for almost a hundred years. This appears to be one more pitiful example, though a minor one, of the ways in which the fruit of the Renaissance was blighted by the violence of its own child, the Reformation. It is certain that the work of Gesner and Simler slumbered until Germany had made her first recovery from the Thirty Years' War, and was then taken up in a fashion inferior to their own. And I do not think we can justly ascribe their failure to found a school of mountaineers to want of energy or of merit. I will endeavour to let the reader judge for himself.

Conrad Gesner of Zürich, renowned among his contem-

poraries as the German Pliny, wrote thus to his friend Vogel of
Glarus (Latinized as Avienus) in the year 1541 :—

Most learned Avienus,—I have resolved for the future, so long
as God grants me life, to ascend divers mountains every year, or
at least one, in the season when vegetation is at its height, partly
for botanical observation, partly for the worthy exercise of the
body and recreation of the mind. What must be the pleasure,
think you, what the delight of a mind rightly touched, to gaze
upon the huge mountain masses for one's show, and, as it were,
lift one's head into the clouds? The soul is strangely rapt with
these astonishing heights, and carried off to the contemplation of
the one supreme Architect. . . . Philosophers will always feast the
eyes of body and mind on the goodly things of this earthly
paradise; and by no means least among these are the abruptly
soaring summits, the trackless steeps, the vast slopes rising to the
sky, the rugged rocks, the shady woods.

It is remarkable to find here, in the words of a Swiss
naturalist, that love of the sublime and picturesque elements in
wild nature which is often supposed to have been born with
Rousseau two centuries later. But this is love of wild scenery
and something more; there is the love of earning the sight by
one's own muscular toil, the genuine mountaineering spirit.
Gesner's employment of a rather artificial Latin phraseology
does him a little injustice in this passage; and we may smile
if we will at his physics, which are surely less rational than
Leonardo's. He claims admiration for mountains, among other
reasons, because they do not sink into the earth by their own
weight, which he treats as a singular wonder of nature. But
his ascent of Pilatus, described by him fourteen years later,
leaves no doubt possible that modern mountaineers ought to
revere him as a true ancestor.

Pilatus, it is true, had a special attraction for bold humanists
apart from the climbing and the view, and whatever might be
found in the way of herbs or crystals. The local tradition of
Luzern, a tradition begotten within comparatively recent times,
it would seem, of the merest verbal confusion, would have it
that the ghost of Pontius Pilate inhabited a certain mere on

THE EARLY HISTORY OF MOUNTAINEERING 11

the mountain, and was sure to do a mischief to the neighbourhood if he were disturbed. Therefore no one was allowed to venture on Pilate's ground save with special permission and in the company of a respectable citizen of Luzern. Not that the ascent was in itself a novelty, for Gesner expressly mentions that there were two known routes,[1] and that names were carved on the rocks at the summit. There were two specially recorded ascents in 1518, one by a company of four Swiss scholars, and one by the exiled duke Ulrich of Wurtemberg.[2] These travellers were not allowed to play any tricks with the enchanted mere, and came down as wise about the mystery of Pilate as they went up. It does not appear that they had any mountaineering interest or pleasure in the expedition. Gesner, on the other hand, went to Pilatus not merely because it was famous as a haunted mountain, but for love of the mountains. In his dedicatory letter he declares his intention of collecting information 'concerning not only this mountain, but others also, specially those of our own pre-eminently mountainous Switzerland,' so that 'a whole book may sooner or later be written of the mountains and the wonders thereof.' Once being started on the ascent, Gesner thoroughly enjoyed the excursion for its own sake. His description is to this day as fresh and sincere a description of the pleasures of mountain walking as can be found anywhere. Nothing is omitted, and everything truly touched.

The party slept out a night on the hay of some huts in the Eigenthal, and enjoyed the various Alpine preparations of milk, for which junket and Devonshire cream are our only English analogies. Gesner saw chamois and thought he saw steinbock;

[1] *Descriptio Montis Fracti iuxta Lucernam, &c.*, Tiguri [1555] p. 45 (the paging is continuous with that of the preceding botanical tract : *De rebus noctu lucentibus et lunariis herbis*) : 'Audio et alia breviore via conscendi sed ea magis acclivi.' Cf. Scheuchzer, *It. Alp.* 399. The letter to Avienus is prefixed to Gesner's *Libellus de lacte et operibus lactariis.*

[2] Peyer, *Geschichte des Reisens i. d. Schweiz*, pp. 24-26. No reference is given to any original authority for Duke Ulrich's ascent. An extract from Vadianus's account in his commentary on Pomponius Mela *De orbis situ* is embodied in Gesner's tract.

not to be seen by the modern traveller, it is needless to say, since Pilatus has been crowned with inns. Next morning a drink of spring water at the first halt is declared to have been the greatest of physical pleasures ; and thereupon Gesner, with the not unpleasing formalism of his time, proves that each and all of the senses receive the highest satisfaction in a mountain walk. For feeling, there is fresh air after heat and the warmth of sun or shelter after exposure to cold. As to sight :—

The sight is delighted with a rare prospect of summits, mountain chains, rocks, forests, valleys, rivers, springs, and meadows. As concerning colour, the greater part of the scene is in full verdure and flower. As concerning the forms of objects, there is strange and wonderful variety thereof in cliffs, rocks, ravines, and otherwise ; and these are no less admirable for their greatness and height. If it please you to extend your views, gazing forth far and wide and all around, you shall not want for spaces and heights to make you seem as one going with his head in the clouds. If you will gather it in smaller compass, you shall behold meadows and green woods ; if you will be yet more confined, then shady valleys, overshadowing rocks, darkened caverns may be your scene. In all things change and variety are pleasant, but chiefly in things of sense. And nowhere else is so much variety found, and that in so little space, as in the mountains ; wherein, not to speak of other things, you may in one day see and have part in the four seasons of the year: summer, autumn, spring, and winter. From the high mountain crests, moreover, the whole hemisphere of our sky will be freely open to your gaze ; you may without hindrance note the rising and setting of the stars, and observe the sun setting far later, and rising sooner.

Hearing is gratified by pleasant talk and social mirth, by the song of birds in the wooded region, and in a manner by the very silence of the higher wastes. Here are no jarring sounds of cities with their crowd and strife ; here in the solemn quiet of the heights a man may almost deem that he listens to the very music of the spheres. Then for smell there are the sweet Alpine herbs and flowers, more sweet and potent than the same kinds as they grow in the plain ; and the air is free and wholesome, not charged with gross vapours or the infection of crowded dwellings.

For taste, the delight of pure water is again called to mind ;

and Gesner adds, with perfect good sense, but contrary to the superstition that still prevails in many parts of the Alps, that there is not the least harm in drinking cold water during active exercise. On the whole, then, Gesner concludes, there is no choicer pleasure than the mountaineer's.

Give me a man of reasonably good complexion in mind and body, of liberal nurture, not the slave of indolence, luxury, or passion ; I would have him likewise a curious admirer of nature, so that by beholding and admiring the mighty works of the Master-workman, and the variety displayed in one mass among the mountains, delight of the mind should be added to the harmonious delight of all the senses ; what entertainment, I ask, can you find in this world so high, so worthy, and in every respect so perfect?

Gesner proceeds to repel with scorn the objections that may be raised by any croaker or idler on account of the hardships to be encountered. The walking is toilsome ; there are hard places and precipices ; food and lodging are rough. What of that? The toil and peril become pleasant in remembrance. And the immediate pleasure and benefit of rest are for a healthy man all the greater in proportion to the work done. 'For as we walk, or at times jump, every part of the body is exercised. We work, and stretch every sinew and muscle, some in going up, others in coming down ; and with variety in each of these kinds too, if (as is the case in mountain walking) the course is alternately straightforward and slanting.' Experience confirms what was already laid down by Aristotle, that the changes of broken ground make the day's work less fatiguing than the same amount of walking on the level or on uniform gradients. Travellers who are subject to giddiness or in any other way unfitted for hard climbing must no doubt limit their excursions accordingly. The lack of town food is amply supplied by those Alpine delicacies—'lactaria illa opera '— which Alpine exercise makes it safe to enjoy. But if you want other food you can have it carried for you. As to lodging, Gesner's eloquence must speak for itself.

There are no beds, no mattresses, no feathers, no pillows. Luxurious and effeminate wretch! hay shall serve you for all; soft and fragrant hay, compounded of the most wholesome grasses and flowers. You shall sleep more sweetly and healthfully than ever before, with hay for a pillow under your head, for a mattress under your body, and spread over you for a blanket.

Gesner concludes his monograph by discussing the Pilate legend, and dismissing it with the robust disbelief of a man who feared God and did not fear devils, as another good man of Zürich (whom he quotes with an apology for his unclassical Latin) had done a century before him. For our present purpose it is more material to note that Gesner expressly mentions the alpenstock ('baculis quos alpinos cognominant et mucrone ferreo praepilare solent'), and that the party anticipated the generations of tourists to come by hearing the Alpine horn: only it appears that they blew it for themselves.

And Gesner, though eminent in his love of the mountains, at any rate in his expression of it, was not a solitary enthusiast. We have an ascent of the Stockhorn in 1536, described by Rhellicanus (J. Müller of Rhellikon, canton Zürich) in Latin hexameters intended to be classical.[1] There is not much about the scenery or the view, but it is clear that the party was a party of pleasure. They carried a gun, and shot one ptarmigan (*Steinhünli*). A generation later than this, and just within twenty years of Gesner's ascent of Pilatus, Josias Simler published his treatise on the Alps, in which 'for the first time

---

[1] See Coolidge, p. 125. The verses construe and mostly scan, but are sadly deficient in rhythm; about as good in a modern scholar's eyes as the feat of shooting a ptarmigan sitting, which they celebrate, in a modern sportsman's:

haec [alpestris gallina, interpreted *Steinhünli* in the margin],
    bombardam dum iaculantes
Intendunt simul atque iterum frustra, tamen istic
Haesitat immotim, donec confixa lapillo
Bombardae, scopulo cecidit tum proniter alto.

But modern sportsmen have not tried shooting with a wheel-lock, nor do modern scholars know what it was to write Latin verses without dictionaries or critical editions. Compare the man stalking partridges from behind a fallen tree in Rubens's great landscape in our National Gallery.

sound practical advice is given as to the precautions to be adopted when making excursions above the snowline.'[1] He describes the alpenstock (in almost the same words as Gesner), climbing-irons, and snow-shoes, and, what is more, he mentions the use of the rope on crevassed glaciers, and of dark spectacles as a precaution against snow blindness. I give the most important passages in Mr. Coolidge's excellent translation. After mentioning the narrowness of mountain paths, the risk of giddiness to persons unaccustomed to such places, and the difficulty of crossing the Alps with a train of beasts of burden, not to speak of the baggage and artillery of an army, Simler proceeds :—

Precipitous and rugged places further increase the difficulty of Alpine paths, and particularly if the tracks are covered with ice, for which reason the travellers and the shepherds, as well as the hunters who frequently roam over the highest mountains, provide for their safety by various precautions. To guard against the slipperiness of the ice, these people are accustomed to tie iron shoes, like those of horses, and furnished with three sharp prongs, securely to their feet, so that they may get firm foothold on the ice ; others furnish the thongs, by which the sandals are tied under the foot, in the same way with a very sharp iron spike, and employ other means in order to resist the slipperiness of the ice and to improve their footing. In some places they use staves tipped with an iron point. and, resting their weight on them, are in the habit of ascending and descending steep slopes. These staves they call alpenstocks, and they are chiefly used by shepherds. Sometimes also shepherds and hunters let themselves down steep and almost precipitous places, when there is no other way, by cutting down

[1] Simler (1574), fo. 110 b ; Coolidge, p. 16 (the original Latin is cited in the notes, pp. 164-5). Perhaps I may be allowed to make one comment strictly outside my subject, namely that snow-spectacles are often needlessly dark. I have found a quite moderate tint sufficient even when the sun was bright on fresh snow. But I am aware that this opinion is not orthodox, so perhaps my experience has been inadequate or abnormal. Veils, which seem to be hinted at by Simler's ' aliquid nigri praetendatur,' have been discarded by practical mountaineers for many years. The other passages cited from Simler are at ff. 112 a b, 213 a, 115 a, of ed. 1574 ; on fo. 112 a this reads ' hi se fune cingunt,' not ' hosce,' which Mr. Coolidge takes from ed. 1633. The earlier reading seems right.

branches from the trees (particularly firs), and then sitting on them as though riding a horse, and thus sliding down. When heavy carts are to be let down difficult places like these, they sometimes lower them by great ropes worked by means of cranes and pulleys.

Further, those who wish to walk over deep snow, in places where there are no paths, make use of the following expedient to prevent themselves from sinking in. They tie to their feet small and thin wooden boards or wooden hoops (such as are used in binding wine casks) woven together by a sort of latticework of cord of a foot in diameter. In this fashion, as they make larger steps, they do not sink in and do not go deep down into the snow.

Further, amongst other evils, the intense cold is very troublesome to those making a journey through the Alps, especially as long as the north wind is blowing, in consequence of which the limbs of many men often burn by reason of the extreme cold, while in the case of others their ears or noses, or in the case of some persons their fingers or toes, and even the feet themselves, grow numb by reason of the cold, and die. Many lose their eyes through continually going over snow. Against these evils there are various safeguards. For the eyes something black should be put over them, or what they call glass spectacles (*vitrea conspicilia*). For other parts of the body it is advisable that they should be well protected against the cold by skins and thick clothes. Paper and parchment protect the breast very well from cold winds; but if the feet are benumbed at night when the shoes are taken off, the feet should be bathed in cold water, warm water being gradually poured in; for so they think they are brought round. The best precaution of all is constantly moving about, for it happens sometimes that while ascending a mountain men get heated by their exertions, and think that they feel no cold; but if they sit down in the snow for the purpose of resting themselves, drowsiness soon creeps over them, and then, with scarcely any feeling of pain, they are benumbed and die.

Further, that ancient ice, over which one must sometimes make one's way, has deep chasms in it, three or four feet wide, and often even larger, into which if a man fall, he must, without doubt, perish. It happens also that these chasms may be covered with fresh snow, or by snow blown together by the wind. Hence those who then travel in the Alps hire men who know the place, to go in front as guides. These men gird themselves with a rope, to which some

THE EARLY HISTORY OF MOUNTAINEERING 17

of those who come after also bind themselves; the leading man sounds the way with a long pole, and carefully keeps a look-out in the snow for these chasms; but if he unexpectedly falls into one of them, he is held up and drawn out again by those of his comrades who are tied by the same rope as he is. When there is no snow over these pits there is less danger, but yet they must be crossed by a jump; for there are no bridges here except that sometimes those who lead beasts of burden over such places (which rarely happens) carry wooden planks with them, and lay them before the beasts so that they can cross these chasms by means of a bridge.

It must be admitted that Simler is not wholly free from superstition concerning things above the snowline. He says that an avalanche may be started by very slight causes, such as a bird flying past, or a loud voice, or even the echo of it. He seems also to have conceived of a snow avalanche as a single gigantic snowball. These statements, however, were copied and recopied for more than two centuries by people who had far less excuse for being inaccurate; and what is said about the use of the rope among concealed crevasses could hardly be bettered at this day.

Now I do not think Simler would have given these careful and practical descriptions as a matter of mere curiosity, and they seem to me to justify the inference that a native Swiss school of mountaineering, with exactly the aims and the spirit of modern mountaineers, was on the point of being formed by the scholars of Zürich. It will be remembered that among the English leaders of Alpine exploration in its most active period a notable proportion were either men of science or men who, as University residents or otherwise, maintained a close connection with the scientific and scholarly traditions of the Universities. Mr. Leslie Stephen was in those days a Cambridge tutor, and Mr. H. B. George (as he still is) an Oxford one; while Mr. Tyndall's interest in the Alps was at first purely scientific.

C

Thus the beginning of Alpine exploration, made in the middle of the sixteenth century, was in some ways much like the more fortunate enterprise which, just three centuries later, was crowned with success both rapid and complete.

But the impulse given to mountaineering by the Zürich band of scholars was arrested, as it seems, by the civil and religious troubles of the seventeenth century. There is a barren period till near the end of that century, and when the mountains reappear as an object of scientific interest, it is in a position subordinate to general topography and miscellaneous curiosity, and with a pomp of exaggeration and fabulous accessories which Gesner and Simler would have laughed to scorn. Wagner the naturalist, writing in 1680, has no doubt that there are dragons in the Alps; he knows of a living man, one John Tinner, who not only saw, but shot one twelve years ago. It was full seven feet long, of a blackish grey, its head like a cat's; it had no feet. Of other dragons footed and winged he gives equally veracious particulars, which lose nothing in their repetition by Scheuchzer a generation later. It has been suggested that some of the dragon legends in the Alps arose from glimpses, caught through stormy weather, of serpent-formed glaciers. The illustration will convince the reader that at any rate the resemblance can be made plausible. The dragons in Scheuchzer's book are mere exercises of the engraver's fancy, and sufficiently amusing ones. Scheuchzer flourished in the first quarter of the eighteenth century; his observation is all but swamped in compilation, and he may be taken as the typical author of this time of partial revival. Of actual mountaineering enterprise there is not to my knowledge any evidence whatever. It would hardly be rash to say that practical knowledge of the mountains had positively declined among educated people between 1550 and 1750. The last we hear of the Zürich school, as I venture to call it, is an imaginary dialogue in verse between the Niesen and the Stockhorn, 'two mountains situate in the worshipful Confederation, and in the Bernese territory thereof. This was first published in 1605,

THE EARLY HISTORY OF MOUNTAINEERING 19

and revised in 1620.[1] It also marks the beginning of the period in which pedantic compilation took the place of firsthand inquiry. One might with some plausibility claim for it the distinction of being the most rambling and tedious poem

Wilderwurm Gletscher.

ever published in Europe. The Niesen addresses a kind of sermon in verse to the assembled notables of the canton of

[1] Coolidge, p. 126. Rebman's interminable verses are often quoted by Scheuchzer, and, indeed, show as much knowledge of glaciers as is to be found anywhere before the days of De Saussure. See the descriptions of *Firn*, &c., at p. 151 (repeating the fable about birds' flight starting avalanches), and of the Grindelwald glacier at p. 488.

C 2

Bern. To them enter the Stockhorn, and the two mountains utter a kind of antiphonal gazetteer of the world in general, and mountains in particular. Rather more than a century later we find, indeed, that parties of young men belonging to the well-to-do families of Zürich undertook walking tours in Switzerland under the guidance of a clergyman, who acted as travelling tutor, and usually composed an account of the journey afterwards. Reports of this kind are preserved in manuscript in the libraries of Zürich and Basel. These tours, however, were a matter of general curiosity and instruction, analogous in some ways to the reading parties of modern English undergraduates; it was a mere accident that they took place in hill country, and it does not appear that any pleasure in mountain travelling, artistic or athletic, can be connected with them either as motive or as result. At most an ascent of Pilatus might be included in the round, not for the sake of climbing or view, but for the amusement of throwing stones into the once enchanted mere, and observing that nothing happened.[1] The view of the Grindelwald glaciers, here reduced from Merian's 'Topographia Helvetiæ' (1654), represents the high-water mark of Alpine knowledge for more than a century afterwards. Illustrations of the same scene in eighteenth-century books are for the most part more or less degenerate copies from Merian. Not in this way was a generation of mountaineers to be raised up. Scientific and picturesque interest in the Alps had, so to speak, to be reconstructed by independent efforts in different quarters; and they were to be brought to a head not in the Central but in the Western Alps; not in the Teutonic heart of Switzerland, but in Romanic borderlands. We must shortly trace the working of these distinct influences.

Towards the end of the seventeenth century the glaciers of the Alps began to excite a specific interest in the scientific world. In 1669 a letter addressed to Hooke by a Bernese physician, 'concerning the icy and crystalline mountains of Helvetia,

---

[1] Peyer, *Gesch. des Reisens in der Schweiz*, 108; Osenbrüggen, *Wanderstudien aus der Schweiz*, ii. 285.

THE LOWER GRINDELWALD GLACIER
(From Merian's 'Topographia Helvetiæ,' 1654)

called the Gletscher,' appeared in the 'Philosophical Transactions.' In 1673-4 this was followed by a more particular relation imparted to us from Paris by that worthy and obliging person Monsieur Justel, who had received it from a trusty hand living upon the place ; which place is a little further specified as 'the icy mountain called the Gletscher, in the Canton of Berne, in Helvetia.'[1] The lower Grindelwald glacier, and the peaks visible around and beyond it, are figured in a rough engraving referred to as 'the Scheme.' Later, in 1708, William Burnet, son of the Bishop, found his way to Grindelwald, and sent a yet fuller account to Hans Sloane, which in due course appeared in the 'Transactions.' As I am not writing the history of glacier theories, I need not dwell on the physical notions exhibited in these early reports ; the assumption that the mountains are not merely snow-covered, but made of ice all through, the belief that the crystals found in the neighbourhood of glaciers are ice hardened by lapse of time, and so forth. But it must be recorded how the 'worthy and obliging' M. Justel adds that 'there is such another mountain near Geneva, and upon the Alps.' For this must be one of the earliest distinct notices of Mont Blanc in a scientific publication, if not the very earliest. Bishop Burnet had already written to Boyle from Zürich on September 1, 1685, how 'One Hill not far from Geneva, call'd *Maudit*, or Cursed, of which one third is always covered with snow, is two miles of perpendicular height, according to the observation of that incomparable mathematician and philosopher Nicolas Fatio Duilier.' It was full two generations before

[1] These communications are reprinted in *A.J.* xiv. 319. I may mention here that an earlier passage in Ray's travels is misquoted in a recent and learned book, Schwarz's *Erschliessung der Gebirge*. He is made to advise shutting the eyes in a snow-storm. What he does say is, ' A brisk gale of most bitter, cutting wind blew just in our faces, which did so affect my eyes, that I could not open them without pain for three days, nor easily endure to look upon snow for a great while after,' *Travels*, i. 357, ed. 1738. The passage is also wrongly supposed to occur in a book of specially Alpine travel, whereas Ray crossed the Splügen as an ordinary incident of a general tour. Apparently the mistake is copied from Scheuchzer. But it is not worth pursuing.

Windham and Pococke's visit to Chamonix. Long afterwards, however, and even after Haller and Rousseau had set the fashion of picturesque sentiment, many who busied themselves with the higher Alps in the way of topography or other scientific research could still see nothing in their scenery but fearful and disgusting desolation, or at best a sublime horror. The real difficulties that still attended high excursions may account for this in part, but in part only; for we have seen how different was Conrad Gesner's feeling, and Gruner in 1760 was at all events no worse off for the material appliances of mountain travelling and climbing than Gesner had been in 1550. Persons of quality who crossed the Alps in the way of ordinary journeying either thought the mountains positively ugly, or were too much taken up with the supposed dangers and more or less real discomforts of the passage to think of the scenery at all. Among Englishmen Gray shows himself on this point, as on so many others, in advance of his time. He writes of the situation and surroundings of the Grande Chartreuse with genuine admiration; he regrets that the beauties of the Apennines were hidden by mist when he crossed them from Bologna to Florence; and he has nothing worse to say of the Mont Cenis, in November, and when there was still no road practicable for wheels, than that it 'carries the permission mountains have of being frightful rather too far.'

In the second quarter of the century, following upon the awaking of scientific interest in the Alps (an awaking indeed, for it was not then first born, but had only slumbered), there came the romantic movement of interest in wild nature, not as yet by way of purely artistic feeling, but combined with a sort of yearning admiration for the simplicity and dignity of human life among the mountains. This mixed sentiment was nourished by descriptions and imaginations of primitive manners which were not always accurate even when they were supplied by persons who had the means of observation; but they served the purpose of furnishing rallying points for revolt against the conventional standards of taste, and to some extent of conduct. The rebels played at realism, and indeed believed in good faith that they were casting out man's devices and restoring the supremacy of nature. In truth, they opposed the established conventions with new conventions of their own, and it could not be otherwise. Two names are conspicuous in the Alpine scenes of this sentimental revolution, which formed by no means an unimportant part of the whole, those of Albert von Haller, and, more eminently though not more deservedly, of Jean-Jacques Rousseau. Haller's didactic poem, published in 1742, celebrated the beauties of the Swiss Alps and the virtues of the Switzers in vigorous though somewhat lumbering stanzas. It had an immense reputation, went through thirty editions in the author's lifetime, and was translated into French, English, Italian, and Latin. Rousseau followed a generation later with *Emile* and the *Nouvelle Héloïse*. There is no fear of his part being forgotten or underrated, and in regard to the special topic we are rather called upon to see to it that the merit of such a precurser as Haller is not unduly eclipsed by Rousseau's greater and wider fame. Haller was the first writer since Conrad Gesner who boldly proclaimed his admiration of the mountains for their own sake. Neither Rousseau nor Haller, of course, had any pretension to be a mountain climber; they were content to admire in a fashion as vague as it was large. Rousseau could write of Switzerland in general, 'Elle offre à peu près partout les mêmes aspects, des lacs, des prés, des bois, des montagnes.' But that does not deprive Haller and Rousseau of their place in the history of Alpine exploration. Many princes and knights were moved to the quest of the Holy Sepulchre in the middle ages by preachers who were themselves perhaps physically unfit for the journey, and as a rule unable to give more than a very poor account of the situation or topography of Jerusalem. In any case, this period can be credited with one distinct mountaineering event, the first ascent of the Titlis in 1739, which was also the first recorded ascent of any permanent snow peak.

Meanwhile the general movement of voyages and travels in all parts of the world had its share, we may well believe, in educating the world of letters to take some interest in these matters. Now and then real mountain travelling outside Europe was heard of, as when Bouguer and Condamine encamped and took observations in the Cordilleras.

It is possible that Humboldt's partial ascent of Chimborazo in 1802 may have also exerted some indirect influence on mountaineering zeal in Europe. But I am not aware of any evidence that it did, and moreover the cult of Mont Blanc had then been fairly started by De Saussure and Bourrit, and only waited for European peace to spread and convert the nations. Humboldt's and Condamine's expeditions have nothing to tell us in the art of mountaineering beyond the truism (as we now esteem it), that above the snow-line all the energy and good will in the world will not make up for the want of proper training and appliances.[1]

We may pass then to the exploration of the Mont Blanc range from Chamonix, by which the modern practice of mountaineering was formed. Windham and Pococke's visit of 1741 calls only for brief mention here, the rather that Windham's account is easy to consult, being reprinted at large in Albert Smith's book. That account is a very modest and sensible one, and free from the exaggerations of many later travellers, who had things made far easier for them. Without doubt Windham and his companions are entitled to the credit of having first ascertained that the *Glacières* of Chamonix were accessible, and promised to reward a more leisurely and scientific view. They seem to have been the first Alpine party

[1] A. von Humboldt's account of his attempt on Chimborazo, together with pretty full notices and extracts of both earlier and later explorations in the Andes of Ecuador, is to be found in his *Kleinere Schriften*, Stuttgart, 1853, vol. i. (all published). This appears to be the ultimate authority for the particulars given in Mr. Whymper's *Travels amongst the Great Andes of the Equator*, Appendix G, as from Bruhns' *Life of Humboldt*. Mr. Whymper's observations (Appendix A) reduce the height of Chimborazo to 20,545 feet, and the height supposed to have been attained by Humboldt's party is likewise reduced in proportion.

who took both guides and porters : 'We took with us several Peasants, some to be our Guides, and others to carry Wine and Provisions.' When they call the Montanvers 'the top of the Mountain,' it is only the current usage of the time. Invidious distinctions between the greater and lesser eminences of a group were still unknown and unconsidered. Whatever prominent part of the chain was ascended was 'the mountain' for the time being. As for Mont Blanc, or any of its higher peaks, they are not mentioned at all. Windham appears to be the true author of the comparison, now taken as a matter of course, that produced the name of *Mer de Glace*. 'You must imagine your Lake' (the lake of Geneva) 'put in Agitation by a strong Wind, and frozen all at once ; perhaps even that would not produce the same appearance.' Alpenstocks were used, and are described by the periphrasis of 'Sticks with sharp Irons at the End.' At the time of this journey Bourrit had not long begun to walk and talk, and De Saussure was hardly out of his cradle. Bourrit was born in 1739, De Saussure in 1740. The design of conquering Mont Blanc was formed as early as 1760, but there is no distinct record of a serious attempt to put it in execution before 1775, when a party of guides, starting from the Montagne de la Côte on the left bank of the Glacier des Bossons, and thence joining what is now the usual route from the Pierre à l'Echelle on the right bank, appears to have gone as far as the Grand Plateau. Then followed eight years of suspense, during which, however, the curiosity of travellers and men of science about Mont Blanc was increasing. It would be useless to pursue here the details of Bourrit's valiant but uniformly baffled endeavours from 1783 onwards, of Jacques Balmat's and Michel Paccard's first successful attempt in 1786, or of De Saussure's own triumph in the following year. The curious reader is referred to M. Durier's excellent monograph as the best and most complete source of information, short of the contemporary narratives themselves, from which, indeed, M. Durier gives considerable extracts. What now concerns us is the spirit rather than the letter of the history; the growth of the mountaineering habit rather than the topographical advance of ascents and explorations. We may smile at Bourrit's way of expatiating on the 'scènes vraiment sentimentales' afforded by the emotions of strangers visiting the Montanvers, his moral reflections (after Rousseau) on the ennobling influences of mountain scenery, or his discovery of a perfectly formed head of Neptune in a crevasse. But, notwithstanding his singular run of ill-luck on Mont Blanc, he was a genuine and an active mountaineer. Such little touches as his enjoyment of a glissade leave no doubt of his quality, and my impression is that he was a better walker than De Saussure, though a less fortunate explorer. His generous admiration of De Saussure, whom he survived a considerable time, must not be omitted in the count of his virtues; it makes ample atonement, if any be needed, for his harmless vanity in persuading himself that being anywhere on the final snow-cap of Mont Blanc was equivalent to having attained the summit.[1]

The French Revolution interrupted the flow of travellers to Chamonix, and Mont Blanc was ascended only twice during the period of the great war. After the peace of 1815, the revival of traffic was felt at Chamonix as elsewhere, but mountaineering did not at first increase in anything like a rateable proportion to sightseeing. More than a generation was to elapse before ascents of Mont Blanc became at all frequent. Let us pause to see in what state the fairly numerous visitors to Chamonix, and the few climbers among them, found mountain craft in the days when Goethe was yet alive or his memory fresh, and Victor Hugo and Lord Tennyson were young, and what manner of impression the few enterprising ones carried away from Mont Blanc. As regards the

---

[1] ' J'estime qu'on est sur le Mont-Blanc dès qu'on est parvenu à dépasser le dernier grand rocher qui, depuis Chamonix, forme un demi-cercle' (the Rochers Rouges). *Description des Cols, &c.*, Genève, 1803, i. 88. The feeling is natural enough, and is not unknown at the present day among mountains which are still unfamiliar.

climbing tackle, there was not any great difference between the material resources available then and now. We have already seen by the witness of Simler that the use of the rope was no new thing. Travellers and guides had only to learn the importance of using it on settled principles, and being sure of its quality. It is true that the lesson was both a long and a costly one, and has been completed by more than one bitter experience well within the memory of men not yet old. De Saussure's expedition to the Col du Géant was provided with one or more ropes ; but the rope was used only to extricate the leading porter when he fell into a concealed crevasse : after which the party seem to have proceeded unroped as before. Sometimes a traveller walked between two guides holding alpenstocks to form a temporary hand-rail ; a device whose value may be set down as strictly moral. I have myself known an excellent Chamonix guide adopt, by way of compromise, on a place where the rope was not strictly needed, the practice of roping a party two and two. The result, though not grave, was unpleasant enough to prove its folly. Not that I mean to deny that on many occasions two men alone may make an expedition in which the rope is a good and useful thing to have with one, at any rate for the amateur ; I could bear witness to that also. But that depends on the character of the expedition and the men, and in particular the Alpine rule of three should be as inexorable as the rules of arithmetic when the way lies over crevassed snow-fields : a truth which will be plain to every mountaineer, and which the didactic portion of this volume will, I trust, make plain to others. Then climbing-irons were well known to the men of Chamonix, having been invented by chamois hunters for their own purposes. Later they fell out of use in the Western and Central Alps, insomuch that fifteen or twenty years ago one seldom or never saw them; but in the Eastern Alps they have never gone out of fashion, and the present generation of climbers, having studied them there, appears inclined to favour them again. De Saussure devised an improved model with only three points under the heel,[1] and used it with satisfaction for several years. He found that these irons might be kept on without inconvenience even where they were not strictly required. Various kinds of moveable spikes or 'glacier nails' have been tried down to recent times by mountaineers and purveyors of mountain equipment, but have not on the whole been found effective. One prominent instrument in the earlier expeditions is seldom heard of now —that is, a ladder—which was deemed indispensable for the passage of the Glacier des Bossons. It is sufficient to recall the name of the Pierre à l'Echelle. Now it may be that step-cutting went slower in the first half of the century than it does now, and it is probable that travellers were on the average less capable of using the steps when cut, and thus, in any case, made the passage longer. But comparison of the various accounts assures us, after making all reasonable allowance for exaggeration, that the glacier was really, during the first half of the century, and even some years later, much more crevassed and difficult than it has been since. A ladder has been used elsewhere in later times under similar conditions. The classical example, if I may be allowed the phrase, is the first passage of the Jungfraujoch.

Mention of step-cutting brings us to that most characteristic and necessary tool of mountaineering, the ice-axe. From the language of the early accounts, which do not use any constant term, one can only gather that some kind of axe was used by the Chamonix guides, as it had long been used by chamois hunters, to cut steps in ice; and it would be easy to believe that until a comparatively late period this was not a special instrument, but a common wood-axe or hatchet. In fact, it is oftener called a hatchet than anything else. It must

---

[1] De Saussure, *Voyages dans les Alpes*, 1779, &c. i. 480, § 558 and pl. 3. His improved *crampon* may be described as a moveable 'tackling' fitted round the heel of the boot and furnished with spikes. It was secured by straps arranged so as to avoid pressure on the instep. Clissold (1822) gives a quite different description of what was recommended to him as De Saussure's pattern.

'MONSIEUR DESAUSSURE, SON FILS, ET SES GUIDES ARRIVANT AU GLACIER DU TACUL AU GRAND GÉANT, OÙ ILS ONT HABITÉ 17 JOURS, SOUS DES TENTES, EN JUILLET 1788.'

(*From an old print in the possession of the Alpine Club.*)

have been a common axe that Gerard found Himalayan mountain folk using for this purpose in 1821.[1] And it appears that sometimes at Chamonix a short-handled axe was made fast to an alpenstock. 'An axe was also fastened to one of the poles'—only one among the party—'that steps might occasionally be cut in the ice and hardened snow,' wrote F. Clissold, who ascended Mont Blanc in 1822. 'A little axe' —it is not clear whether each guide had one or not—occurs in a printed record as late as 1852. On the other hand, Dr. Hamel mentions an 'Eisbeil' as early as 1820, and Dr. Edmund Clark, who made the ascent with Sherwill in 1825, speaks of the ice-axe by its present name as if it were a well-known and distinct thing. His narrative is one of the most careful, and the vagueness of other people's language proves nothing. Indeed, it was many years before the term 'ice-axe' was firmly established. It is not fully recognised even in 'Peaks, Passes, and Glaciers' (1859), where no less a mountaineer than John Ball speaks of a *pole-axe*. And in Mr. Tyndall's 'Glaciers of the Alps,' published in 1860, the word used is still generally 'axe,' without addition. Bennen is once described as cutting steps with his *mattock*, possibly to mark the difference, to be presently mentioned, between the Chamonix and the Oberland pattern. Conclusive evidence, however, is furnished by the contemporary drawing of De Saussure and his following on the march to the Col du Géant, where we see clearly figured an ice-axe that could be nothing but an ice-axe, the regular Chamonix *piolet*[2] of the form used till 1860 or later. Therefore a true ice-axe was known from early days, though it seems that a makeshift of one sort or another sometimes did duty for it, and it also seems that step-cutting was regarded as an extraordinary aid in difficult places rather than a normal mode of progress above the snow-line. Certainly

[1] Lloyd and Gerard's *Narrative, &c.*, Lond. 1840, ii. 32, 58.
[2] Ice-axe = piolet = Eisbeil or -pickel = piccozza. As to hunters carrying an axe, see De Saussure, *Voyages dans les Alpes*, ii. 126 § 736. *Hache* appears to have been the official term at Chamonix as late as 1855 (Hudson and Kennedy).

the practice was to carry only two or three axes, or even only one, for a whole party; so far was it from being a matter of course that every guide should have his ice-axe. It does not appear to have ever occurred to the travellers of those days to prefer the axe to the alpenstock : and, in fact, it is not much more than twenty years, if I remember right, since this habit, a convenient and prudent one in all cases, and oftentimes necessary in modern mountaineering, became general enough for a commencing, mountaineer to feel quite at his ease in adopting it. But, again, I must not trespass on the dogmatic office of my colleagues.

Another point that excites the wonder of a modern climber is the enormous load of clothing carried by his predecessors from the time of De Saussure to that of Albert Smith. H. M. Atkins gives this inventory of what he wore on Mont Blanc in August 1837 : 'I had on a good pair of lamb's wool stockings, two pairs of gaiters, two pairs of cloth trousers, two shirts, two waistcoats, a shooting coat, and, over all, a blue woollen smock frock, a nightcap, three handkerchiefs round my neck, two pairs of woollen gloves, and a straw hat, from which hung a green hood. For my eyes, a pair of spectacles, and a green gauze veil.' No wonder that the pace of men so hindered, and going with an overgrown train of guides and porters, was slow according to the modern standard. We must remember that they had to sleep out at the Grands Mulets with very poor shelter; and the 'woollen smock frock,' doubtless a common fisherman's frock, would be a quite practical addition to the walking clothes for that purpose. I have not used one in the Alps, but I have found it an excellent thing for camping out on a Canadian river, where hot days were constantly followed by cold nights. But still one does not understand why some of these warmth-preserving garments were not left at the Grands Mulets. Yet more remarkable is the almost uniform record of distressing sensations, chiefly of great lassitude, caused by the rarity of the air at great heights. Trouble of this kind is now a most exceptional drawback, not only at

Alpine altitudes, but at those of the Caucasus. Dr. Clark describes his party, in 1825, as 'pretty nearly worn out with fatigue, and nausea, and pain.' Auldjo, in 1827, attained the summit only by hauling himself along on a rope held fast by two guides in front. Compare this with the experience of 1868 on Elbruz, when 'at a height of 18,500 feet no single man out of a party of six was in any way affected;' or with that of M. Durier on Mont Blanc itself in 1869, when his guide reproved him for levity in smoking his pipe up to the very top. Yet Mr. Whymper's experience in the Andes has shown that some two thousand feet more may give a serious check, though not an insuperable one, to the best of European mountaineers.

Another variety of 'mountain sickness' is often caused by the joint effects of heat, glare, and fatigue in high valleys, and in monotonous walks over snow-fields. This was set down by De Saussure's generation, and even later, to some peculiar malignant quality of the 'stagnant air' supposed to lie in such places. I cannot avoid concluding that the average health and condition of pedestrians travelling for their own pleasure, if not their measurable bodily strength, must have greatly increased in the last two generations. Clissold, apparently a better walker than most of his contemporaries (for he knew Snowdon well, and had traversed it in winter), makes a curious little disclosure when he remarks how 'the cold surface' of the snow-fields 'prevented those inflammatory effects on the legs which are experienced when walking upon common ground.' Any such effect of 'common ground' is, to the best of my belief, wholly unknown among English walkers of the present day; nor do I remember to have seen any warning against them in any of the ordinary modern guide-books. Doubtless many persons even now cause themselves to be conveyed to the summit of Mont Blanc who are at least as ill prepared for the continuous exertion as the travellers of sixty or seventy years ago. But these have ceased, as a rule, to publish their experience.

We can hardly doubt that the fatigue and unusual sensations of amateur climbers in this archaic period were actually and physically heightened by an overstrung imagination of the difficulties and dangers. While real danger was still not seldom overlooked, the ordinary incidents of mountain walking were hugely magnified. It was thought hardly decent to come down from Mont Blanc without moralizing on the question whether the ascent were justifiable. I will take a quite late example, the reflections of Mr. Philips, of Christ Church, Oxford, whose party joined Albert Smith in 1851. 'I strongly recommend any one who may feel ambitious of ascending Mont Blanc to consider well before he attempts an expedition which cannot be productive of any good to himself or others, and which is attended with fearful risk, not only to himself, but to those persons who, allured by the desire of gain, endanger their lives in his service.' And these are the sentiments of an active young man, after an ascent which does not appear to have offered any unusual incident. Lest I be misconstrued, I hasten to say that in mountaineering, as in war, it is far better to take things too seriously than too lightly. Nevertheless philosophy reproves manifest excess even in the more laudable direction. This Oxford party used one precaution that is not now common. As soon as, being at Ouchy, they had formed the project of ascending Mont Blanc, they solemnly went into training on stale bread and beef-steaks. Over-training may possibly go some way to account for the generally depressed tone of Mr. Philips' account. His little book is otherwise interesting as an independent confirmation of Albert Smith's much better known story. For example : 'Albert Smith was utterly exhausted, and it was only by the most undaunted courage that he ever reached the summit.' The guides must have strained a point to give Albert Smith the flourishing certificate which during the past winter (1891-2) was on view at the Victorian Exhibition in London ; its terms might be read as asserting, as literal matter of fact, only that the travellers walked and were not carried ; but any reader without other

THE EARLY HISTORY OF MOUNTAINEERING 33

knowledge of the facts would naturally infer that they walked like mountaineers. There is one curious passage in Mr. Philips' narrative, immediately preceding the moral, which gives rise to difficulty. 'We all commenced the ascent with the idea of finding the fatigue and danger much exaggerated. Albert Smith, in particular, fully intended to expose the whole affair as an imposture, fancying the guides were leagued together in representing it as much more hazardous than would prove, on trial, to be the case.' Now such intentions or beliefs on Albert Smith's part are not consistent with the words or tone of his own account, or with his previous knowledge of Chamonix and personal acquaintance with the guides. On the other hand, there is no reason to doubt Mr. Philips' good faith. I can only suppose that Albert Smith thought his companions a little too confident in their own power, and could tno resist the temptation of a hoax. If this were so, Albert Smith was the first and true inventor of the part played by Bompard in 'Tartarin sur les Alpes,' and M. Daudet has builded better history than he knew.

We must now turn our attention to what was doing in other parts of the Alps. In Placidus a Spescha of Disentis (1752-1835) De Saussure had a worthy, nay a brilliant follower. Had he not fallen on the evil times of the revolutionary campaigns, Father Placidus would have done for the region of the Upper Rhine as much as his contemporaries did for Chamonix, perhaps more. As it was, his ascents, and still more his prospecting expeditions in the Bündner Oberland, were such as amply to entitle him, in Mr. Freshfield's words, to 'a high place among the founders of mountaineering;'[1] but his work was sadly interrupted, and had to be completed by others long afterwards. When Spescha was an old man climbing for climbing's sake was still far, very far, from the comprehension of tourists in the Central Alps, if we may judge of their current opinions by the anonymous and singularly illiterate author of 'The

[1] *Placidus a Spescha and Early Mountaineering in the Bündner Oberland*, by Douglas W. Freshfield, *Alpine Journal*, x. 289-313.

D

Diary of a Traveller over Alps and Appenines' (*sic*), published in 1824, who reflected thus on the arrival of a few walkers at Grindelwald :—

These pedestrian excursions are by no means uncommon in Switzerland, and it is most extraordinary, that they appear sometimes to be undertaken by persons to whom œconomy need not be an object. The travellers move with a napsack upon their backs, and a long pole with a spike at the lower end of it in their hands. The last appendage is extremely necessary for their security upon glaciers and precipices. It is a system of peregrination that must be pursued with a relinquishment of the most essential comforts of life. . . . At hotels, too, the walking itinerant will meet with second-rate attention. The meanness of his style will prevent him from being regarded as a person of condition.

The same person wonders that 'men and women living almost without the pale of society'—to wit, in the Bernese Oberland—should be of decent appearance and behaviour, and when he gets into Germany plaintively records that 'as the language of the people is German, it is not easy to gain any information from them.' He is an exaggerated specimen of a type that still survives, abstaining however from print, and consoling itself, as Mr. Conway has noted, with being sick of the sight of ice-axes. We are not to suppose, however, that no progress was being made in this period. A limited number of Swiss explorers, among whom the foremost names are those of the Meyers and the Studers, laid siege in a slow but methodical fashion to the strongholds of the Oberland, and formed a generation of guides who only waited to be urged to a series of final assaults by the example, comradeship, and emulation of the Chamonix leaders. This chapter, however, does not deal with the statistics of first ascents, but with the growth of that body and spirit of enterprise which within our own memory has profoundly changed the meaning of the words mountaineer and mountaineering. Readers who seek the details in English may find them in the late Mr. Longman's narrative, edited and enriched by Mr. D. W. Freshfield, and

THE EARLY HISTORY OF MOUNTAINEERING 35

published with the eighth volume of the 'Alpine Journal.'[1] It may suffice here to mention that the Jungfrau was ascended in 1811, the Finsteraarhorn in 1812 (the doubts entertained for some time as to the reality of the latter ascent are now known to have been due, not to any defect in Meyer's own account, but simply to an editor's perverse alterations) ; and the principal glaciers of the Oberland, and most of the passes of general practical utility, became fairly well known, though hardly frequented. The old superstitious dread of the ice-world was dispelled as effectually as at Chamonix, although it long remained a tradition of the elders that, other things being equal, a Chamonix man preferred ice, and an Oberland man preferred rocks.

Yet the normal difficulties and fatigue of climbing were estimated on a scale which we must now consider excessive. A fair test case is afforded by the Strahleck, a pass known to most travellers who have made any glacier expeditions in the Bernese Oberland. Writing about 1840, Gottlieb Studer said that this route from the Grimsel to Grindelwald must always present extraordinary fatigue and danger even to practised Alpine tourists. It is true that his own experience was specially unlucky. The party were baffled by mist, failed to hit off the true pass, and were driven back to the Grimsel. And it is ever to be remembered that in one sense there are no easy expeditions above the snow-line. All high expeditions may become difficult or dangerous, or both, in bad weather. Some take worse conditions to make them so than others ; there is no total exemption. At the same time Studer wrote with full knowledge that other parties had done much better since ; for he tells us so himself. Now let us hear what John Ball said in 1859 : 'This is one of the most interesting of glacier excursions, and in fine weather offers no serious difficulty to a moderately good mountaineer.' Moreover, it is 'one

[1] *Modern Mountaineering, and the History of the Alpine Club*, by William Longman, 94 pp. Not separately published.

of the least fatiguing of the high glacier passes.' These are the words of a most careful and discreet author, one whose general knowledge of the Alps and Alpine travelling was almost unique, and who has never been accused of lightly depreciating real difficulties. Indeed, Ball's own account of the final ridge of Monte Rosa, for many years conspicuous in Murray's Guide-book, must be pronounced somewhat overwrought according to the modern standard.[1]

On the whole, then, the work of the first half of this century was to prepare the way for modern mountaineering; the critical impulse was yet to come. It came within a few years in a series of independent adventures undertaken by Englishmen. Among these, the ascent of the Wetterhorn from Grindelwald by Mr. Justice Wills in 1854, the final conquest of the highest peak of Monte Rosa in 1855, and the ascent of Mont Blanc without guides in 1856, hold the foremost place. Mr. Alfred Wills, as he then was, made his attack on the Wetterhorn with two Chamonix and two Oberland guides; and for this cause he became the first chronicler of the modern 'ice-axe.' 'The sticks the Oberland men carried were admirably suited for their work. They were stout pieces of undressed wood, with the bark and knots still upon them, about four feet long, shod with a strong iron point at one end, and fixed at the other into a heavy iron head, about four inches long each way; one arm being a sharp spike, with which to hew out the ice, where needed, the other wrought into a flat blade with a broad point, something like a glazier's knife. . . . This kind of alpenstock is hardly ever seen at Chamouni . . . the great utility of the Oberland implement called forth repeated expressions of admiration from the Chamouni men, to whom it was new.'[2] Mr. Justice Wills's brilliant Alpine descriptions, both in his 'Wanderings' and in his later volume, 'The Eagle's Nest,'

---

[1] G. Studer, *Topographische Mittheilungen aus dem Alpengebirge*, 2nd ed. 1844, i. 36; *Peaks, Passes, and Glaciers*, i. 255, 279. As to Monte Rosa, cf. Coolidge, *Swiss Travel, &c.* 310.

[2] Alfred Wills, *Wanderings among the High Alps*, 2nd ed. 1858, 280.

were doubtless potent in turning the energies of English vacation tourists in this direction. At the same time Hudson and Kennedy's concise and workmanlike account of their triumph over professional routine on Mont Blanc put an end to many vulgar errors, and Hudson and Smyth's unaided discovery of the true route to the Dufourspitze of Monte Rosa completed the proof that the traveller's part was not one of mere mechanical effort. Unaided, I say, for though there were guides, and one of them, Ulrich Lauener, a first-rate one, it was the travellers who led the guides. Henceforth the amateur of mountaineering was to have an active and directing share in Alpine enterprise, and at need he might be independent. Climbing was no longer a laborious curiosity; it was a new form of active life and pastime, a sport embracing the noblest forms of outdoor pleasure, and exercising the powers of mind as well as body. In one word, the spirit of Conrad Gesner lived again. Hudson and Kennedy were likewise among the first, if not the very first, to destroy the superstition about 'mountain sickness' in some form being inevitable at great heights. Their party had read and heard of the 'nausea, vomiting, and drowsiness,' which everyone ought to suffer from, more or less, on Mont Blanc, and 'which are sometimes accompanied by bleeding at the nose, eyes, or ears, and by an utter prostration of strength.' None of them, however, felt anything of the kind; so they boldly set down their 'belief that these symptoms proceed chiefly from fatigue, though they may be increased by the rarity of the air.' In these latter days the practice of climbing without guides, which Hudson and Kennedy tried as a necessity in finding a new way up Mont Blanc, has been taken up for pleasure. Its development is too recent to fall within the scope of this chapter.

We are now on the threshold of the modern period. There followed in quick succession the explorations and adventures of which the history is written in 'Peaks, Passes, and Glaciers,' in Mr. Tyndall's 'Glaciers of the Alps,' in the early volumes of the 'Alpine Journal,' and in G. Studer's 'Ueber Eis und Schnee.' For many years extraordinary ignorance of the whole matter was still to be met with among professed instructors of public opinion, and not a few prudent British parents hesitated to admit Alpine climbing as a legitimate form of recreation for their sons. Nevertheless the early history of mountaineering was closed. The precursors had done their work, and another generation entered on the full career of conquest which their labours had made possible.[1]

[1] References and citations might easily have been multiplied in every part of this chapter. I have endeavoured to restrict myself to such as are curious and little known, or else may be useful as clues for the guidance of those who wish to pursue the subject farther. My thanks are due to Mr. W. M. Conway, Mr. W. A. B. Coolidge, and Mr. D. W. Freshfield, for information and suggestions, and to Mr. Coolidge also for the loan of some books not in the possession of the Alpine Club.

Training

CHAPTER II

EQUIPMENT AND OUTFIT

By C. T. Dent

Hamlet iii. 1, 81.

FOR an ordinary tour in the frequented Alps little special outfit is needed, but it is important that every article should be of the kind that experience has shown to be best adapted for the purpose. Many, however, now seek new fields for mountaineering, and for such a more elaborate equipment is needed. It has been thought better to enter at some length into the details of outfit that would be required by men who are compelled, necessarily, to combine travel with their mountaineering. This chapter does not pretend to give a complete list of all that may be needed for expeditions beyond the Alps. Opinions on points of detail vary so much that it is no easy matter to give a digest of the collective experience of many. At the same time the following remarks are the outcome of very numerous inquiries, and can fairly claim to represent the advice of a large number of mountaineers qualified in the highest degree to form a valuable opinion.

Clothing.—The best material is of medium thickness and of strong texture. Tweeds answer excellently well, or any woollen fabric in which the fibres are cross-woven. Materials such as serge and flannel are too apt to tear. The colour of the suit should be medium, of a 'subfusk' hue as the old Oxford University statutes phrase it; if too light every stain will show, while if too dark it will absorb heat. The mistake is commonly made of having the clothes too heavy. It is much better to carry a spare flannel shirt in case of great cold, than to be condemned to wear through the hot summer months a suit fitted for an Arctic voyage. Very thick clothes, too, take much longer to dry, and are intolerably heavy when soaked with rain. It is best not to select too pronounced a pattern, for in case of an accidental rent and consequent repair, the resources of the village tailor may not be equal to supplying a good match. It is not a bad precaution when ordering a suit of clothes to take also a piece, say one foot square, which may be used if necessary to patch with. In the Alps the costume adopted by our countrymen appears to be generally agreed upon. It consists of a Norfolk jacket, waistcoat and knickerbockers. The advantages claimed for the Norfolk jacket are, that it is loose, warm and comfortable, and can be worn with or without a waistcoat. In the writer's opinion, and speaking rather from a medical point of view, it is much wiser to have a coat that is not too thick to wear with a waistcoat. In tropical climates a man would be safer with a waistcoat and no coat, than a coat and no waistcoat. A very good form of coat, although no longer fashionable, is that made in the shape of a double-breasted pilot jacket. Whatever the actual cut, the coat should be fairly loose, especially in the armholes, and capable of being buttoned up close round the neck. It should be well provided with pockets, one or more being lined with mackintosh to contain small articles, such as matches or a knife, that might be damaged by wet. A 'game' pocket large enough to carry maps is convenient. All the pockets should button up close, the buttons being on the main material, and the button-holes on the pocket edge. Flaps are also required to all the pockets. If these precautions are not taken, half the loose articles will be lost when the coat is taken off and thrown over the arm. Put a number of loose coins in each pocket, button them up, turn the coat upside down and shake it; if none fall out the pockets are safe. The loop inside the collar, made to hang up the coat, should be very strong and well sewn on. On a hot day it is very comfortable to take the coat off and let it hang over the shoulder, threading the axe through the loop. A tab of doeskin stitched on linen is recommended. The waistcoat pockets also should button up and have flaps. The coat, waistcoat, and knickerbockers can be lined with flannel. The waistcoat certainly ought to be lined with this material. It is a mistake to line any of the pockets with washleather, as this gets very hard and rots if frequently wetted.

Knickerbockers are now generally in vogue, and have many advantages; they allow free movement of the knee, and do not get so draggled and dirty as trousers. A disadvantage is that they entail wearing long stockings, which take up more room in the knapsack. In ordinary knickerbockers a cloth continuation is provided, which fits close to the upper part of the leg; if this continuation is made to turn up, instead of down, as usually worn, the upper part of the stocking can be folded over it, and is kept up perfectly. The continuation is turned up inside, and is concealed from view. This device entirely dispenses with the necessity for any form of garter, and does away with the objection that many have to wear anything tight round the leg just below the knee. A strong strap and buckle are more convenient than braces. Some advise that the knickerbockers should be double-seated. The necessity for this depends on the climber's habitual mode of progression when descending rocks. The precaution ought to be unnecessary. All woollen clothes will be found to resist wet much more effectually if they have been dipped in a solution of alum.

Hat.—The majority of mountaineers adopt a light-coloured felt hat with a tolerably wide brim. A kind of material which is extremely light, owing to the fur being loosely 'felted,' is sometimes known as French felt. Some have a very high crown to the hat, and claim that it is cooler. On a very hot day, it is not a bad plan to make a little depression on the crown of the hat, and to carry in it a small lump of snow. Small ventilating openings on each side of the hat are a distinct advantage, and so is a piece of flannel between the lining band and the felt; a piece of blotting-paper is not a bad substitute, and some use oiled silk. It is more important for the brim to shade the nape of the neck than the face. An ordinary tall white hat is considered an eccentricity on the mountains, but as a matter of fact it is not so uncomfortable as it appears. Hats that fit close to the head are unsuitable. So are solar 'topees' or cloth caps, though some adopt this latter form of headgear with a white canvas covering which can be removed. The brim, when turned down and tied under the chin, should be wide enough to cover the ears, a useful precaution when there is a cold wind blowing. A hat-guard is required.

Boots.—Perhaps the most important part of the clothing equipment is the boots. It is certain that an ill-fitting pair of boots may spoil the whole pleasure of a tour, and that a sore foot may become much more than a very troublesome annoyance, and may even prove a source of risk; the man who constantly endeavours to spare his foot will neglect his foothold. The ordinary shooting boot is not the best form for mountain climbing. The soles should project at least half an inch all round beyond the uppers when the boots are first made, and it is in the writer's opinion equally important that the heel should also so project. After a few days of walking, it will be found that the uppers will be pressed down till they are nearly flush with the soles. The uppers should extend well above the ankle, and must be absolutely waterproof and made of the very best leather obtainable, for good leather is not only much softer, but much more durable. Laced boots are now invariably adopted. A weak place in many otherwise well made boots is to be found at the junction of the tongue with the upper leather : water can often get in at this point. Some advise that the seam crossing over the instep should be done away with, as the foot is liable to be chafed at this point. It is a difficult matter, however, to cut the upper leather in one piece so as to insure a perfect fit. To get over this difficulty, it has been recommended that the boots should lace nearly down to the toes as in a running shoe. This plan does not appear to the writer to offer any advantages. The places where a perfect fit are most important are the heel and the instep. The instep is the part most tried in descending. If the boot is too long, the foot will inevitably shift in it when descending, and if too short, the toes are bent up and get rubbed on their upper surface. In ascending, the tendons just in front of the ankle start up under the skin at each step, and if the boot is laced at all tightly the foot will be chafed at this point. If the boot is at all new, or in the beginning of the tour when the feet are still soft, it is a very good plan to vary the pressure by lacing firmly at the upper part in descending, and relaxing this somewhat in ascending so as to give the 'extensor' tendons free play ; but if the boot is allowed to become so loose that the heel rises in it, a sore is very apt to form just over the heelbone.

The tags are best made of leather, and they should be firmly sewn into the upper and extend down the whole depth of the boot. A great strain has to be thrown on the tags in endeavouring to drag on the boots when they are half-frozen, and the spectacle of a man in the early morning trying to pull on his boots with benumbed fingers by means of a broken tag is more instructive to others than pleasing to himself. The soles need not be of any very great thickness, only sufficient indeed to prevent the points of the nails from getting through ;

the heel should be two thicknesses of leather deeper than the soles. Too low a heel throws a considerable strain on the tendons of the back of the leg in ascending, and renders the boot less secure when going down steep soft snow slopes or loose scree.

The *boot-nails* are a most important part of the equipment. Bootmakers in England almost invariably use cast-iron nails. Wrought-iron nails, such as are generally used in Switzerland, are better. Various patterns are adopted in arranging the nails on the sole. A method that will be found to answer well is shown in the figure. One row of nails (A) with heads shaped something like the caps worn by coal-heavers, are required for the very edge of the sole, and should pass entirely round it from the waist of the boot forward. If their edges overlap,

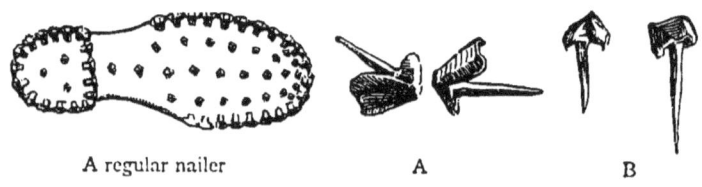

A regular nailer　　　　A　　　　B

they will be found far less liable to knock out. These are the most important nails in the boot, and with a row round the edge a climber can go anywhere in perfect safety, but it is better to add others (B) as shown in the drawing. The heel of the boot should be similarly garnished, a row of nails (often omitted) extending across the front part. Some add a few light nails in the waist. With English-made boots the nails are very readily knocked out in rock climbing, and still more in walking over loose moraine or scree. More commonly this is due to bad walking, but sometimes it is accounted for by the manner in which the nails are driven into the boot. The leather ought to be perfectly hard and dry when the boots are nailed. If the sole is thoroughly well hammered before any nails at all are put in, they will be found to hold much better. Another plan is to bore a small hole in the boot and to put a drop of water in it, and then drive in the nail, which rusts and holds more firmly. Some bootmakers render the nails secure by driving them through the bottom layer of the sole, and then clinching the points. The nailed sole is then fixed in the usual way. There must be plenty of leather always between the nail points and the foot. Yet another plan is to drive the nail in obliquely with a hammer, and then to finally clinch it with two or three vertical blows. The head of the nail for the sole (B) should be bent down a little, and have a sharp edge and points which can thus be driven a little way into the leather. A mountaineer will give as much thought to his boots as to his ice-axe, and if he wants to get the maximum amount of service out of them, he will wear them once or twice just after they are made and then put them away for at least twelve months. A well-made pair of boots so preserved can then be made to last three or four seasons. When in use in the mountains the leather should be well greased every day. Various ointments have been specially devised for this purpose. Vaseline, mutton fat, or common yellow soap answer perfectly well. In camp, after a wet walk, it is a good plan to stuff the boots with hay or dried grass at night. A greasy boot is not a very agreeable article to pack, and it is a good plan to take a couple of stout linen bags for the purpose. Do not give your boots to the guides to pack, or they will generally stray into the provision bag. In the crowded hotels of Switzerland the method adopted by the officials of the boot department of chalking the number of the room on the sole does not always answer or prevent confusion. The owner's name may advantageously be stamped conspicuously inside the boot. Warm slippers, as a change, are more than a comfort; they are a necessity.

Stockings should be hand-knitted and as thick as possible. Those made of undyed wool are much the best. Undyed stuffs are often considerably lighter than the same materials dyed. Dyes, even in these sanitary times, occasionally irritate the skin, and will worry any little excoriation. The elaborate patterns, after a long wet day, are apt to leave a more or less permanent impression of their design on the legs and feet. People whose skin abrades easily will often do wisely to wear two pairs of stockings even at the risk of masking the symmetry of their legs. A thin sock worn underneath the thick stocking will often save sore feet; a little play and rub is allowed between the two. If a wrinkle in the stocking, or knots (and such things should never be in a properly hand-knitted stocking),

The apparel oft proclaims the man

chafe the foot at all, it is a good plan to change the stockings at once, putting the right stocking on the left leg and *vice versa*. Stockings, just as much as boots, should fit accurately. The usual fault is to have the foot too short. The toes are bent up, and abrasions follow.

Garter.—A simple piece of Berlin wool wound three or four times round the leg is the best form of garter.

Shirts should always be of flannel and made to fit loosely round the neck, even though this method be temporarily

EQUIPMENT AND OUTFIT

opposed to the dictates of fashion. A flannel shirt without a collar will look untidy, even though worn by a bishop, but if provided with a collar of fine-wove linen, the wearer can always look respectable without having any discomfort. In the Alps, at any rate, some concessions must be made to the possible prejudices of others, and no man has a right to go about looking like an ostler unless he happens to be one. A shirt of white silk is extremely comfortable, and silk underclothing generally, as a change, is the most convenient, as it packs up into an extremely small space. *Pyjama sleeping suits* are on many grounds the best to take, made either of thin flannel or silk. They render the wearer safe to a great extent from the terrors of damp bed-linen, and, at a pinch, will serve as a change of clothes. Pyjamas and a sleeping bag have obvious advantages in some inns. A woollen knitted jersey, large enough to go over the waistcoat and under the coat, is a useful thing in a bivouac or for sleeping out. Shetland wool has the advantage of being extremely soft, light, and warm, and packs up in a small space. 'Cardigan' jackets are warm but bulky. A most valuable addition to the clothing store is the ordinary flannel cholera belt, such as is commonly used in tropical climates. Excellent for sleeping in at nights, it is invaluable for the slight derangements that are apt to follow a chill. A woollen comforter is a more bulky and less efficient substitute.

A *waterproof coat* is unhappily a necessity in the changeful climate of the mountain regions. It should be made very loose, and the sleeves should be long. If the smooth surface extends for an inch or two inside the sleeve, or if the edge of the sleeve be tucked inwards for an inch, the wet will be found to drip off. A little waterproof cape, made very loose, with sleeves very wide at the armholes and only just long enough to cover the coat, will be found useful in climbing on the high mountains; a longer garment hampers the movements. A wrist-strap or elastic band is required. The arm can be drawn back through the wide sleeves, and the pockets are thus accessible without unbuttoning the coat. In order to preserve a waterproof cloak, it should be spread out until it is entirely dried, or it will be found to crease and the indiarubber will perish. The silk waterproofs made nowadays are more expensive and last longer.

An excellent form of waterproof coat, especially serviceable in countries such as the Caucasus, where mountaineering implies travel also, is a large waterproof sheet made after the fashion of a poncho. This consists of a square sheet of mackintosh, measuring say 8 feet by 5 feet, with a hole in the centre large enough to allow the head to pass through. The opening can be closed up by a flap which buttons over it. On a very wet day the poncho will be found less irksome than a closely fitting coat. A poncho, too, keeps the traveller on horseback perfectly dry down to his boots, and also protects the saddle-bags. The garment can be used as the ground sheet of a tent, and is useful to protect any baggage left outside the bivouac.

A knitted woollen Balaclava (Templar) cap is useful for sleeping out, and is worn by some for rock climbing. Unless loose, however, it is apt to cause headache, and at the best is a poor protection against the sun. Where great cold is to be provided against, as in winter climbing, the 'Dundee whaler's' cap may be required. This is made of leather, lined with wool, and has a fold that can be turned down over the ears and back of the neck, together with a peak. The cap is rather heavy, but very warm.

Thick gloves are required on the mountain-side. A perfect glove for mountain purposes is still a desideratum. The closely knitted woollen gloves usually adopted are warm so long as they are dry, but they hold the wet terribly. They are improved by a piece of flannel or box cloth sewn into the back. Leather gloves get hard when cold. Waterproof gloves have been devised, but are not very satisfactory. Gloves in any case should fit loosely, and they are much warmer when the fingers are all together, the thumb being in a separate pod. They ought to come well up over the wrist, or even be loose enough to encircle the coat-sleeve. Some find a pair of old kid gloves, with a moderately thick woollen glove worn over them, a very warm covering, and gloves have been suggested with a separate indiarubber case to go on over the whole. Fur-lined gloves are bulky, and too thick to enable the wearer to take hold with his hand. In no form of glove that has yet been devised sufficiently warm for protection against great cold can the climber get a really good grip on the axe. On rocks the handhold is never so safe when gloves are worn, and yet it is often necessary to wear them when rocks are very cold, for when the fingers are numb the hand-hold is still more insecure. If gloves are worn with fingers, care should be taken that no pressure is made on the fingers just at the web. A wet glove is almost as bad as no glove at all, if the cold is extreme and the wind high. If it be necessary to plunge the hands into deep snow when on a high ridge, it is best to take the gloves off for a moment, so as to be able to replace the hand into a dry covering afterwards. It is often worth while to carry a second pair.

Gaiters are needed for walking in deep snow, and they are of use when walking on fine scree or moraine, to keep stones out of the boots. They should come well up to the knee, though not above it, and fit closely over the foot and round the heel. The strap or string usually provided to pass under the waist of the boot and keep the gaiters in place is useless, as it gets cut to pieces at once on rock. Only a well-made chain will stand the wear. Coarse woollen stuff or box cloth is generally used. Box cloth is the better material, but rather hot. The portion of the gaiter that covers the boot is best made of box cloth. The gaiter may either be made to button (horn buttons are best), or strap, or lace at the side ; but, whatever the pattern adopted, it should always be possible to put the gaiters on without taking off the boot. Some use gaiters of thin leather. They can be kept soft by rubbing with yellow soap. It is as necessary that the gaiters should fit as that the boot should fit, and it is never wise to trust to the chance purchase of a suitable pair. The fore part of the gaiter has especially to be so cut as to lie close to the boot over the instep, for it is at this point that the snow will work in, and if the snow once gets under the gaiter it will soon find its way over the top of the boot. If the gaiters really fit well, it should be possible to walk for hours in deep snow without getting the feet wet, and as long as the feet are dry there is little danger of frost-bite. For walking on loose rubble or over an ordinary glacier high spats answer perfectly well, and are considered at present to give an elegant appearance to the leg. *Knitted or canvas anklets*, which just cover the top of the boot and extend an inch or two up the leg, will effectively prevent small stones from getting into the boot, and will also keep out a good deal of wet. These anklets are often worn under gaiters, and in deep snow the double protection will be found an advantage. 'Putties,' i.e. list bandages such as are worn in India, make effective snow gaiters.

Climbing outfit for ladies.[1]—Women who climb should, like men, dress in such a manner that they are protected from extremes of either heat or cold. Every garment should be of wool, and the softer and lighter the material the better. The only exception to this latter point should be the skirt, and this will be found most serviceable if made of cloth, rough in texture and as thick as the wearer can get, provided it is not clumsy. A closely woven tweed is suitable. A small check pattern mends neatly if torn. The skirt should be made perfectly plain, except for a deep border of stitching. The pockets should be large, plentiful, and in definite places; one can be nearly in front, two others quite at the back. All the pockets should be outside, and have flaps to button down. Several buttons are desirable on each flap, so that nothing may slip out when sleeping in huts. Three yards round the hem will be found a good width for a skirt, which can be of an ordinary walking length. When climbing the skirt must, whatever its length, be looped up, and therefore it is easy to have a skirt which, in the valleys or towns, does

[1] For this section I am indebted to Mrs. Main and Miss Richardson.

not look conspicuous. For looping up the skirt, the following simple plan is effective. An extra belt of strong ribbon is put on over the skirt which is then pinned to it in fishwife style. The length is arranged according to the requirements of the occasion. One safety pin attaching the two sides and another fastening the back, the hem being pinned on to the outer belt, do the work. A belt with pockets attached might be the best arrangement of all for a woman, as it would enable her to have her skirt of a neater and smarter cut. Some employ loops and buttons for shortening the skirt. This arrangement is objected to by others, on the grounds that the weight is not distributed so well as by the belt arrangement, and that pieces of the material are apt to be torn out. Dark blue or dark grey are both suitable colours for a climbing skirt.

Under the skirt, a short under-skirt reaching just to the knees, of the same colour, but of a lighter material, can be worn. On neither skirt is a bordering of mackintosh practically of the slightest use. A rough cloth coat lined throughout with silk may be taken in case of cold.

Knickerbockers of waterproofed cloth lined with flannel will probably be found best, and, though a little heavy, are a great safeguard against a chill when sleeping out.

The bodice is an important part of the outfit. A soft grey flannel blouse, high in the neck, long in the sleeves, and loose, is the best for both heat and cold. A short jacket of thick cloth, and lined with the lightest possible fur (squirrel is the best), with the collar rising high all round the neck, is, though unusual, extremely comfortable. The fur should not come to the edges of the jacket; it is thus less liable to get wet.

A light grey felt hat is cool; each wearer must decide on the shape she prefers. A knitted helmet, which can be pulled over the whole head and face, the eyes only being uncovered, is a necessity in very cold weather. A large silk handkerchief is useful to tie the hat on in a high wind.

Woollen stockings (one pair on, another pair in the knapsack), thick, water-tight, nailed mountain boots, and cloth gaiters to button or to pull on in the Chamonix style (hooks and laces are apt to catch in a skirt) are all essential. Gloves should invariably be of wool, and of the shape worn by babies, the fingers being enclosed in a bag, and the thumb only having a separate casing. Let the gloves come well up the arms, and take at the very least two pairs with you on an expedition. A very fine woollen mask to protect the face is much pleasanter to wear than one of linen.

A more extensive outfit is required on a tour when access cannot be had to heavy luggage for several days. The climber may have to spend a few nights in the more civilised of the Alpine centres, or perhaps twelve hours may even be passed in such places as Geneva or Turin. It is necessary to be provided against such contingencies, and if a little thought and trouble are given to the matter, neither the weight nor the bulk of the extra garments necessary need be great.

Silk (only to be worn when not climbing) can be substituted for wool for the under-clothing, of which two complete changes are desirable, not including what is worn. A dark blue or grey silk blouse can be worn with the climbing skirt in the evenings, and a small dark felt hat, which will fold flat, and a pair of *gants de suède* will help to do away with the stamp of the climber. Leather soles, without any heel whatsoever, put to a pair of neat black laced shoes, will pack flat and take up very little room. The whole weight of the bundle (which can be tied up in a large silk handkerchief) need not exceed 4½ lbs., including such essentials as soap, a comb, pocket-handkerchiefs, and other small things, which the experience of each climber will suggest.

Knapsack.—The old form of *knapsack* is now almost entirely superseded by the *rücksack*, which is not only a much more convenient article for carrying, but has many additional advantages. There is no need here to describe the thousand-and-one forms of knapsack that have been invented. All are open to the same objections. The weight is carried too high,

AN ICE SLOPE

EQUIPMENT AND OUTFIT

so that the balance is interfered with, while the strain falls uncomfortably, and the shoulder-blades being pressed upon have not the free play which is required in climbing. For an ordinary pedestrian tour, implying little more than tramping along high roads, the knapsack may serve, but even for this purpose it has no advantages and is far less capable of keeping its contents dry. In countries where the mountains have to be approached by riding, the rücksacks are most convenient, for they can be slung in pairs across the saddle. The writer has known them, when fording rivers, submerged under the water, and yet the clothes inside were kept perfectly dry. No knapsack that has ever yet been made could have stood this test. All knapsacks have many straps liable to be torn away from their attachments. The rücksack has one, to close the neck, on which no strain is thrown, and even that can be replaced by a cord if desired.

The rücksack is a form of pack that has for a long time been generally employed in the Tyrol. It consists of a simple bag which can be closed by a string at the mouth, and which is carried much lower down than the knapsack. The greater part of the weight lies in the hollow of the back, and thus, some vertical support being provided, part of the strain is taken off the shoulders. The carrying straps are connected to the neck of the bag by a swivel and ring, or they can be attached by the ends of the cord closing the sack. Thus the position of the bag on the back can be readily shifted a little. The skin over the lower part of the back perspires less freely than that over the shoulders, and this gives an additional advantage in carrying. Moreover, as the weight of a rücksack falls chiefly at the lower part of the back, it is much easier to keep the balance. In a rapid descent the flapping of the knapsack on the shoulders is very uncomfortable. The rücksack, being soft, is moulded more or less to the shape of the body, and not liable to this drawback. Three different sizes have been recommended. The first measures twenty inches in length and seventeen inches in width, and has two outside pockets;

the second is twenty-four inches in length and twenty-two in width, and has two inside pockets; the third is twenty-eight inches long and twenty-two inches wide, with two inside pockets. The first form, the smallest, suffices for carrying the clothes and such articles as a custom-house officer once described to the writer as 'les petits nécessaires de voyage,' a conveniently elastic phrase. Many climbers content themselves with little more than this for a tour. The second is rather more capacious, but suitable for the same purposes, or may be used as a provision bag; while the third is useful as a provision bag for a large party, and is adapted to more extensive mountaineering expeditions than are usually undertaken in the Alps. A change of clothes for three men can be carried in this largest size. The small size is made of canvas with a loose waterproof lining, the second and third of the green Willesden canvas. This material is only waterproof to a limited extent, and these bags, if required for clothes or for articles which it is necessary to keep dry, had better also be provided with a mackintosh lining. Locking straps to close the bags, or padlocks (which should all have the same lock, each member of the party being provided with a key), are useful additions. It is difficult for even a most skilful packer to wedge a pair of greasy wet mountain boots into the interior of the sack without damaging the other contents, and accordingly pockets are provided on the sides of the small sacks each capable of holding a boot, or, if such an article be required, a bottle. Guides and porters in the Alpine districts prefer the rücksack to any other kind of pack. In other countries too the bearers have not been slow to recognise its advantages. It is not so easy to find any required article in a rücksack as in a knapsack. This drawback can be fully compensated for by a regular method of packing, the same articles being always stowed away in the same place.

For carrying a few small articles, such as sketching materials or scientific apparatus, the *havresac* is even more convenient than the rücksack. The havresac is slung by a single loop of leather or webbing worn on the yoke principle ; the arms are passed through the loop as in the action of putting on a coat. The pack will then be found to lie close against the back, and will not flap about. Articles such as wine bottles, field glasses, and the like are best carried on this principle, and they will be found less liable to damage in rock climbing than if carried in the rücksack.

Provision bags.—A number of coarse linen bags of various sizes are a great convenience for packing provisions. In mountainous countries, such as the Caucasus, such bags are indispensable. Small oiled silk bags are useful for salt and sugar. Saccharine tabloids can be carried in a little box in the waistcoat pocket. Horn or celluloid boxes are good for butter.

A mountaineer who uses a tent or camps out in the open will find a small linen or mackintosh bag useful, to contain at night all the loose articles he usually carries in his pockets. If he is not so provided, all small articles are best stowed away in one of the trouser pockets, and secured by a string or elastic band.

The climber will exercise his own discretion as to the kinds of *preserved foods* he carries. 'Self-cooking' tins are very portable and convenient. They furnish a hot meal in about ten minutes, without the necessity of making any fire. With one of these tins in the pocket, an opener to reveal the contents, and a spoon to consume them, a climber may go out for a day without any other provision save a little bread. By boring a number of holes close together through the cover, the tin can be opened even when the proper opener is not to hand.

Bottles are too heavy and too liable to get broken to be conveniently carried. Sybarites, whose tastes or constitutional ailments require champagne or wines that cannot be decantered into gourds, will find the outside pockets of the rücksack the best place for the bottles : the more they are in view the less likely they are to be broken, and a guide or porter will always be extremely careful of a bottle committed to his charge. Wine bags made of indiarubber, sufficiently large to carry two or three bottles, may be slung over the shoulder or carried yoke fashion. They are not very easily cleaned, and it is difficult to pour the wine out of them, but they get less bulky as the contents are consumed. Various kinds of gourds are employed. The flat tin 'flacons' that can be procured in Alpine villages are not much in favour. Some gourds are made with a stopper in two pieces, so arranged that when the top part is removed the bottle cannot be opened. Spirits of wine for cooking, or brandy, cannot be carried in leathern or indiarubber wine bags. A flask is best for brandy or liqueur. Vulcanite gourds covered with felt, made in various sizes, are excellent. A strap passes through the base of the drinking cup, which is placed upside down over the cork, and secured by a small padlock. The wine is thus locked up and the drinking cup can never be forgotten. By damping the felt and exposing it to the sun or wind, the contents can be kept cold. If warm liquor is put in the bottle the heat will be retained for a long time.

Leathern drinking cups folding up flat are very convenient, but they flavour their contents rather strongly when new, and should be soaked for some days in constant changes of water before use. Large collapsible drinking cups, sufficient to hold a brew of wine and glacier water, which, when flavoured with lemon, sugar, and a dash of brandy, is known on the mountains as punch, are sometimes carried. With mutual forbearance enough can be made in one of these for a party of five or six. Metal drinking cups which fold up are far pleasanter but more bulky.

Snow spectacles are indispensable. The glass should not be green or blue, but of a tint which is described usually as neutral or smoked. The latter are most efficient in relieving the eyes from the glare of snow, and are less opaque; there is also less loss of colour in objects seen through them. The right tint is shown by the colour of the shadow of smoke

EQUIPMENT AND OUTFIT

thrown on to a piece of white paper in a ray of sunlight. These spectacles are of the shape known as goggles, made with fine wire gauze. The piece of elastic usually supplied to fit over the bridge of the nose is apt to press and cause discomfort; velvet is better. The edges of the wire gauze cage must be covered also with velvet. See that the spectacles fit properly, and do not pick up the first pair you come across. The goggles are fastened on by a tape or elastic tied round the head. If a glass is broken an efficient substitute is a piece of card or wood in which a fine horizontal slit is cut. This is the form of protection adopted by the Esquimaux. The Tartars of Ladakh plait eye protectors from the tail hairs of the yâk; or pull out wisps of this hair and fasten them loosely across the eyes when traversing snow. Blackening the eyelids, and the skin included in the ring of bone encircling the eye, prevents the worst consequences of snow ophthalmia. Burnt cork would answer the purpose. If the glass of one goggle is broken, shift the sound glass every ten minutes to the eye that has been exposed. If a pair of goggles is lost on the climb, each member should take his turn to do without them; but no man should be without glasses for more than a quarter of an hour, and he must keep his eyes well off the snow while unprotected. It is cruel to let the leader go without spectacles, as he above all men must keep his eyes on the snow. Neither is it right to let a porter go unprotected. Either he will star-gaze and go into a hole, or else keep too close to the man in front of him, so as to be in shadow. Even plain glass is much better than no glass, and will cut off many of the irritating rays. A bit of horn or talc cut from the lantern to replace a broken glass is a tolerable makeshift; with a strip of a handkerchief, with two holes cut in it and two discs of talc, a temporary pair of goggles can be fashioned, but the slit card will answer better.

A *knife* is, of course, an essential item; it should be provided with a corkscrew, a button-hook, a screw-driver and a leather punch, in addition to a tin-opener (which must lock), and two blades (which had better lock). The best form of tin-opener has the fulcrum on the same vertical plane as the cutting part, so that the two resemble a partially opened pair of scissors. The cheap knives that can be bought at any village store will open tins. Highly tempered steel breaks readily. If the knife has a loop and chain, it is less likely to be left behind after a meal.

A *whistle* for signalling to a guide or companion is often valuable. By adopting the Morse code it is easily possible with a little practice to communicate by means of the whistle when the voice cannot be heard.

A *field-glass* is more bulky than a telescope, but gives a better idea of the modelling of a mountain. Aluminium glasses are extremely light, but expensive, and are easily damaged. The cap should be fastened to the telescope, or the guides, who are perpetually in search of chamois (for which purpose they constantly borrow the instrument), will soon lose it.

Matches.—The best kind of matches for lighting in a wind are the sulphur matches usually sold on the Continent. With wax matches, however, a light can always be obtained. If a strong wind is blowing, a wax match will be found more efficient if one or two fibres are separated and twisted loosely round just below the head. If you fail to light the first match do not throw it away, but split it up into its component fibres, crumble them up, and twist them round another match just below the head. Tinder paper can be made by rubbing some tobacco ash into a piece of soft paper. The piece is then twisted up. It is worth while to recollect that a light can always be obtained by crumpling up a piece of newspaper and lighting a match inside the folds. A burning glass will always provide a light for a pipe or cigar if the sun is out. Chinese joss-stick can be easily started by a burning-glass, but it requires considerable knack to wave the stick properly so as to get 'a head on it,' and then, by a very gentle quick puff, make it burst into flame. Flint and steel also requires practice. The sparks fly upwards, proverbially. The lens of a telescope can always be used as a burning-glass.

EQUIPMENT AND OUTFIT

Needles, thread, and buttons have their obvious uses. A small pair of pliers, combined with wire-nippers and a piece of copper wire, can be turned to a thousand-and-one uses in making repairs. Spare boot nails should be carried in journeys in remote mountain regions, but are not needed in the Alps. Nails with four-pointed heads to screw into the boots are of little use, but are sometimes taken.

The few medical stores which may be necessary are described elsewhere.

Scientific apparatus.—For ordinary climbing purposes this part of the equipment need comprise but very few articles. The most essential is a compass; the card compasses answer best, one half of the dial being black. Luminous dials are useful. Each party should be provided with at least two compasses. There should be a catch provided so that the needle can be locked, or it will get out of order during a rough descent. When endeavouring to trace the way through mist the compass has perpetually to be called into requisition, and in very cold weather, or if rain is falling heavily, people as a rule are too indolent to consult it as often as they ought. It is very hard in thick weather to keep a true line for more than a few yards on a great undulating snowfield. A good plan is to have a compass fitted into a leather bracelet worn on the wrist, such as ladies use for watches. The card can be made to lock by turning a milled edge on the top of the compass. The card turns on an agate centre, and the north point is of pearl and can be readily seen at night. A prismatic compass is far the best kind to carry, and can be used for many more purposes than merely determining the right line. Very little practice is required to enable the observer to use it accurately. A prismatic compass and a clinometer can conveniently be carried together in one small case slung on the shoulder. Care must, of course, be taken in using the compass not to consult the instrument when holding it close to the axe-head. Occasionally magnetic rocks occur—as, for instance, on the well-known Riffelhorn and some of the hills in Skye—and will cause deviation of the needle. All compasses ought to be tested beforehand, and compared from time to time with standard instruments.

Clinometers are not of much value in the Alps, where the mountain regions have been most minutely surveyed. In less known districts, very good rough observations can be made as to comparative heights.

A circular clinometer will give very accurate readings. A popular little instrument, known as Abney's clinometer, is rather more difficult to work with, but will repay the trouble taken to learn its use.

Apparatus such as the theodolite, plane-table, hypsometrical instruments, and the like, need not be here described, for the average climber who pursues mountaineering chiefly as a sport is only concerned with the use of such instruments as may assist his way, or enable him to give an intelligent record likely to be helpful to others.

Aneroids, graduated up to 18,000 or 20,000 feet, are useful instruments, but only under certain conditions. In the Alps the heights have been determined by other and more accurate means so fully that it is not worth while to carry an aneroid merely to ascertain how, even with the best instruments, the reading will differ from the true height. Mr. Whymper[1] has made an elaborate inquiry into the behaviour of aneroids under diminished pressure, and finds that, if subjected to low pressures for long periods, the readings will diminish, and the ordinary test employed for determining the errors is of little or no value, as sufficient length of time is not allowed. A very rough idea, however, of the height gained in a given space of time suffices for the ordinary climber, and this a good pocket aneroid will give him. The less the time that intervenes between the readings at the higher and the lower stations, or the lower and the higher stations (as the case may be), the better will be the result.

---

[1] *How to Use the Aneroid Barometer.* By Edward Whymper. London: John Murray. 1891.

EQUIPMENT AND OUTFIT

Those who mountaineer in regions where the heights are undetermined must not depend on aneroids alone. Mercurial (Fortin) barometers must be taken as well. Still, Mr. Whymper is of opinion that aneroids are of real value. His work must be carefully studied by those who plan a tour in unsurveyed mountain regions. The mountaineer who is unable to give, with some approach to accuracy, the heights of new mountains which he has ascended will be thought little of. That much is always expected of him ; but the information cannot be acquired by merely buying a mountain aneroid at a shop, however first-rate the maker, and quoting the readings noted, without making somewhat elaborate deductions for errors. An aneroid should be held horizontally when the reading is taken. Ascending and descending readings should be noted down. Better results are likely to be obtained from the means of ascending and descending readings, than from ascending or descending ones alone. The readings of a single instrument in a prolonged tour are hardly likely to be worth recording. Frequent intercomparison with other aneroids and with the mercurial barometer is needed.

A small *pocket divider*, a rule graduated in inches and centimètres, and a horn *protractor* are of great use in working out distances on the map.

Maps are best cut up into small sections and mounted, and these should fold up into a size that can be conveniently carried in the pocket. When engaged in exploring minutely a limited district, it is a good plan to have all the sections separate, each being the size of a sheet of note-paper ; these can be numbered, and then stored in an oil-skin envelope like a pack of cards. The required sheet is placed uppermost, and in wet weather the climber can make out what he wants through the oil-skin without damaging his map. Maps have been printed on gutta-percha, but there is no special advantage in this, and they weigh more. A piece of tracing paper included in the map case enables corrections and notes to be made readily.

Tents are now rarely necessary in the Alps, but for those who wish to render themselves independent of huts they are requisite, and in remote mountain districts, such as the Caucasus or the Selkirks, essential. The best pattern for Alpine purposes is that devised by Mr. Whymper, and described in his 'Scrambles amongst the Alps.' The main features are that the floor and the sides are continuous, and that a single rope and two tent pegs suffice. No ridge pole is used. The tent can be pitched on snow if this is well stamped down over the pegs. The guy rope may be secured to a big stone or to an axe driven obliquely into the ground. If fastened to an axe the rope should be attached by a double clove hitch, or it will slip up. In the shallow soil found in high mountain regions the 'dáterâm' principle (*cf.* Galton's 'Art of Travel') is inapplicable. The best material of which the tent can be made is Willesden canvas, which stands wet well and is rot-proof. In fine weather a ground sheet is unnecessary if the tent is made of this canvas: still it is a luxury and conduces to warmth. A poncho answers perfectly for the purpose. Large pockets in the inside of the tent are a great convenience for stowing away small articles at night. A small loop fixed to the centre of the roof is useful for suspending a lantern.

The poles are usually made of ash, but bamboo is lighter and answers admirably. The poles are jointed in the middle for convenience in packing. Both ends forming the joint should be brass shod. This adds weight, but if wet, the wood will swell so much that it is often a matter of the greatest difficulty to separate the pieces. The joints should fit easily and all the pieces be interchangeable. There is no need to have long spikes; an inch and a half is quite sufficient. A tent having a floor area of 6 feet square will just accommodate four at a pinch. If the night is very wet the position of the two outside men is the least enviable, for the wet will come through if any pressure is made on the sides. Luggage stored in the tent must not be pressed up against the sides with the object of giving more space inside if rain is expected. A tent with a floor area of 7 feet square will accommodate four, and a 6-foot tent will really only suffice for three if the party proposes to spend several days under canvas. The extra 12 inches adds little to the weight and much to comfort, but it is not so easy to keep the sides properly stretched. The tops of the poles at each end are fastened together by screws, which are apt to get lost, and no good makeshift can readily be substituted. A good device is to make a small pocket at the bottom of the window-flap to hold three screws, one being a spare screw. The pocket can be tied up, and the weight keeps the flap down when it is desired to close the window. A 7-foot Willesden canvas tent with bamboo poles, packed in a bag measuring about 44 inches in length, weighs, complete with spare pegs, &c., about 28 lbs. The bag containing a 6-foot tent measures about 42 inches, and weighs but little less. A thick ground sheet weighs about 8 lbs., but a lighter material will answer.

The 'Mummery' Tent

Mr. A. F. Mummery has devised a small tent which is extremely ingenious and portable, weighing only $3\frac{1}{2}$ lbs. No poles are necessary, two ice-axes furnishing all that is required.

The area covered by the tent is 6 feet long by 4 feet wide. A ground sheet is required. The tent gives shelter, but is necessarily only the height of an ice-axe. A thin mackintosh ground sheet suitable for the tent weighs 1 lb. Mr. Mummery used this tent during a high mountaineering tour in the Caucasus, and found it to answer perfectly.

Sleeping bags in fine weather render the use of a tent almost unnecessary. A simple bag 6 feet long, made of striped Austrian blanket having a circumference of 90 inches, will suffice, with a string round the neck to draw it up close round the head. These bags, however, are very bulky when rolled up, awkward to carry, and in case of wet are useless. The best form of sleeping bag is that devised by Mr. Tuckett (figured in Galton's 'Art of Travel'). It is made partly of mackintosh lined with blanket felt, but Jaeger's woollen material answers much better as a lining, as it is lighter and packs in a smaller compass. The chief advantage of the Tuckett bag is that it folds up and can be carried as a knapsack, and that when folded the mackintosh only shows, so that the lining does not get wet. The bag weighs about $7\frac{1}{2}$ pounds. The original measurements given are rather short even for a man of average height. In default of a sleeping bag the ordinary shepherd's plaid is much more portable, much warmer, and adapted to many more purposes than the miserable blankets the innkeepers of Switzerland are prone to hire out to travellers. It is desirable to take a lesson from some well-informed person of the various ways in which a shepherd's plaid can be put on.

Some luxurious climbers carry inflating mattresses. Perfect comfort can be obtained by a circular air pillow 20 inches in diameter, with a depression in the centre to admit the hip bone. The air cushion should be put inside the sleeping bag. An air cushion makes a very uncomfortable pillow, and a coat rolled up or even something much harder will answer better.

Canteen.—In the Alps it is usually possible to borrow (for a consideration) any cooking apparatus that may be required. A great variety of canteens are made, most of which contain far more than is necessary. The 'Rob Roy' cooking apparatus, though small, is a very admirable one. One of the early parties to the Caucasus took no other, and found that it was quite sufficient. The canteens are usually packed in round cases which are very awkward to carry. The writer has found that a wicker basket, flat at one side, is convenient for packing, and it will stand a good deal of knocking about. The cooking appliances should fit easily into this. The makers have a detestable trick of constructing the outside case in such a way that it is a perfect puzzle to fit the various pieces in. A list of the various articles, fastened to the inside of the lid, enables the contents to be checked off every time the canteen is packed or unpacked. The frying-pan and stew-pan should have a handle at least 13 inches in length, made with a joint in the middle : with a handle of the length usually supplied, equal only to the diameter of the pan, it is impossible to cook over a wood fire without burning the hand.

Lantern.—The requisites for a mountain lantern are, that it should be strong and capable of being folded up small when not in use. Block tin is the material generally employed, and the transparent portions are made of plates of talc or horn. The socket should be adapted to any size of candle. If the lantern is of square shape—and this is the form usually made —at least three of the sides should be transparent. The candles should be of the hardest obtainable variety; those known as carriage candles answer well. In case of loss or damage it is well to carry two or three spare talc plates. The best pattern is probably the 'Excelsior Mountain Lantern,' adopted by the Italian Alpine Club. Lanterns specially made in Vienna for mountaineering purposes, known as the Austrian pattern, are very light and portable, but not strong enough for rough work. Both of these lanterns fold up flat and will go into the breast pocket. For travelling in more remote countries a stronger instrument is required. The pattern known as the 'Beresford' is good but heavy. The case into which the lantern fits may be made large enough to take a

F

few spare candles. A good substitute for a lantern is a clear glass bottle, of which the bottom has been knocked out. To do this neatly, fill the bottle with water to the depth of an inch or so and set it on hot embers. The glass will crack at the level of the water. The bottle is turned upside down and slung by a string, the candle being inserted into the neck. For early starts a lantern is indispensable, and if the expedition is likely to be a long one or soft snow is expected, the more that is done in the early hours by lantern light the better. If there is only one lantern to a party of three, it is usually carried by the second man. Four or five persons will go much faster over rough ground if provided with two lanterns. If any member of the party is unable to see well in the dull early light, he had better carry the lantern. The next best place for him is immediately in front of the man who carries it. It is often desirable to rope while the lantern is in use, as time is saved.

The best *Alpenstocks* are made of well-seasoned ash. Although these aids are now almost out of fashion, and it is customary for everyone in the mountains to carry an axe for all purposes, they are yet most serviceable articles for an ordinary morning walk, in which the traveller has no ambition to do more than keep to some kind of track. The alpenstock or 'helping stick,' as it was once ingeniously termed by someone not very conversant with the German language, should taper slightly from base to summit. A piece of wood should be selected in which the grain is straight and free from knots. The top may expand into a slight knob. For a person of ordinary height a stick 4 feet 6 inches in length including the spike is a good size. The top of the stick may have a diameter of an inch, and the thickest part of the stick, at the bottom, of an inch and a quarter. Such a staff will easily support a weight of twelve stone on the middle when the two ends are supported. The spike, made of a single piece the same shape as that recommended for the ice-axe, is riveted through the wood. The mistake is commonly made of having the spike of the alpenstock too sharp, and the shoulder between the spike proper and the hollow socket into which the end of the stick is inserted should not be cut square. Neither should the spike taper from base to point. In many of the early expeditions across snow passes—the first passage of the Schwarz Thor, for example—and up high peaks in the Alps alpenstocks only were employed, and when steps were required in ice or hard snow these were laboriously made with the point of the alpenstock by a series of little prods. This is a very insecure and tedious proceeding. Alpenstocks are sometimes provided with a sort of chisel point to facilitate step-making, but this has little advantage over the plain spike. No proper step can be quickly made with an alpenstock. In any expedition in which it is the least likely a step will be required, an axe should be carried by the party. The various uses to which the alpenstock can be applied will be described later on. It is a fact that everything which can be done with the alpenstock can be done also and better with the axe.

In the early days of mountaineering in Switzerland, a man who flourished about on a mule path or on a high-road carrying rope and axe was looked upon much as a new boy at Eton would be if he appeared in 'Sixpenny' clad in white flannels; but the times have changed, and little children are seen now playing about the hotels with toy axes. The Alpine traveller's ice-axe is as familiar an object as the Londoner's umbrella. This being the state of things, no apology is needed for dwelling a little on the proper use of axe and rope.

The *ice-axe* is as important an article of the mountaineer's special equipment as the rope. Axes are made on many different patterns, but these hardly differ more than the so called variations of routes up a popular mountain in a climbing centre. Some prefer a round stick, others one that has rather an oval section. Some have the pick formed in the shape of one curve and some of another. The cutting edge is sometimes shaped like an adze, sometimes like a chisel. The power of an axe depends mainly on its balance. An axe that is light in the scales may, if properly used, cut out a step quite as easily as a much heavier instrument in which the weight is in the wrong place. Guides generally carry rather heavy axes, and amateurs when they wish to emulate guides imitate them in this respect. Some of the best step-cutters, however—Melchior Anderegg for example—use axes no heavier than those carried by the average amateur. In any case it is best to learn with a light axe, certainly not exceeding 3 lbs.

It will be sufficient here if we recommend a form of axe which will be found generally useful. It is better for the axe to be too short rather than too long. The whole length need not exceed 44 or 45 inches. Such a length is quite sufficient for a 6-foot man. The early patterns of ice-axe were prodigious halberds, and 5 feet was no uncommon length. It was impossible to deal an effective stroke with good aim with such a weapon when holding it at the end, and it was most awkward to deal any blow on the ice when holding it anywhere else. Forty-four inches is an average length for the golfer's driver—that is to say, the longest club with which he finds he can hit accurately. Accuracy of blow is as important in step-cutting as in golf.

The stick is made of seasoned ash straight in the grain. The circumference of the axe just below the iron of the head is $4\frac{3}{8}$ inches. Six or seven inches from the top of the axe the circumference is 4 inches, and the section at both these places is oval with rather flat sides; the axe gradually tapers and becomes more rounded, until 12 inches from the extreme point the section is circular. The thinnest part of the axe just above the collar of the spike is $3\frac{3}{8}$ inches in circumference. The head and the braces by means of which the head is riveted on to the upper part of the axe are made out of a single piece of iron. From the point of the pick to the cutting edge is $10\frac{7}{8}$ inches. From the extremity of the pick to the centre of the top of the axe measures $6\frac{3}{8}$, and from the cutting edge to the same point measures $4\frac{1}{2}$ inches. The pick is square on section, each side of the square measuring $\frac{3}{8}$ or $\frac{1}{2}$ an inch two inches from the centre of the head; it gradually tapers to a point. It matters little whether the edges are turned upwards and downwards or whether one of the flat surfaces lies uppermost—i.e. whether the pick gives the appearance of being square or diamond shaped. The pick forms part of a circle whose radius is the whole length of the axe—a circle, therefore, about 22 feet in circumference. The cutting blade may be made slightly more curved. The radius line runs through the central axis of the pick, so that the force of the impact in a truly delivered blow passes through the metal. To the ice-man, the sensation of a blow delivered with good swing on to ice is as pleasant as a well-timed square-leg hit is to the cricketer. The surface of the top of the axe-head should be perfectly level, and any projecting nails or fasteners are objectionable. The braces are riveted through the wood in two or three places, and the ash is slightly grooved so that they may fit in without causing any projection. The spike is made of a single piece riveted on in the same manner as the spike of an alpenstock : an inch and a half or two inches of the wood may be sunk into the collar.

The axe should balance at a point from $12\frac{1}{2}$ to $14\frac{1}{2}$ inches below the top. This is a matter of very great importance, and too much care cannot be taken to ensure that the balancing point shall come at about the level mentioned. If the balancing point were too low, the lower part of the shaft might be tapered off a little so as to ensure the proper distribution of weight, even at the risk of spoiling somewhat the symmetrical appearance of the weapon. But an axe that does not come out right at first can rarely be tinkered into a good one by alterations. A badly balanced axe, or one that has the weight too much in the lower part, entails undue labour in step-cutting ; on the other hand, an axe with too much weight in the head is not only awkward and uncomfortable to carry, but 'runs away' with the stroke. The upper, oval part of the stick is that which is grasped in ordinary walking, but in step-cutting, when the axe is held at the lower part, the circular portion comes into the grasp. Some, who prefer a flat surface to grasp, advise for this reason that the stick should preserve the oval shape throughout. Each one must try for himself the form of stick that he finds most convenient to his own handhold. As a rule, the spike, the pick, and the edges are, when first made, far too sharp. There is no need to have them with a point and an edge like a knife. They do not cut a step any better, and they are apt to inflict injury on the clothes or person. Soft metal should be employed if the head is made of steel; if of iron the pick and blade should be steel tipped, and the hardening must only extend a very short distance, as it tends to brittleness. The metal should be soft enough to be cut by an ordinary penknife. An axe-tip, if too hard, can be softened without removing it from the wood, by placing a wet rag over the head before putting the tip in a blacksmith's fire to soften. Greater heat has to be applied if the metal requires hardening, and the head then must be removed from the handle. The spike point may be hardened like the edge of the pick if desired. It is better to harden in oil than in water. It is a mistake to keep the axe during the winter in a warm room or suspended over the fireplace as a decoration. The wood shrinks and the iron bands are left projecting from the surface, or the spike may get loose. Axes should be kept in any room in which no fire is lighted, and should be from time to time oiled, and at least have as much attention paid to their keep as a favourite cricket bat. The uses to which an axe may be put, and the manner in which its efficiency as a step-cutter may be tested, will be described under the heading of 'Snowcraft.' It will suffice here to point out that a man who desires to learn the difficult art of step-cutting cannot do so properly unless he is provided with an axe possessing a good balance and a properly curved pick. There is no need to have any teeth cut on the under surface of the pick.

In rock-climbing the axe is often a temporary encumbrance, and a sling is advisable. This can be made of a piece of strong string (an old silk necktie answers well) fastened round the shaft just under the head by means of a clove hitch; the loop should be sufficiently large to admit the arm up to the shoulder. Leathern slings are also made, and are more comfortable and convenient. These can be readily slipped on and off the axe. Some climbers use a long strap fastened to the head and to the lower part of the shaft, so that the axe can be slung on the back like a rifle. This plan is useful for the chamois-hunter, and would save the danger of the axe clanking on the rocks and disturbing the game. The device is too bulky for ordinary use. Some makers cover up the upper six or eight inches of the shaft of the axe with a leathern collar, which is said to be warmer to the hand. If the axe is really used for step-cutting, this leathern collar will be torn to pieces in half-an-hour's work, and at the best seems to the writer a useless addition. A red rubber tube, such as is used for lawn-tennis racquets and bat handles, is comfortable, and is said to stand the wear of step-cutting. Twelve or fifteen inches above the spike of the axe a ledge is sometimes made, and an indiarubber ring, like an umbrella ring, is placed above it, or a leathern collar, the object of which is to arrest the hand if it tends to slip over the shaft. There is no great harm in this addition, but it appears to the writer unnecessary. No one, happily, has yet succeeded in solving the problem of how to make an axe with a removable head. The ingenuity of makers is still expended in this direction, but as the most important part of the axe is the head, it appears rather waste of labour to devise an elaborate means of abolishing it; if you want an axe without a head, use an alpenstock.

A stout *walking-stick*, with a steel crutch handle made in the form of an axe-head, is a useful companion on a ramble where it may be necessary to cross a glacier or cut a step or two on a moraine.

Rope.—There is no part of the Alpine equipment for those who intend to go above the snow line, and, it may be added, for a great number of expeditions below it, more important than the rope. Of the methods of using it much is said in other parts of this book, and we need here only consider it as an essential article of equipment. The 'Alpine Club' rope, originally recommended by a special committee of the Alpine Club in 1864, and manufactured by Messrs. Buckingham, is still the best that can be procured, and, in fact, it seems hardly possible to improve upon it. This rope has a circumference of 1¼ inch, and the 'lay' is gentle. It is made of pure Manilla hemp, and between the strands a red thread is inserted as a distinguishing mark. Unfortunately, such a mark is easily, and possibly often, imitated. When new, the pure hemp has a glistening white appearance which should only be faintly tinged with yellowish-brown; the whiter the hemp the better the quality. The fibres, if picked out, should be long and of a uniform strength throughout. By looking at a section of the rope a good idea can be obtained as to the uniformity of its quality. When burnt, Manilla hemp produces an ash of a dull, greyish-black colour. Sisal hemp, of which the price is only half that of Manilla hemp, when burnt, leaves an ash of a whitish-grey colour. If the Manilla hemp is mixed with the Sisal hemp—a common adulteration—the resulting ash is of a grizzly white and black, which has been compared to a man's beard when turning from black to grey. Sisal hemp is stated to have only about half the strength and one-third the durability of Manilla hemp. The ordinary Russian hemp is very much cheaper, but is quite unsuited for Alpine purposes. The fibres are neither strong nor hard, and will not stand wet or wear on rocks. Russian hemp is much darker than either Manilla or Sisal, being of a *café-au-lait* tint. Silk ropes have occasionally been used; they are very strong, very light, and very expensive, but probably not as durable as Manilla hemp ropes, and for ordinary mountain work have no advantages which compensate for the extra cost. Alpine Club rope weighs about 1 lb. per 20 feet.

Belts, with loops attached to which the rope can be secured, are sometimes used, but are not to be recommended. It is better to make a loop in the rope. The various kinds of knots are described elsewhere (p. 101).

Crampons, or *climbing irons*, do not find much favour with English mountaineers, and have been spoken of contemptuously on many occasions. They are sometimes branded as artificial aids, a vague term, but implying great disrespect. Among German and Austrian climbers, however, they find much favour. Those who have really learnt how to use them speak warmly in their favour. They are made with six, eight, or more spikes. Whatever the exact pattern, the irons should be specially made to fit the boots. No doubt they have often been condemned by those who have only tried them casually, merely because they did not fit. Unless the crampons are really part of the boot they give no security. A good skater can do little when his skates are of an unsuitable size, and it is the same with climbing irons. A good pattern of climbing iron is one made of the best steel, in two parts, hinged together under the arch of the foot and provided with ten spikes. They strap on like the old-fashioned skate. The irons fold up at the hinge, and are conveniently carried in a separate leathern case. Climbing irons are especially serviceable on dry glacier when hard and slippery in the early morning, and on hard snow slopes. It requires a good deal of practice to use them on rock, but experts will readily stand firmly on smooth slabs where a man shod with ordinary hobnailed boots finds a good deal of difficulty. On rocks, however, a man who has a really good balance is perhaps safer than if he wore crampons. Undoubtedly on hard snow and on slippery ice an immense amount of time can be saved by the use of these irons. With two or three days' practice a man can acquire enough proficiency and confidence to enable him to dispense with steps in many places where they would otherwise be needed, and to trust to his crampons alone. They are, of course, of no use unless every member of the party is provided with them. The beginner will, however, do well to acquire a good balance before he tries crampons. Swiss guides for the most part have no experience in the use of crampons, and never employ them. In the Tyrol, however, they are commonly used. In an old engraved portrait of Jacques Balmat a hatchet and a pair of crampons are represented beneath the figure, and no doubt he employed them in his chamois-hunting expeditions.

Snow-shoes do not seem to have had the attention paid to them that they deserve. Much labour could be saved in deep snow on a mountain like Mont Blanc or Elbruz, for instance, by the use of snow-shoes, which can be made very small and light. Dr. Nansen found that the 'ski' used by the Norwegians were of great service to him in his journey across Greenland; but they would be of no use in the mountains. Good and practical snow-shoes which can be used with little preliminary practice are obtainable in the Engadine and the Tyrol. The Engadine pattern consists of a hoop of wood strung coarsely, like a racquet, with rope. The diameter is about 12 inches, and the shoes are very light, and would be useful in the deep snow met with in winter climbing. At great heights the snow will commonly be powdery and yielding. Those with experience of the Himalaya and the Andes can bear witness to this. Crampons and snow-shoes would probably more than compensate for the extra weight they would entail in the outfit, and should find a place in the stores of any mountaineer visiting a new country, where the length of a given expedition is usually quite uncertain, and where any device for saving time is of value.

The nightly task

CHAPTER III

MOUNTAINEERING AND HEALTH

BY C. T. DENT

A chamber scene

HEALTHY as the pursuit of mountaineering may be, it is not on physical grounds alone suitable for all sorts and conditions of men. 'It is hard,' says Lord Bacon, 'to distinguish that which is generally held good and wholesome from that which is good particularly, and fit for thine own body.' At the same time, even high mountaineering does not demand any exceptional physical requirements of its votaries. Broadly speaking, any youth who is sufficiently sound to pass the medical examination for entrance into the army is fit for mountaineering, while minor defects, such as short sight or slight varicose veins, need not debar him. Men climb, as trainers say horses run, in all shapes. Short, thick-set and muscular men, lean, flat and wiry persons, may make equally good mountaineers. On the whole, the best type physically for climbing purposes is the wiry man of average height and of a weight proportionate to his stature. But something more than the *corpus sanum* is wanted. In all forms of active exercise, as the 'Autocrat of the Breakfast-table' has noted, there are three powers simultaneously in action, the will, the muscles, and the intellect; and in mountaineering the first and the last are by no means the least important.

As in other sports, there is an immense advantage in beginning early. But it is the greatest mistake for the very young to indulge in high mountaineering. Before the age of, say, eighteen expeditions involving more than eight hours' walking are altogether unwise. The spirit of competition is strong in early life. Fascinated by novelty and spurred by emulation, youths in their keenness are apt to misjudge the strength of the machine which has to be driven. The big peaks will attract, of course, and the beginner is anxious to make what in Alpine slang is called a 'book'; but an Alpine 'book' is one that requires a long and careful preface. Set any promising member of the rising generation of mountaineers to work with a veteran guide, and not with a young man who has only just emerged from porterhood. The older man will be the better teacher, for he will have learnt to set more store on precision than on mere activity.

There is no need for any man who leads an average healthy life to submit himself to any special training for mountaineering, but, at the same time, those whose avocations force them to a sedentary life at home must not suppose, even if they are old hands, that they are fit to go up a snow mountain directly they come in sight of one. Nor should they, as is too commonly the case, imagine that a single training walk— perhaps a few hours' ramble on a glacier or a hard pound up a steep, hot mule-path—is sufficient in itself to overcome the effects of eight or ten months of mountaineering inaction. Mountaineers too often, in the desire to make the most of a holiday, ruin their pleasure for the greater part of their tour by over-exertion at the outset. Rowing, especially on a sliding seat, is the best possible form of preparatory exercise for a man to take who wishes to get himself into condition for mountaineering.

A certain amount of time is required to enable anyone to accustom himself to change of diet and surroundings. The diminution of atmospherical pressure, too, exerts a sensible, though perhaps not much recognised, effect even at elevations of six or seven thousand feet. The matter of diet may be almost summed up in the advice to get the very best you can and to take as much of it as you feel inclined, and this applies to actual climbing as well as to life in valleys. The question of alcohol or no alcohol has of course been debated warmly with regard to mountaineering. On the whole, the less taken on mountain expeditions the better, particularly in ascending. The majority of the Swiss guides, it is true, do not abjure wine when engaged in their work, especially when it is provided for them free of cost; but then they are in their own country, and accustomed to the native wines from their youth up. Many guides hold to the tradition that on the mountains white wine is bad, inasmuch as it 'cuts the legs.' Each one must find out the kind of drink that suits him best, and it need only be noted that in case of fatigue alcoholic stimulants do harm rather than good, and that the thin and sour bottled mixtures which do duty as wine in Alpine resorts are more prone to derange digestion than restore the strength. There is a prejudice in the minds of some against drinking glacier water. It is of course unwise, when much heated, to drink largely of any cold water just before a rest; but while on the march waterdrinking, in moderation, will not do the slightest harm. The water from the glacier pools is particularly good. Streams coming from melted snow patches need not be avoided unless in trickling over the surface the water has acquired some ingredients injurious to health. This will not be the case in high regions. It will be prudent to abstain from drinking water that courses near chalets or in the neighbourhood of herds of cattle. It is best to avoid water when any nettles are seen growing close by the stream. In the valleys, and especially in the villages, of countries such as the Caucasus, it is not wise to trust the water even when it has been filtered. All doubtful water should be boiled before drinking.

Fatigue is a condition that the mountaineer cannot wholly escape. To a great extent he can prepare himself by carefully regulating his pace and the length of his expeditions, allowing due intervals of rest, until he has brought himself into good condition. Without entering into any physiological details, it may be pointed out that fatigue is a complex condition involving something more than the mere tiring of the muscles. Ordinary walking is an action so familiar that it becomes automatic, and the mind is not conscious of making any effort; but directly extra work is entailed, as in walking up hill or climbing, or wading through deep snow, the automatic mechanism ceases to be sufficient, and an effort has to be made; the will, in short, has to be called in to stimulate the muscles to do their work. At first the mind simply becomes conscious of the muscles; later on this consciousness amounts to discomfort, and then to a sensation of pain. In other words, the unduly exerted muscle becomes recognised, and the tired muscle becomes painful. Additional exertion—that is, additional will power—is required to overcome the sensation and to stimulate the activity. This expenditure of energy involves waste, and in a measure explains fatigue.

That the mental factor is a strong element may be seen any day on the mountain-side. A man who seems utterly tired out, if placed in sudden peril or confronted with something unexpected, becomes immediately capable of great exertion; when all are getting a little tired the breakdown of one member of the party has an astonishingly beneficial effect on the condition of the rest. The practical outcome is that the keen and determined men are able to make their muscles do more work than the apathetic, and the men who on the mountain-side find constant employment for their minds, whether in attending to the details of the route, enjoying the beauties of the scenery, or in pursuing some scientific aim, will go better than those who treat a mountain as a treadmill.

The expenditure of energy has to be made good in two ways—by taking in oxygen through the lungs, and by the ingestion of food. The natural inclination of the tired man is to stimulants, or at any rate to drink, for fluids are rapidly absorbed, and the relief, therefore, is brought about more quickly. Unfortunately the benefit is only transitory. The great point with a tired man is to feed him. When you have a weak man in the party, it is well to feed him early, while he can still eat, and feed him often. If a man is utterly tired out, it is better to let him rest till he can eat, however little, than to attempt to stimulate that which is incapable of response. One of the worst possible things to give a man when in this condition is brandy, though it seems to be considered a universal panacea. A little champagne, however, will often provoke an appetite in an exhausted man. Thirty to sixty drops of 'sal volatile' in a little water answers almost equally well.

Many of the slight intestinal derangements that travellers in the mountains experience might be obviated by the adoption of simple precautions. The chilling due to rapid evaporation from the surface of the skin is one of the commonest causes of these troubles. From anatomical reasons, chill of the surface of the abdomen provokes very directly derangements of the viscera beneath. There is no better preventive of trouble than a cholera-belt. Immediately on arriving at an hotel after a walk, it is wise to bathe in tepid water and to change the damp clothes. If at a bivouac or hut, a vigorous dry rub, with a change at least of flannel shirt and stockings, will serve almost as well.

It cannot be expected that the digestive powers will be in perfect order after severe exertion, and the traveller who arrives at an inn somewhat tired, at night, will do best to take very light food and abstain from wine altogether. When the entire body is in need of rest before anything else, it is injudicious to throw on it the labour of digesting a heavy meal. On the other hand, if no food at all be taken, that best of restoratives, sleep, will keep aloof. Weak tea, for those who can take it, or soup of not too rich a nature (and many of the tinned soups are extremely rich), will probably prove more efficient in inducing sleep than a meal of meat. Hot bread and milk is an excellent light supper.

There is no need to carry much of a medicine-chest, and certainly the less medicine that is taken the better. For the mild diarrhœa that often attacks mountain travellers chlorodyne is in most cases a simple and efficient remedy, but it does not suit all. It is best taken in small doses of ten or fifteen drops, repeated two or three times at intervals of an hour. The heartburn that often occurs at night after a long walk can be controlled by bicarbonate of soda lozenges.

It is always unwise to start early in the morning on a perfectly empty stomach. When solid food is distasteful, milk, if procurable, will carry a traveller a long way. Some take a flask of rum with them, and a few drops of this in the milk is perhaps less objectionable before an ascent than any other spirit : rum is a strong respiratory stimulant. Warm food is best ; tins of chocolate and milk can be bought which form an excellent breakfast. It is well always to carry some chocolate in the pocket. Kola chocolate or biscuits answer well.

The special discomforts to which mountaineers are subject are sunburn, snow-blindness, and frost-bite. Under certain conditions marked symptoms due to the rarity of the air occur, which are spoken of, collectively, as 'mountain-sickness.'

The inflammation of the exposed parts of the skin consequent on the intensity of the sun's rays is a very familiar trouble. The affection is essentially the same as that experienced in milder degrees by those who work with intense electric illuminants, or are exposed much to the sun's rays reflected off water. In the case of arc electric lights the direct rays set up the trouble, but on snow the rays reflected from the surface are the chief agents. A high cold wind aggravates the trouble by producing dryness of the skin. Freshly fallen snow at great elevations is the most powerful reflector of the irritant rays. It seems probable that the effect is produced largely by the 'chemical' rays. The burning is felt at the time to some extent, but the worst of the discomfort comes on after a few hours, often at night. Fair people suffer more than the dark-complexioned. Sometimes a considerable degree of inflammation is set up. A tolerably acute condition of eczema may be produced if adequate precautions are not taken. The cracks and blisters that ensue are not only painful, but rather unsightly, and in the Alps, at any rate, mountaineers should, from motives of regard for others' feelings if not for their own comfort, take measures to mitigate sunburn. Lady mountaineers are especially concerned, for the effects, like those of tattoo marks, may be somewhat permanent, and have to be repented of at leisure. There is no application better than the substance sold as 'Toilet Lanoline,' which can be purchased in small tubes. A little of this smeared from time to time on the parts most likely to be affected will altogether prevent any trouble. The application should be renewed every two or three hours. Cold cream or zinc ointment is recommended by some, and answers almost as well. Glycerine is useless. Whatever is used should be employed as a preventive. The effect of powdering the face with starch-powder over some ointment renders the mountaineer rather like a 'Pierrot,' but is extremely efficient.

The skin may be as severely burnt on a foggy, cloudy day as when the air is clear. For, as pointed out by Professor Bonney, a very large portion of the light which reaches us does not come directly from the sun, but is reflected to us by the vapours of the atmosphere.

Snow-blindness is a more serious affection. The commonest form is essentially an ophthalmia—that is to say, inflammation of the mucous covering of the eye and inner lining of the eyelids. The eyes become greatly bloodshot and very sensitive to light; there is a free watery discharge which gums the margins of the eyelids together; the slightest endeavour to use the eyes causes a copious flow of tears. At great elevations this affection will occur more readily than at low heights. The trouble usually subsides after a day or two, though sometimes the eyes remain weak and sensitive for days, or even for months. A more serious form of snow-blindness is an affection of the deeper parts of the eye. Here there is much less superficial inflammation, but extreme intolerance of light. The symptoms are much more grave, and the effects pass away much more slowly. Both forms are tolerably familiar to those engaged in electric-lighting work. As in the case of sunburn, vaporous misty days do not render the mountaineer exempt from snow-blindness. Proper snow spectacles are most efficient preventives. They should be put on *before* the glare begins to be felt. A certain amount of discomfort may be expected at great elevations even when spectacles are worn. A five per cent. solution of cocaine dissolved in rose-water, and with a little boric acid added, acts like a charm in snow-ophthalmia. It is not easy, unless the right method is adopted, to introduce the fluid into the eye, for directly the lids are separated a gush of tears ensues and washes out all the lotion. The sufferer should be directed to lie down with the back of his head to the light and with the eyes closed ; a few drops of the solution are then poured into the little depression which is above the inner angle of the eyelids by the side of the nose. If the eye is then covered, and the sufferer directed to blink the lids a few times, the fluid will be drawn in. Cold compresses give a good deal of relief. For the more serious snow-blindness prolonged rest of the eye is really the only means of cure.

Frost-bite is a trouble that the mountaineer is less likely to meet with if he recognises the necessity for taking precautions against it. The mildest form is merely a deadness or numbness of the fingers or toes. The part affected becomes white, the skin a little wrinkled, and sensation is lost. As the symptoms pass off, the part swells a little and becomes of a bluish-violet colour. Return of sensation is characterised by itching and tingling. The more quickly artificial warmth is induced the more pronounced are the after effects. It follows that in the treatment of slight, or, indeed, any degree of frost-bite the temperature of the part should be very gradually restored. Gentle friction with snow will restore vitality, or the hand or foot may be immersed in cold water, of which the temperature is only to be raised very gradually indeed to a warmth of 50 to 60° F. Avoid, especially, warming the affected part at the fire. After rubbing, wrap up the frost-bitten part in cotton-wool. A more severe degree of frost-bite is associated with the formation of blisters after a few days. Here the after consequences may be serious as regards the frost-bitten extremities. The blisters may be pricked and the fluid gently pressed out. It is often impossible in the early stages to judge of the extent of frost-bite. If too energetic friction is made the superficial skin may be rubbed off and troublesome sores form. After a severe frost-bite complete sensation will often not return for weeks or months. On the other hand, a part that appears gravely frost-bitten will frequently recover entirely, or perhaps with the loss of a nail or two, after days or weeks. The most profound stage of frost-bite is a complete freezing or stiffening of the whole body. The sufferer loses consciousness and falls into a state of suspended animation. This third degree is especially likely to occur when a man is overcome by fatigue and cold and lies down. The most gradual raising of the temperature must be resorted to. Gentle friction of the limbs towards the heart may be made for several hours. Artificial respiration may be necessary. So long as there is the slightest trace of heart-beat the efforts should be continued. If the frost-bitten parts become very painful as they recover vitality, it is a sign that the warmth has been applied too rapidly, and they may then be wrapped in cloths dipped in cold water. In applying any remedies to a hand that has been attacked, take care not to make any pressure on the sides or webs of the fingers. Such pressure is very likely to be made if the fingers are wrapped up separately and a bandage placed over the whole hand. The fingers and toes and tips of the ears are most liable to frost-bite. The point of the nose may be attacked. Tight boots especially favour frost-bite of the feet. Cold wind and moist cold are the most to be dreaded. In still, dry air there is much less likelihood of the trouble. It is essential, therefore, to keep dry the parts that are likely to be attacked. The utmost attention must be given for days afterwards to any part that has been attacked by even the mildest degree of frost-bite. In mountaineering the fingers are more often affected than any other part, especially after climbing on rocks in bad weather.

'*Mountain-sickness.*'—This term has long been in use to denote the effects that rarefied air has on the body. The affection is not by any means peculiar to the mountains. Aëronauts are familar with the symptoms, and M. Paul Bert has shown that they can be produced in the laboratory. The recent publication of Mr. Whymper's experience in the high Andes has drawn fresh attention to the subject. The matter is one of profound interest to mountaineers, for it involves the possibility of ascending on foot the highest points of the earth's surface. The symptoms which are collectively known as 'mountain-sickness' consist in lassitude, which may amount to an utter incapacity for the slightest exertion, headache, difficulty of respiration, feverishness, quickening of the heart's action, and great disinclination for food. Nausea and vomiting may occur, and also bleeding from the nose and ears. The more permanent of these symptoms are the lack of muscular energy and difficulty of respiration. But even the more transitory effects subside slowly if the travellers remain at the same altitude. Thus Dr. Frankland found, in his own case, that at the summit of Mont Blanc (15,782 feet), after six hours perfect rest and sleep his pulse was about double its ordinary rate. Mr. Whymper at his second camp (16,664 feet) on Chimborazo suffered from intense headache, which did not disappear for three days, during which time he remained at the same elevation.

In this chapter little more can be done than to put forth some arguments for and against the possibility of ascending on foot the highest summits of the earth. In the first place, existence can be maintained at a greater height than that of the highest mountain, assuming this to be about 30,000 feet. Messrs. Coxwell and Glaisher in one of their balloon ascents believed that they reached an elevation of 37,000 feet. The estimate was probably a rather exaggerated one. The ascent was made from Wolverhampton, and the highest altitude reached in less than an hour. M. Paul Bert, in an experiment lasting 89 minutes, submitted himself to rapidly diminishing pressures and remained for a few minutes at a pressure about equal to that which would be experienced on the top of Mount Everest. During this experiment he inhaled oxygen.[1] In the case of aëronauts the rapid diminution of pressure is not without risk. Messrs. Crocé-Spinelli, Sivel, and Tissandier rose in two hours in a balloon to a height of 26,000 feet, and for two hours more remained at an elevation of from 26,000 to 28,000 feet. M. Tissandier was the sole survivor of this adventure. Both his companions were suffocated.

Mountaineers seem now able to attain to great heights with less difficulty than formerly. In the early days of mountaineering symptoms due to rarefied air were the rule on mountains such as Mont Blanc. De Saussure and many others described them. The ascent is now made almost daily in the summer by numbers, and few experience any symptoms whatever. Mr. Whymper, as we have said, on Chimborazo, suffered severely at a height of 16,664 feet, but was able a few days later to walk from his third camp (17,285 feet) to the foot of the southern walls of Chimborazo (18,528 feet) at the rate of 14·6 feet per minute. Mr. W. W. Graham, in an ascent of the Himalayan peak Kabru, which he estimated at a little under 24,000 feet, suffered 'no inconvenience whatever.' Even assuming that the height of this mountain was over-estimated, it is probable that this is as high an ascent as any yet made on foot, and yet no symptoms due to the rarity of the air were experienced.[2]

Too much importance must not be ascribed to the experience of aëronauts and scientific experimenters working in laboratories, for in such cases the fatigue due to muscular

---

[1] The inhalation of oxygen as a remedy is not practicable on the mountains, any more than the injection into the blood stream of permanganate of potash solution. Eating chlorate of potash lozenges has been advised; the benefit is rather fanciful.

[2] See also the reference to Mr. Johnson's ascent, p. 287, *supra*.

exertion of the limbs is wholly absent. But their observations show that 'mountain-sickness' is a real malady, and that the symptoms may not be explained away by assuming that they are due to fatigue, indigestion, or want of condition.

One marked effect of very low pressure (as also of greatly increased pressure) is to diminish the flow of blood to the lower limbs, and thus seriously affect their nutrition.

Additional demands, too, are made on the muscles concerned with inspiration owing to the prolonged necessity for extraordinary efforts to inhale the requisite *volume* of air. At the sea level, without exertion, thirty-five cubic inches is computed to be the volume necessary to supply an adult male with the essential oxygen. M. Vallot estimates that on the summit of Mont Blanc the amount of oxygen is reduced by one half as compared with the sea level. It is by no means certain that in the case of mountains the proportion of oxygen diminishes proportionately at still greater elevations; a man, therefore, on a mountain may be better off in this respect than an aëronaut. The element of time is a most important consideration. Travellers in balloons have undergone great diminution of pressure in extremely short periods of time. The mountaineer, who must necessarily ascend extremely slowly, has an advantage in this respect.

Diminished pressure affects different individuals in very varying degrees, but it is clear that all who are accustomed to long sojourns at great heights improve to a greater or less extent in their resisting power.

The extent to which, by special training, men can overcome the physical obstacles incidental to exertion at great heights is unknown; but there are good reasons for supposing that *after a long time* the capacity for exertion of the limbs and of the muscular respiratory mechanism will largely increase. M. Vallot's observations on the top of Mont Blanc point to the same conclusions. Mr. Whymper, who speaks guardedly, says that he became *somewhat* habituated to low pressures. But even if men could breathe on the top of Mount Everest, could they walk there? Mr. Whymper himself seems disposed to doubt the possibility, and, in a passage likely to be often quoted, he remarks :—'From the effects on respiration none can escape. In every country and at all times they will impose limitations upon the range of man ; and those persons in the future who may strive to reach the loftiest summits on the earth will find themselves confronted by augmenting difficulties, which they will have to meet with constantly diminishing powers.' Beyond question, the ascent, if it be possible at all, of a mountain 30,000 feet high would require very exceptional men, exceptionally favourable circumstances, phenomenal perseverance, immense expense, and a great length of time.

The party could not consist of less than three, and certain physical and mental attributes would be required in each one. The men had best be between thirty-five and forty years of age, with long experience of mountain climbing in regions besides those of the Alps. All should be possessed of confidence, determination, and the scientific spirit which, having a definite end in view, brushes aside minor obstacles. Physically they should be thoroughly sound, under rather than over the medium height, with a large 'vital capacity,'[1] and with a slow, full pulse. To attain the greatest possible elevation, it will not suffice for a man, however well qualified in other respects, to go straight out from England. His chances of success will be far greater if he has spent twelve months or so at considerable elevations. If Swiss guides are taken, select the intelligent rather than the simply strong. There will be little real chance of exercising mountain craft. Prolonged step-cutting or difficult rock-climbing will render success hopeless. The porters and carriers would really determine the success of the expedition. Supplies would have to be carried up, at the lowest estimate, to a camp at about 25,000 feet, and very light loads indeed would have to be apportioned to each man. These natives would

---

[1] The vital capacity is indicated by the quantity of air a person can expel from his lungs by the most powerful expiration, after the deepest possible inspiration.—

have to be trained even as the leading members of the party, or no efficient work would be got out of them. It is very unlikely that even the most carefully selected and trained carriers would go higher than 20,000 feet. The rest of the carrying work would have to be done by the party themselves. Depôts of provisions would have to be formed high up.

The more gradual ascent characteristic of the great Himalayan peaks might enable beasts of burden to be utilised to some extent. Sir Douglas Forsyth, in his journey to Kashgar, took a large train of yâks and ponies over the Changlung Barma La (a pass of 19,280 feet). All the animals suffered severely, and many were lost. The yâks were far more efficient than the ponies. In a further journey to Kashgar, Sir Douglas Forsyth started on September 29 from Leh, and arrived on October 15 at Suget. During this journey he crossed the Khardong (16,757 feet), Sasser (17,277 feet), Karakorum (18,550 feet), and Engel (16,937 feet) Passes. The expedition consisted of 350 persons and 550 beasts of burden. Only eight of the baggage ponies were lost. Possibly the effects of rarefied air are not the same in different climates, and they may be less in the Himalaya than elsewhere.

Great uncertainty exists as to the precise conditions that determine the onset of 'mountain-sickness.' The symptoms may come on with great suddenness and in the most unexpected —in the present state of our knowledge, almost in a capricious manner. In the Alps at elevations no higher than 10,000 feet marked symptoms have been noticed. It is impossible, therefore, at present to state with any certainty the precise atmospheric conditions likely to prove favourable or the reverse. An explanation may possibly be found in the varying temperatures of different strata of the atmosphere.

It is quite certain that nothing short of a regular siege would enable a party to ascend a mountain of 30,000 feet. The ascent from one camp to another would have to be very gradual. Probably many descents would have to be made before a point could be reached from which the final assault

MOUNTAINEERING AND HEALTH 89

could be delivered. It is not likely that on the final day any height greater than 5,000 feet could be accomplished.

So far the abstract possibility of ascending to the highest point of the earth's surface has alone been discussed, and the question whether any work of value might be done on the summit of Mount Everest, or whether any useful result would come of it, has been ignored. Time perhaps will show. All that need be here pointed out is that the upward limit has assuredly not yet been reached; and, taking all the evidence into consideration, it seems to the writer premature to fix it at a point even below that of Mount Everest. The problem is one to be worked out practically on the mountain-side and nowhere else.

An awkward drop

CHAPTER IV

THE PRINCIPLES OF MOUNTAINEERING

By C. T. Dent

Les Tricoteuses

OUNTAINEERING is a many-sided sport of which climbing is but a single, though a very important, branch. The part should not be mistaken for the whole. A man may be an active or even a good climber and yet a very poor mountaineer. Ability to run rapidly up the rigging does not qualify a man in seamanship. A proper balance must be observed between the various departments by anyone who wishes to excel; and to mountaineer well means to mountaineer safely. In that best of all training-grounds, the Alps, far too much importance is attached to a man's purely gymnastic prowess on rock or snow. Not only do people apply a purely climbing test in gauging their own proficiency as well as that of others, but they even go further and imagine that the traditional difficulty and the height of mountains climbed constitute the real measure of excellence. A far better criterion than a long list of peaks and passes successfully overcome, is the manner in which a reputedly easy mountain has been accomplished. The skilled mountaineer has always, as it were, a large balance in hand. It is in the avoidance of risks that his experience tells. The man who is a climber and nothing else while on the mountains has very much less to spare. Strong limbs count for much, but they will always be outweighed by average activity *plus* intelligence fortified by experience. On the elementary principles which should guide a mountaineer all are pretty well agreed. The misfortune is that more people appear to be able to quote the rules than to follow them.

Walking up hill is a tolerably obvious necessity in all mountaineering. Yet, even here, there is a right and a wrong way, and a particular style adapted to the mountains has to be acquired. In most cases the method comes intuitively and without much difficulty up to a certain point. The great object, of course, is to ascend with as little exertion as possible, and in even so simple a matter as walking up hill, there is so much that may be acquired that a man may go on improving for years. Now, this is an extremely convenient fact. The pedestrian athlete has but a short career. He soon finds that his elasticity begins to fail. Even the professional runner can only hope to pursue his calling for a few years, and almost from the first looks to the public-house as his real goal. The mountaineer, too, with advancing years, loses suppleness and perhaps increases in weight : but he is not therefore debarred from his favourite sport. On the contrary, he will find that by practice he has gained knack enough to fully compensate for these drawbacks. Many instances could be quoted of guides and of amateurs upwards of sixty years old who can ascend high mountains with little more exertion than when they were quite young. Yet more remarkable is the instructive fact that they will find that their pace for long distances diminishes but slightly as they grow older.

The essence of walking up hill, whether on snow, rock, or path, is the cultivation of a quiet swing and a methodical rhythm of gait. A certain slight degree of roll, or, to put it more scientifically, of swinging of the pelvis or hips, is of the first importance. The front leg should, as it were, drag up the hinder one. The beginner usually walks up hill with far too lively an action. In ascending, he raises his weight by a spring from the toe of the lower foot. This method throws a great strain on a small group of muscles (those of the calf of the leg), which have, for the moment, to support nearly the whole weight of the body. Watch any beginner, if he is suitably long-stockinged, and you will notice that the muscles of the calf contract powerfully at every step. In the older man, or in the one who has acquired the proper gait, they contract but slightly; that is to say, other muscles, acting at a better mechanical advantage, contribute to do the work of raising the weight. The upward movement is effected partly by the swing and partly by the muscles about the hip. However powerful a man may be, he will soon tire a small group of muscles if he throws a constant strain on them, whereas, if he calls many into play, he exerts none to anything like the full extent, and is enabled to last much longer. In ordinary quiet walking on the level, as a step forward is made the disengaged leg swings forward, almost passively, by a pendulum-like movement. Where rapidity is required, the oscillation is supplemented by muscular contraction, which is necessary to draw the leg quickly

Ein Junger

THE PRINCIPLES OF MOUNTAINEERING 93

forward. Most of the muscles which effect this movement act at a great mechanical disadvantage, and thus entail considerable exertion. Even on a steep hill a certain amount of pendulum action is possible, but to gain this advantage slowness of pace is essential. There are proverbial sayings in every language enforcing the necessity of a slow, steady pace when walking up hill. 'Plus doucement on monte, plus vite on arrive au sommet,' say the French guides, and the Italians have a parallel maxim, 'Chi va piano va sano, chi va sano va lontano.' Most beginners walk too fast. Pace is an all-important consideration, for, however good a man's action may be, the exertion in ascending must be great if he exceeds the rate suitable to the gradient. A good test, because applicable to the condition in which a man finds himself, is the ease and comfort with which he can talk as he ascends. It is better, no doubt, to be silent as much as possible while ascending. A small pebble held in the mouth does not contribute to comfort solely by keeping the tongue moist : the chatterer runs a risk of swallowing the object. If a man has something to say and finds that he catches his breath, or slackens his speed, or stops altogether in order to get out a few words, it is a very sure sign that he is going too fast. If he finds a sense of admiration for the surrounding scenery come upon him every half-hour, he is going too fast ; and if he finds

All very well at the Curragh

it necessary to readjust a bootlace that is perfectly in order, he is going much too fast.

Regularity of pace is of prime importance, and the rhythm of the footfall is of more consequence than the length of the step. It is better to lose a few inches in the stride than to interrupt the rhythm. It is the amount of effort put forth at each step which should be rhythmical, and not the mere number of steps made in a given period of time. Varying the speed unwisely will soon cause a man to realise that he is tired. In a long walk up a valley or mule path, or up steep snow slopes, the duties of a leader are much like those of the stroke of an eight : he sets the stroke best suited to those behind him, and if he desires to get over the ground quickly, must keep to that pace, neither spurting nor slowing down abruptly. It is better to halt for a minute or two than to slacken the speed for a long stretch. It is generally better to halt for a minute or two when it is desired to increase the pace ; the leader may then start off at a quicker step, and the change will hardly be noticed. The exertion can be rhythmical still on slopes of varying gradients. If a slope suddenly becomes a little steeper, the pace should slacken a trifle ; if this consideration is not observed, the pedestrian is really trying to ascend faster when the slope becomes more steep. If, on the other hand, it becomes gentle or nearly level, it is best not to quicken the pace.

Unless a man walks up hill with a stride which comes natural to him, he will soon get tired and his legs will ache, whether his stride be too long or too short. A proof of this can easily be seen in ascending a flight of stairs in which the tread is of unaccustomed height. In walking up hill the zigzag system cannot be too generally adopted. On a mule path, the turns should be taken as wide as possible, and when there is a long straight piece up hill the pedestrian should cross from side to side as often as may be necessary to ensure that his track shall be as far possible at one uniform angle of inclination. The experienced man has a very great advantage over the beginner in the power of choosing with certainty the proper place for his foot. The selection is made with so little conscious effort and with such celerity that it seems mechanical. Yet he may frequently step on a loose stone which slips away as he strikes off from that foot. Still he goes placidly on and the rhythm is not broken. The beginner puts his foot on the same stone and it gives way in the same manner. He strikes out with his leg as if he were skating, and is brought up stationary. He has spent energy to no purpose and has to start afresh, finding that he has lost perhaps ten yards or more by his stumble and has to make up the distance by quickening his pace. The secret is, in all hill-walking, to accept any slip you may make, whether it is one of a few inches on a mule path or whether it is a more serious one, such as the giving way of a loose bit of crag in rock-climbing. The foothold that has proved insecure must, so to speak, be discarded as worthless, and the slip checked by other means. Disregard the hold that has given way, and turn attention to those that are secure. Accept a slight slip and you will not make a serious one. Before everything preserve the rhythm of the footfall.

To ascend a vertical height of 1,000 feet in an hour is good and quick walking. At the beginning of a tour, the pace had better be less than this. On a good mule path 1,500 feet in an hour is quick walking. To ascend through a height of 2,000 feet in the hour implies that a man is a fast goer and in good condition. Of course this assumes that the inclination of the slope is tolerably uniform, and that the climber has no long journey before him. A man who has ascended some 4,000 feet in two hours without a halt will not be fit for much at the end of it. At the beginning of an expedition in which the ascent will probably occupy six to eight hours, 800 to 900 feet per hour on a good path is a fair speed to adopt. On snow slopes in good order, at a height of 8,000 to 10,000 feet, twelve to fifteen feet per minute, vertical height, is not bad going, with a brief halt at the end of every hour. To ascend 3,000 feet in four hours, including halts, would be average work. This assumes that no step-cutting is necessary. Men in good condition, starting from a height of, say, 7,000 feet, and ascending a mountain of 14,000 feet, will often average 1,000 feet an hour if conditions are favourable. At great elevations, such as 15,000 to 20,000 feet, the low pressures tell greatly and the pace will slacken proportionately.[1]

As a broad example the ascent of Mont Blanc (15,782 feet) may be cited. The summit is six miles south of Chamonix (3,445 feet) The ascent from that village involves about three hours' walking on a mule path and good track, two hours over glacier and the lower *névé*, and the remainder on snow slopes. There will not be a moment's hesitation on the part of the guides as to the best route from first to last, and not a single step may have to be cut. An exceptional walker in good condition would require everything in his favour to ascend through the height (12,337 feet) in a continuous walk of twelve hours. Excluding fractions, this would be at a rate, for 720 minutes, of just over seventeen feet per minute.

The whole foot should be well planted on the ground at each step. Even in climbing up very grass-covered slopes, where there is no path at all, it is quite possible to preserve the rhythm. The irregularity of the steps makes this kind of climbing very tiring, and the beginner on grass slopes is extremely prone to forget the value of zig-zags. Tall people have rather an advantage in ascending, but this is more than counterbalanced by the drawback they find from their height in descending. Guide-books as a rule warn the pedestrian against attempting short cuts on the hill-side, and, generally speaking, this advice is good; but there is no objection to selecting a track which cuts off a long détour, provided the climber alters his pace and the length of his stride proportionately to the steepness of the slope.

Descending requires more practice than ascending, and a

---

[1] The effects of low pressures on rates of speed are dealt with fully in Mr. Whymper's *Travels among the Great Andes of the Equator*. London, 1892.

THE PRINCIPLES OF MOUNTAINEERING 97

good deal of experience is necessary before a man acquires just the right action and balance that enable him to go rapidly and safely down hill. The beginner here exaggerates the mistake that he makes in walking up hill, viz. putting too much exertion into each step. In descending a fairly steep slope, or even a path inclined at a moderate angle, the movement, however slow, should be continuous. To remain completely stationary, even for the briefest possible moment, not only necessitates a fresh start, but demands an adjustment of balance which implies an unnecessary outlay of muscular effort. The knee is almost constantly just a little bent, and the period for which it is actually straightened is extremely minute. Now, it requires much greater exertion to maintain the equilibrium when the knee of the supporting leg is a little bent than when the limb is straight, and were it not that the degree of bending is constantly changing as the weight descends, and is, incidentally, being shifted from one leg to the other, fatigue would rapidly ensue. What the beginner has to learn is to overcome his natural tendency to bend the knee by an active effort. With practice, the bending of the knee becomes a result of relaxation of muscular effort. The moment this is done the weight tends to fall forward. When the movement is effected, as it should be, by an effort of relaxation rather than of exertion, there is a momentary condition of instability. Some say that a certain looseness of the joints should be cultivated, and that then a man is able to descend rapidly without exertion. This is only another way of expressing the same point. Inasmuch as there is a momentary period of instability, a balance has to be acquired and all the muscles have to be kept on the alert to counteract any slip or disturbance of the proper equilibrium. The ordinary action of walking, merely allowing the hinder leg to swing forwards, is not sufficient in descending ; for, as this limb is more bent than it would be in walking on the level, the foot would naturally tend to catch on the slope. To overcome this, a certain, though a very slight, amount of spring is necessary. Practically, it is almost sufficient

H

to keep the body, and with it the chief part of the weight, at the proper angle of inclination, so that the force of gravity lifts, as it were, the disengaged leg from off the ground.

Beginners are in constant dread of spraining their ankle or knee when they descend rapidly, and it is this fear which chiefly accounts for the difficulty they have in learning a proper and easy action. But a man will never sprain his ankle when he expects that he may do so at any moment, and he is perfectly secure from any such accident if he is in a state of perpetual motion so long as he is descending. The steeper the slope the greater amount of spring required. On very steep slopes the best form of gait more resembles a canter than a walk. The footsteps occur in pairs, the feet falling almost side by side, and one immediately after the other. If one foot is placed on an insecure spot, the other is ready to back it up. On a suitable slope the interval between each pair of steps may be eight to twelve feet. The more loose and shifting the foothold, for instance when descending the side of a grass mountain strewn with small loose stones, the more necessary it is to preserve the continuity of movement. It does not at all follow from this that the writer is urging that the steeper the slope the greater should be the pace. The very reverse is generally desirable. While descending the stick is grasped by both hands, one in front of the body about the level of the hip, and the other rather below this level and behind, the hands being well separated. The point of the axe directed backwards rests on or touches from time to time the slope, so that when both feet are touching the ground, the climber really supports himself on a tripod. The tendency is always to fall backwards if the foot slips. But the man who is always prepared to slide will rarely slip.

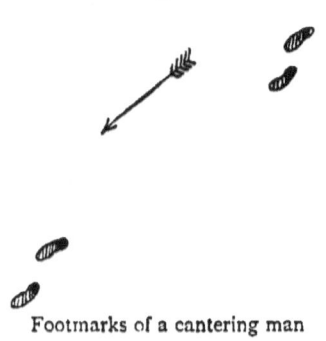

Footmarks of a cantering man

THE PRINCIPLES OF MOUNTAINEERING 99

The action of actually running down hill is a very fatiguing one, and, when employed, is usually associated involuntarily with a good deal of sliding action, not always performed on the feet. The most rapid way of descending, as anybody may try for himself on a flight of stairs, is to turn the body sideways. It has the advantage that the length of the foot can be planted fairly on to any hold that offers itself; and though the whole width of the sole of the foot may not be utilised, the edge, with its nails, gives ample bite. If a man descends with his back to the slope, in the majority of steps he must needs extend his feet or point the toe on hard ground; if he keeps the toe up and trusts to his heels the jar is considerable; this method is only applicable to soft snow or fine scree. Moreover, in descending sideways, the foot is not thrust forward in the boot at each step, and there is less liability of rubbing the skin over the instep or the upper surface of the toes.

There is, perhaps, no place more difficult to descend properly, and none on which care and deliberation are more necessary, than a steep and slippery grass slope. If a slip is made in such a place, the ordinary method of digging in the heels and throwing all the weight possible on to the axe, a method which is quite sufficient to arrest the progress sharply under other conditions, will often prove useless. The best plan is to turn completely round with the face to the slope and to try to anchor the pick of the axe into the grass, endeavouring to do so at about the level of the knee, or even lower. If a climber has no axe, but only a stick or alpenstock, he had better discard it at once, turn his face to the slope, and try for a handhold low down. To catch at a handhold at the level of the head is worse than useless. The arms would be extended, and as the momentum allows no time to get a firm grasp, the effort to stop with a jerk will involve too much strain. Some brake must be put on first.

Lastly, it is better if the place allows the choice, to avoid descending one below the other. Each man should choose his own line. It is never justifiable to race down short cuts, and run the risk of starting loose stones, when there is a zig-zag, or indeed any path, in a straight line below. A good many perils have been described in the Alps, but one usually omitted is the 'perils of our own countrymen.'

Use of the rope.—A 60-foot rope is a good length for three men, and on simple rock-mountains will suffice for four. On névé the real width of crevasses cannot be certainly known, and an allowance of 20 feet per man is judicious. About 4 ft. 6 in. is required for each waist-loop. The intervals, with three men, should be 15 feet; with a 60-foot length this would leave 16 ft. 6 in. of spare rope, weighing less than 1 lb. It is always better to carry too much rope than too little, even though some is coiled up and not in use.

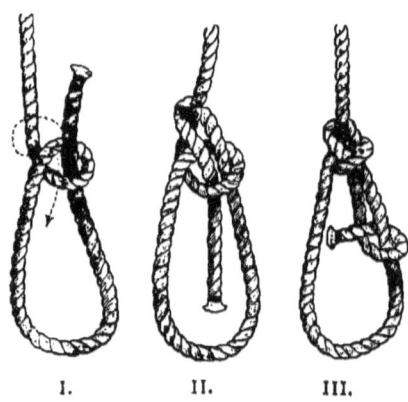

FIG. 1.—Simple bowline knot

A 60-foot rope with four men on it will take 18 feet for the loops, leaving 14 feet intervals, a minimum length even if the rope is always kept properly taut. On difficult rock-mountains the length may well be increased to allow intervals of 20 feet or more, and a spare rope should be taken. It is a mistake to lengthen the rope because there are inexperienced men in the party; a 15-foot interval is a good average length to allow. Avoid joining two lengths of rope whenever possible. The joining-knot is apt to catch on rocks. A 40-foot length can be taken as a spare rope to be left behind in places of difficulty. The ends of the rope should be securely bound with waxed twine to prevent unravelling of the strands. The tendency of new Manilla rope to kink can be overcome by thoroughly stretching it. A friendly tug-of-war is not a bad way of doing it. A wet rope also is likely to kink; after use it should always

THE PRINCIPLES OF MOUNTAINEERING 101

be well stretched out and allowed to dry in the sun. For carrying purposes the rope can be wound in a convenient circle, a little larger than a horse-collar, by winding it round under the foot and over the bent knee. If the coils are not made in sailor-like fashion, the rope will be much twisted in rolling it up in this manner, and will lose its suppleness.

Few knots are required, but these should be practised until they can be tied properly and with certainty. Knots should always be tied 'with the lay' of the rope. The loop should be fastened round the body at the waist. The end loop may be made with a simple bowline knot (fig. 1), or that shown in fig. 2 (in the diagram, A' is the free end of the rope). The latter, in the writer's opinion, is the better of the two. Guides generally employ the knot shown in fig. 3 (the simple overhand knot), and in the case of the Alpine Club rope, which is very strong, it answers the purpose perfectly well.

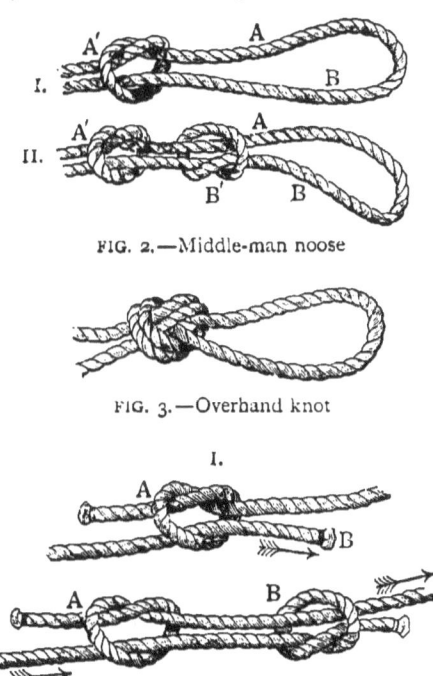

FIG. 2.—Middle-man noose

FIG. 3.—Overhand knot

FIG. 4.—Fisherman's bend

The Committee of the Alpine Club, appointed in 1864 to consider the subject of ropes and ice-axes, spoke strongly against the use of this knot, and advised the 'middle-man noose,' but only imperfectly figured it in their report. The best knot for the central members of the party is the 'middleman noose,' or 'running knot to hold' (fig. 2). Essentially this is the 'fisherman's bend' (fig. 4). A little practice is needed to make the loop the right size, and few Swiss guides know how to tie it. It is possible by manipulation of the knots to convert this into a slip-knot, but only with much difficulty. Practically, the noose is perfectly safe.[1] The trick of making a clove hitch, and also the double clove hitch, should be learnt. After a long climb on a difficult rock-mountain, inspect carefully every inch of the rope, and if any portion is badly frayed, cut the damaged piece out and join the ends together by means of the knot shown in fig. 4, or a 'figure-of-eight tie.'

Care should be taken to fasten *fixed ropes* in such a manner that they are not likely to be cut on any sharp projecting rock, and if a wedge or stanchion is driven into a cleft and the rope secured to it, the utmost attention should be given to the fastening. Many ropes in the Alps have been fixed for years without renewal, but no cord ought be much trusted to after it has been exposed for a year. The cords may serve to indicate the proper line, but should be handled gingerly. Too often the fixed ropes are made of inferior material, presumably from motives of economy, but these are precisely the ropes which should be of the best procurable material. They are certain to have a great strain thrown on them; they lie exposed to all weather, and, in addition, must suffer by friction against the rock. If a party takes down a rope which has been fixed temporarily to assist in the descent of a difficult place, they will do good service to any who may follow by leaving a string marked with rags, like a kite-tail, to indicate the proper route,

---

[1] Messrs. L. A. Legros and O. Eckenstein have made careful experiments to test the strength of various ropes used for Alpine purposes. In plain tension Buckingham's Alpine Club rope gave way only under a load of 2,200 lbs. Knots are, of course, the weak points. The 'fisherman's bend,' used as a middleman noose for instance, has only 65 per cent. of the strength of the rope. The simple overhand knot would show a worse result. But an enormous margin is provided *if good Alpine Club is used.* A simple overhand knot would certainly stand a load of 1,000 lbs. Practically, in the writer's opinion, it is perfectly safe, for this is a greater pull than any man on the mountain-side could withstand. Nevertheless, it is not a good knot.

THE PRINCIPLES OF MOUNTAINEERING 103

but the string must be so thin as not to tempt any climber to trust his weight to it.

The great use of the rope is on snowfields, especially and imperatively where there is any suspicion of crevasses. Now crevasses may lurk unsuspected even on well-worn routes. New crevasses form principally in the summer months, when the glacier motion is fastest, the very time when mountaineering is chiefly practised. On a snow expedition the rope should always be put on the moment the dry glacier is quitted. The rope is unnecessary when the path lies along old winter or avalanche snow that has fallen by the sides of the glacier between the lateral moraine and the bank. The more level the snowfield the greater will be the vertical depth of the crevasses. Consequently on the easiest-looking places the rope is most necessary. It must be remembered that the rope practically abolishes all risk on open snowfields, if properly used with a sufficient number in the party, such number not being less than three. It is not to be put on merely because the snow and ice look formidable, as in an icefall. As a matter of fact, it is far less essential on the icefall than on the level part immediately above or below it. The rope is a security against hidden danger more than against that which is apparent, and acts better as a preventive than a cure. On a pass like the Theodul, it is not less necessary than on one where there is complicated ice-work, such as the Col delle Loccie, or the Jungfraujoch. By experience alone can a man get into the habit of using a rope properly, and no opportunity should be lost of adding to this experience.

Inspect every loop, and see that the knots are properly made when the rope is first put on. The knot is worn on the left side for the middle members of the party, so as to leave the right hand free. The loop is best put on over the coat. The great point in the use of the rope is to keep it lightly stretched between the various members of the party. A trailing rope is a source rather of weakness than of strength. A party of five or six good mountaineers on one rope will cross long stretches of snowfield, and the rope will be almost as dry at the end as it was at the beginning, never having been allowed to drag in the snow.

Unless the party actually keep step, which is not practicable, the tension on the rope must vary a little. It is customary, therefore, in order to obviate any undue drag, to gather up a short loop of a few inches and hold it in the left hand, so that a little play is allowed, and an even tension maintained. This hold is let go when a bad crevasse or treacherous snow has to be crossed. The rope is then held taut by keeping the bodies at their fullest available distance apart, and should stretch from one waist-loop to the next without any sag at all. Those who are on sound ground lean away from the crevasse. The man actually on the bad place leans forwards, as, in case of a slip or breaking through into a crevasse, he can then more quickly recover himself.

The rope cannot be kept evenly stretched unless the pace is carefully regulated so as to ensure uniformity. The leader is chiefly concerned, but the functions of the second man on the rope are, in the matter of pace, hardly less important. The second man occupies much the same position as No. 7 in an eight-oar. The leader requires both hands free, and has to look about him a good deal for the best route in detail. The second man's duty is to watch the rope carefully. The attention is chiefly directed to the loop of rope in front, but the length behind must not be neglected. An Alpine party on a snowfield should not proceed in a jerky, spasmodic manner, like a loosely-coupled train; nor should they perpetually vary their relative distances. The pace has to be lowered to that of the slowest man. If any attempt is made to exceed this speed, the slow man will become an exasperated source of danger, and all the party will be miserable—or angry. It is of no use to be annoyed with a slow man. If a practical person, he will conduce to the general equanimity by tying the flaps of his hat over his ears. The delinquent cannot be approached, lest the lecturer and the lectured both descend into a hole together. Forcible

SERVE HIM RIGHT

THE PRINCIPLES OF MOUNTAINEERING 105

remarks, too, lose much value when a dryness of the tongue and throat mars their distinctness, and it is almost useless to throw much expression into the eye when it is hidden behind a smoked goggle. In mountaineering, as in other sports, loss of temper is a serious handicap. The hasty man may find much relief in dealing a furious blow at the snow with his axe when his feelings overpower him. He will sober down while getting the weapon out of the snow again. However good the stroke set, the boat will not travel smoothly and well unless the crew 'pick it up.' A division of labour, which makes the leader lead, and the second man follow the leader and lead the followers, is a good one for comfort and security of walking.

A man who is not unprepared can support a great weight if it does not fall on him with a jerk. As long as the rope is in use each man should expect a pull to come at any moment. At first this close and constant attention to the rope allows little time for looking about, but the habit is very soon acquired. When the rope can be used properly and unconsciously, the mountaineer will find as much time to look about him as if he were walking along the high road. Mountaineering would be but a poor amusement if there were few opportunities afforded of appreciating the ever-varying beauties of the surrounding scenery. There are times, however, when the security of the party is of more consequence than the æsthetic gratification of the individual, and a man who is perpetually star-gazing when on a rope is but a dangerous companion.

Amateurs often think that they go much faster when not roped. The reverse is the case, especially when a guide is leading. If a man is becoming a little tired and beginning to lag, it is often wise to put on the rope earlier than is really necessary. He will generally be found to walk much better when he is responsible in some measure for the safety and comfort of the party. Some men cannot go well on an empty stomach, and lag early in the morning, when steady progress is most desirable. It is better then to feed than to rope them. It is not perhaps out of place to observe that, while the rope should be put on early on snow expeditions, it should never be taken off until the ground in front is absolutely, not merely relatively, safe. Young mountaineers, when the chief difficulties of an expedition are over, are prone to seek too soon the kind of freedom they enjoy in walking without a rope. Some, as may unfortunately be seen almost any day in the Alps, appear to consider that putting on the rope on a pass such as the Tschingel or the Theodul is a ridiculous precaution, and they seem half-ashamed if they meet a party who are foolish enough to go without one. Second-rate guides are often to blame for this unwise and slovenly mountaineering. They dislike the trouble of unwinding and fastening the rope, and suggest that, as the pass or mountain is 'easy,' there is no need of it. It must be reiterated that the rope is not to be put on because a place is difficult, but because it is or may be dangerous. No precaution can ever be ridiculous. It is the impossibility of employing the rope, which is by far the greatest safeguard known to the mountaineer, that renders solitary climbing so reprehensible. No one, who has the interests of mountaineering at heart, should countenance so dangerous a practice.

On rocks, where a fall might entail any serious consequences, the use of the rope is not less imperative, although the security it affords is undoubtedly less than on snow. On snow a rope gives absolute safety as regards crevasses, and it is extremely efficient for checking a slip. On rocks the rope can only check a slip; but where the rocks are rotten, slips are, of course, far more frequent than they would be on snow, and the consequences more serious. On the other hand, a good deal more assistance can be given by a rope on a rock than on snow, and the rope is often employed as an aid rather than as a safeguard. The rope is much more difficult, however, to use on rock—that is to say, it requires a good deal more care. It is not sufficient merely to attend to the length in front, and to see that it does not catch as the climber

THE PRINCIPLES OF MOUNTAINEERING 107

ascends, but attention must also be paid to that behind ; in fact, each length of rope requires the watchfulness of the two men on it, the attention being principally devoted to the length in

Player in hand

front. A long, difficult rock descent, say of six or eight hours, is in itself very trying to the patience of some. The perpetual worry of the rope, which hooks here, catches there, dislodges small stones, and is always twining itself round the legs, becomes annoying, to put the case mildly. But these worries do not occur with good mountaineers. A climber who, while always ready to give a pull when wanted, never gives a pull unnecessarily, never allows the length of rope in front or behind him to catch and get jambed in a crack, or to hook round and detach a loose stone, is a real expert in his art. And here a word may be put in against a common misuse of the rope in case of a slip, which applies equally to snow and rock work : a word of warning which is at least as applicable to the ordinary guide as it is to any amateur. A man may slip a dozen times on rock and, if a fairly good climber, will check himself almost instantly. There is an end of it. He has made a false step, and at once corrected it. The climber is rather pleased with his own promptness. But his elation is often short-lived. Some, especially second-rate guides, have a detestable habit of giving a violent pull at the rope, not when a man stumbles but after he has checked himself, with the result over and over again that a second slip is the consequence. A pull, to be really efficacious, should be really almost contemporaneous with the slip. The rope should have been taut at the time. The violent jerk is often little more than an apology for having allowed the rope to fall loose ; and apologies are generally mistakes. If a man has slipped, has recovered himself and got into a secure position, leave him there ; be ready to pull, but do not jerk violently at it when safety, momentarily imperilled, has been regained. That is the principle adopted by the London cabmen, who jerk savagely at their horse's mouth when they have steered the beast badly round a corner ; the result is prejudicial to the animal's temper, but does not replace the paint that has been grazed off. Too much zeal is as much out of place here as in any other pursuit. Some guides have been known, in order to prove their strength and watchfulness, to jerk their employers clean out of their steps in order to show with what ease they can hold them. It is of little use being angry with these officious creatures. It is more to the point to

On the Tit-für-tatt Berg

change the order of the party, and place the offender where you can have a turn at him; there is nothing, let it be said in a parenthesis, that a guide hates more than to be dragged at by an amateur. No man will ever learn to climb properly unless a certain amount of responsibility is thrown on him. If he says that he can climb up a place, let him try it; do not assist him unnecessarily, but be prepared to aid him at any moment. The writer has seen rock-climbers who were but poor performers suddenly begin to go well and safely, merely because they found themselves with a good guide who gave them a fair chance to exercise their powers. But if a man is allowed this freedom of action, he must not be above asking for help when he really wants it. If you cannot trust a climber to do this, pull and drag at him without remorse if you think it makes the party any safer. There is no more dangerous companion than the inflated climber who thinks it undignified to ask for aid though in sore need of it. Such a one is for ever helping others when they do not want assistance. Many lady climbers go well enough, and would go far better if they could persuade their guides that gallantry does not consist in squeezing their waists with the rope unnecessarily. As a rule they get far more assistance, if it may be so called, than is at all necessary for their comfort or security; and they find it hard to persuade the guides that they are anything more than bundles, which have to be hauled up to the top merely for the sake of being let down again on the other side.

On a wet day knots in a rope are not very easy to untie, and it is a good plan to put a little pad into each knot; a small piece of rope or rag will answer the purpose. The rag is pulled out and the knot can then be easily untied. After using a rope untie all the knots completely. It is not wise to make the hitches always in the same place. A well-kept rope will serve perfectly well for two seasons, and often for longer.

Use of the axe and alpenstock.—The value of the stick in ascending is obvious enough. As the disengaged leg is swung forward in making the step up, part of the weight of the body is thrown vertically down through the stick, and the strain on the employed leg is relieved. Neither the axe nor the alpen-

THE PRINCIPLES OF MOUNTAINEERING 111

stock should, as a rule, be kept in front in descending. Timid persons are apt to advance their alpenstock in front of them and then to descend down to it. If they lean back and slip, the staff will not help them in the least. If they slip forward, as they sometimes do, they impinge painfully on the top of the stick. The proper way is to hold the stick behind and nearly horizontally, so that at any time the weight can be thrown back

The axe fiend

on to it, and the base of support converted into a tripod. An axe may be more than an assistance and a source of security to the climber; it may also be a source of danger to his companions. We have heard in this country of the umbrella fiend; the axe fiend is not less well known in the Alps. It must be admitted that he is prone to do himself injury, but his companions are not exempt. The climber should carry his axe with due consideration for others' comfort, especially when rock-climbing. The use of the axe cannot be acquired without practice. It cannot even be handled properly without practice, and it is as easy to tell an experienced mountaineer by the way he takes up an axe as it is to tell a tennis-player by the way he grasps his racket, or a fencer by the way he holds the foil. Practise, therefore; practise diligently; admit that there is something to be learnt, learn it of the right persons, and work till you acquire the knack. A glacier is not the only place where the axe can be wielded. A gravel-pit will do; what is a bunker to the golf-player may be an admirable practice ground to the mountaineer. Do not imagine that the use of the axe, or the alpenstock for that matter, will come intuitively when you find yourself on the snow. As well imagine that you will be able to swim the first time that you are thrown into the water.

Guides.—Except in the case of expert and experienced mountaineers, for whom this chapter is not primarily intended, professional aid, in what may properly be termed mountaineering expeditions, is a *sine quâ non*. Immediate safety demands that guides—not merely men to guide—should lead the party; and, as has been repeatedly urged, it is of the guides that the art can best be learned. But the amount of guiding power that may be necessary for any given climb; the number of guides or porters to be taken; and the occasions when a porter may be substituted for a second guide (a more expensive article) are questions which require careful consideration. The practice generally is to leave matters pretty well in the hands of the principal guide engaged. It is not always wise to do so unless the climber has been fortunate enough to secure a leader on whose judgment and disinterestedness absolute reliance may be placed. The desire to do a good turn to a friend, to take advantage of an opportunity of giving practice to a son or relation, or the necessity, which too often occurs in the Alps, of obeying the imperial mandates of the hotel porter, may all combine to influence the guide's advice, sometimes to the disadvantage of his employer. For expeditions of moderate

THE PRINCIPLES OF MOUNTAINEERING 113

difficulty, in fine weather, this may matter little. Yet the climber should be capable of exercising his own judgment on a point which so materially affects the pleasure, and it may be the safety, of his expedition. As to the number of professionals to be taken, one rule may be laid down of a tolerably comprehensive nature. A sufficient number of professionals should be engaged to ensure that two beginners shall not be roped together. A guide and a porter, for a simple climb under favourable conditions, will suffice for two amateurs; but then one of the amateurs must go last on the rope in ascending and first in descending. On an ordinary climb an experienced amateur will be quite as fit to precede or follow the beginner as a heavily-laden porter. If the party consists only of three, one guide, an expert amateur and a novice, the expert must be prepared to do some of the carrying. As a rule, he will prefer to engage a porter.

At Chamonix the Corporation of Guides is regulated by the State, and very precise rules are laid down. Thus for the ascent of Mont Blanc a single traveller is compelled to take two guides; two travellers, three guides; beyond the number of two travellers, one guide is deemed sufficient for each traveller; and for 'courses extraordinaires' one guide for each traveller is prescribed as absolutely indispensable. Certain exemptions from the rules are granted. Unhappily the system that still obtains at Chamonix is far from being a model, and appears framed in many respects to worry and fleece the traveller to the utmost possible extent. If safety in this matter were to be found purely in numbers, there would be little to say against the regulations. But, at Chamonix in these days, quantity is more conspicuous than quality. The first-rate Chamonix men are scarcely ever to be found waiting for their 'tour de rôle,' but are engaged for the season elsewhere.

The advice of an experienced mountaineer can usually be obtained as to taking a second guide or a porter, and the number of each required. The value of the entries in testimonial books (which all guides carry) can best be appreciated by those who have most often had occasion to write them. Failing an amateur of judgment, the hotelkeeper is the best person to consult, especially if he has had experience of mountain expeditions. But beware of the hotel porter's counsel.

The proper order of a party, when roped, is not a matter to be decided at haphazard. Under different conditions the order may best be varied from time to time. Without pretending to give a complete set of instructions, the following suggestions may serve to indicate the general principles. The best number for a party for an ordinary rock or snow mountain is three. A traveller of some proficiency, with a good guide and a porter who is really up to the rank of a second-class guide, can go fast and safely on a snow mountain. There is no objection whatever to adding a fourth; two travellers with a guide and a porter can go almost as fast and perhaps even more safely than a party of three. For a snow expedition—that is, one in which snow and ice work will probably form the chief difficulties—the numbers of a party may be largely increased even to eight or ten without finding much drawback from the number, except that the progress will be slower. It must always be remembered that the pace of a party is really regulated by that of its weakest member, and it is not easy as a rule to make up a large party all of whom are equally expert mountaineers. When the party is larger than seven or eight, it is best to divide into two sections: two parties of three or four, who keep together, will go better

Vingt-et-un

than one long caravan. Each section should of course be properly provided with professional aid. On rock expeditions four is, as a rule, quite enough on a rope. On a difficult rock mountain, such a party may require to extend over one hundred feet. Progress in any difficult place will be rather slow, and danger from falling stones dislodged by the uppermost climbers considerable.

The order that the party may best assume, expressed, for convenience, in tabular form, is somewhat as follows :—

With three climbers, let A be the leading guide, B a second guide or good porter, and C an amateur. The order on the rope for all expeditions would be A, C, B in ascending, B, C, A in descending.

A second amateur D is added. The same order cannot now be adhered to throughout, but varies with the conditions, and the difficulties; D is assumed to be a better mountaineer than C. For ascending, on a straightforward snow or rock expedition, A, C, B, D. For the most part the stronger of the two amateurs should go last.

For descending the whole order is simply reversed thus, D, B, C, A.

In ascending difficult rocks where a passage occurs just within the powers of the best of the amateurs, he had best go first. The order will then be D, A, C, B. It makes little difference whether B follows C or not, and it may save time to leave C last. The leading guide usually in such places shortens the rope and follows closely his 'Herr.' On rocks he can do so with safety. This order is advised by Melchior Anderegg. When the best of the amateurs is but a weak vessel, the order has usually to be A, D, B, C.

In descending the same passage, the best order would be D, B, C, A; D going first.

In ascending difficult snow or ice, or where step-cutting is involved, the order again is A, D, B, C. The need of a third professional to follow C is tolerably obvious if the amateurs are clumsy, nervous, or beginners.

The same principles can be adopted if the party consists of more than five. No doubt it is annoying to spend time in changing the order on the rope, but it is prudent to do so when the occasion really arises.

On rocks there is less objection to placing two amateurs together on the rope, but still it is a practice that should be avoided. Beginners get into dreadful difficulties with the rope on rocks, which tangles up in a manner marvellous to behold. Parties have been seen, through telescopes, in a condition resembling that of Laocoon and his family. Apart wholly from the fear of criticism and on the much more valid grounds of safety and comfort, it is best to do everything in the mountains, as far as lies in the power, in the right way.

As a point of mountaineering etiquette, it may be mentioned here that great caution must be observed if two independent parties find themselves on a mountain together. Every consideration should be shown to the slower party by the other. If a fast party catches up a slow one in ascending on a rock mountain, they have often no right to go on ahead, as there is always a chance of sending down loose stones. In descending, if the faster party choose to go ahead they do so at their own risk, but they have no right to expect that the slower party must necessarily halt altogether when the others chance to be below them, and in the line of fire. It is much better for all to keep together on rocks, and also on snow mountains, when any gully is being ascended.

An *early start* for an expedition is often half the battle The time of starting may in most cases be settled better by the traveller than by his guides. It is a good and a common plan to discuss during dinner overnight the time at which the party should start the next morning; it is better, but less common, to write the hour decided on down at once; it is best, but extremely rare, to act rigidly on the decision. Start at the earliest possible moment consistent with being able to progress continuously up to sunrise. To make a start too early, however, when there is difficult work a little way ahead which cannot be managed in the dark, is bad policy. To sit down for an hour or two in the cold is depressing, and takes the courage and strength alike out of a man. Do not merely adopt the estimate that the hotelkeeper, or guide, or guide-book tells you is needed for an expedition, and arrange your hour of starting accordingly, but get every minute of sunlight you can out of the day. A great point is to arrive fresh at the most difficult or the most interesting part of the expedition, and the man who feels that he is an hour or two earlier than he really need be, and has the day well before him, will go strongly and well ; while the man who feels that he is an hour or two late, becomes hurried, anxious and easily discouraged. There is nothing more delightful, after the fullest time has been enjoyed on the top of a pass, than to feel that there is ample time during the descent to wander a little, to vary the route, or trace out new expeditions ; and every expedition is a new one, as has been well said, to a man who has never undertaken it before. Let every sort and kind of preparation be made overnight. The prudent traveller at an hotel gets his bill in overnight, or at least endeavours to do so ; for it must be admitted that the worldly-wise landlord generally omits to oblige him in this respect. But the moral remains the same. It is often very difficult to get the provisions in and packed overnight, but this can be done if insisted on. In attacking a mountain new to the party, early starts are of absolutely vital importance. It is impossible to foresee all that may happen during the day, but it is tolerably certain that, in the majority of instances, the best weather is found in the early morning. Near the Italian frontier of the Alps, the clouds rise up about midday with almost clockwork regularity, and this condition will occur, too, day after day in the Caucasus. Travellers should be very peremptory with their guides and porters in seeing that their orders as to early starts are properly carried out.

Recollect, when starting from hotels, that the whole of the visitors are not necessarily interested in your departure ; they are only too anxious, as a matter of fact, to speed your going.

A start from an hotel ought to be effected as noiselessly as possible. There are some tourists whose waking is not brought about without a regular tattoo on their door. Being roused up, they relieve their feelings with a short outburst of song. They then begin dressing at the wrong end, and put on their heavy boots first. Soon, becoming concerned about the proceedings of their companions, they stamp about the passages and hail them noisily: they ring bells, and they whistle going downstairs, or drop tinned-meat cases on the top landings, and allow them to roll down with a plaintive clatter. They shout for their guides at the back doors, and run upstairs again to fetch something they have forgotten, and they jodel, or try to, before they are clear of the hotel. A well-conducted mountaineer always puts on his boots when close to the front door of his hotel, and slips away in a few minutes without a sound. He cannot possibly do better than take as his model a professional burglar skilled in all the details of his profession. But in truth the etiquette of mountaineering extends far beyond early starts from an hotel. People are apt to forget that when mountaineering or travelling they are in the position of guests, and that something is due to their hosts. Too many of our fellow-countrymen forget that the Alps are not merely for the climber. The mountains are, in fact, as many-sided in attraction as they are in architecture, and different people may enjoy them in different ways. There must be some give and take, and one great value of travel, in whatever form it is enjoyed, is to teach this fact. Not that the climbers are always to blame. Live and let live. Mountaineering 'shop' may be in bad taste unless all are equally inclined to talk it ; and so may any other form of 'shop.' Wherefore, let the little *salon* and the smoking-room be free of access to the readers and the smokers even on Saturday evening in the Alps.

The last words to be said before a start, whether at the beginning of an expedition or after a halt, should always be, 'Has anything been forgotten?' and a short list of the things always required on a mountain expedition, such as knife, drinking-cup, spectacles, compass, gloves, &c., may well be carried in the pocket-book and regularly referred to. If one man does this the others will do it. The forgetfulness of the average guide in small matters is something phenomenal.

Emergencies may arise of serious moment on the mountains. For example, a man may be hit with a stone or taken ill, and it is often difficult in such cases to decide promptly on the best line of action. One thing is at least certain, that no man who is ill or injured should ever be left alone. If the party consists of only three, one must remain behind and one must go on for help, even though he be exposed to the inevitable risk of having to cross snowfields unroped. It is not proper to leave a man alone in a hut if he is taken ill. In these days in the Alps, where in fine weather the mountains are tolerably crowded, where the hotels are always provided with telescopes, and where the huts are common objects of telescopic interest, the propriety of having some recognised signal of distress, known to the innkeepers and guides in all parts where climbing is practised, is worth considering. Such a plan, if adopted generally, might often save much valuable time spent in searching for assistance. The hoisting of a white or red flag on a stick on the roof of a hut would in nine cases out of ten probably be perceived within a few minutes ; and a simple well-marked light signal, say a red light, could easily be agreed to for use at night. Such signals could be left in every hut, for use if required, and could be extemporised without difficulty. A party travelling in an unexplored country should always agree upon some such signal. The matter is well worth the concerted attention of the various Alpine Clubs.

Some means of communication will often be found very convenient on the mountains when a party is widely separated, though it should be a rule for a party always to stick together throughout, and to arrive, as they start, together. The Morse code of signalling, by flag or other device, can be learnt without much difficulty, especially if the alphabet is carried in the pocket for reference. It requires much practice, however, to read a message accurately. Reading is much easier when a good pause is made between each letter signalled, but it is not advisable to make the movements indicating a letter too slowly. With a bit of looking-glass properly fixed, messages can be signalled at an enormous distance on the mountains. Reading a message is much easier when one person keeps his eye on the flag and another writes down the letters as they are recognised and called out. An axe with a handkerchief tied to it makes a sufficiently good flag. The movements must be made very sharply and decidedly, and a strong distinction observed between the longs and the shorts. To read a message at a distance of a mile would require good glasses. But on the mountains two parties might be unable to reach each other for some hours, though only a short distance intervenes. A simpler method of flag-signalling, and one easily learnt but very much slower, was described not long ago in a newspaper letter by a military authority. The movements in this system are always made from the central vertical line of the body. The signaller and reader face each other. The first half of the alphabet consists of decided movements to the signaller's right, and the second half of movements to the signaller's left. There are no longs or shorts. One wag to the right is A, two B, and so on. One wag to the left will be N, three will be P, and so on. The only point to be observed in reading is that the observer must not count the number of wags, but must for every letter run through the alphabet as far as necessary. Thus if the signaller gives five wags to the right the observer must repeat A, B, C, D, E. This method of signalling is more rapid if the alphabet is divided into three sections. Thus the wags to the right go from A—H inclusive; those to the left from I—P inclusive; for the remaining letters the flag is held above the head and waved to and fro, beginning a little on the right and going through a quarter of a circle to the left of the middle line; each set of movement describes a quarter of a circle.

The writer has employed this method, and found, after

THE PRINCIPLES OF MOUNTAINEERING 121

a day or two of practice, that messages can be sent with a very fair rapidity and be readily understood.

Weather signs.—Mountaineering is unfortunately a sport that is largely dependent on the weather for success. This is especially true for those who seek their pleasure in high places. Certain weather signs are tolerably constant in all mountain districts. It is, generally speaking, as true in the mountains as elsewhere that a red sunset predicts fine weather for the morrow, and that a red sunrise is of evil omen. The French proverb runs

> Le rouge soir et blanc matin
> Font réjouir le pélerin ;

and in most languages the saying has its parallel. A northerly wind in the Alps, for the most part, signifies fine and settled weather. During the summer months in the Mid-European Alps, there is generally a point or two of east in the north wind. This is a cold wind. The southerly or south-west wind in the Alps brings up the moisture from the Mediterranean, which condenses on the snow mountains and produces clouds and rain.

The ' Föhnwind' in the Alps often blows with tremendous force ; it is a warm south-east wind, most common in the spring and the autumn. The melting of the winter snow in the Alps is chiefly due to this wind. In some valleys it is rare, and in some very common ; for instance, at Grindelwald, in the neighbouring Haslithal (the Meiringen valley) and the Reuss valley, it is very common. In the Kanderthal, in which Kandersteg is situated, which is at no very great distance from the valleys mentioned, it is very infrequent, this valley being altogether protected on the south side. Little mountaineering can be done when the Föhnwind blows, but it does not often occur in summer. A true south wind will constantly bring a day or two of extremely fine weather, but it is apt to break suddenly and end in rain. A single day may often be snatched when the wind is nearly due south. Rocks will be found warm and the snow very soft.

To estimate the true direction of wind when in a mountain

valley, the higher strata of clouds must alone be observed. All kinds of eddying currents are produced, especially when a south wind is blowing, by the variations of temperature in different parts of the country, and at different elevations. Nothing is more uncertain in the summer than the weather that is likely to follow during the day, when there are clouds low down, or even drizzling rain, with little or no wind at one or two o'clock in the morning. Many of these days, however, turn out fine, and the keen mountaineer will very often be rewarded if he perseveres at least till sunrise. Often a thick mist hangs over valleys and the low hills, when the upper stratum of air is cold and clear, and this is constantly a condition that precedes a fine day. If any trace of blue tint can be seen on looking directly upwards through the mist, persevere. Do not be driven back by merely damp weather until at least you have got as far on the upward journey as is compatible with prudence.

Clouds to leeward and not 'in the wind's eye' do not signify bad weather. Long wisps of grey cloud that hang about the mountain sides and rise or fall rather than drift to and fro commonly presage evil weather when the upper sky is overcast. A very eminent authority has noted that an intensely deep blue in the sky often precedes bad weather. Showery weather with light winds often gives the most beautiful cloud effects and need seldom stop an expedition. In summer, thunderstorms that occur with short warning are frequently single, and are followed by a bright and cool atmosphere. But when the weather has been close and threatening for some days, with distant thunder only in the evenings, the storms, when they do break, are likely to occur in a series lasting for some hours. In a mountainous country the topographical meteorology should always be inquired after of those who have the best opportunities of watching it. The shepherds can generally give very trustworthy predictions about the weather. General rules are often misleading as regards a single day in a particular locality. Wind is a much more formidable obstacle and danger than mere rain or snow. The presence of high winds on the heights can often be recognised by an appearance like a bright band of misty cloud clinging to the crest of a high mountain. This is really a thin veil of snow whisked up by the wind. It looks innocent from below, but is of serious portent.

Most mountaineers of experience will admit that they have on many occasions been deceived by the appearance of the weather at one or two in the morning, and got up at seven or eight to find a perfect day wasted, as far as high mountaineering is concerned. Anglers say that the man who keeps his line longest in the water catches most fish, and assuredly the man who always starts, if he has made up his mind to do so overnight, and goes at least as far as he can, will get the most pleasure out of the mountains.

In all expeditions, whatever the number of the party, it is well to appoint the most experienced mountaineer as leader or rather the captain of the expedition. There should be no doubt as to who *is* leader. The arrangement will relieve the leading guide of some responsibility, and will often enable him, in cases of emergency, to give his opinion with more decision. A question that constantly arises, and one on which even first-rate guides are prone to speak with rather an uncertain voice, is the propriety of abandoning an expedition in bad or doubtful weather. The voice of the captain ought to carry as much weight on such occasions on the mountain-side as in the cricket field. He should take counsel with his leading guide, and in the event of obtaining no decided opinion should not hesitate to decide promptly as he thinks best in the interests of safety. Too often the guide but answers, when appealed to, ' Whatever you think best,' or 'What you will.' The abandonment of an expedition may mean loss of money, and this consideration, it is to be feared, has often led to an unwise decision. Now and again mountaineers may even succeed in persuading a guide to persevere really against his better judgment. No one, naturally, likes to turn back when arrived at the point which in Alpine literature is generically termed 300 feet, or 'a stone's throw' (up or down is not particularised) from the top. A decision has to be arrived at promptly. The compromise, so often adopted, of going on for another few minutes and then seeing is commonly unwise, and in the vast majority of instances when the question seriously arises, 'Ought we to turn back or go on?' the answer should be 'We ought to turn back.' On no occasion is the distinction between the climber and the mountaineer brought into more marked prominence. The climber urges 'Go on,' because it is possible to do so ; the mountaineer says, 'It is possible, but it is also unwise and imprudent, and therefore we will turn back.' On the other hand, when the more faint-hearted sometimes question the propriety of persevering, the captain of an expedition may, after consultation with his leading guide, decide that it is safe to persevere. Many an expedition has in this way been brought to a successful issue without any undue risk being run. If a man is known to have the moral courage to decide on retreat when that seems to him the proper course, he has a much better chance of being followed when he determines to advance. Sometimes, in order to retreat in good order, an advance has to be made first. For instance, when bad weather arises on a pass, the best course may be to make a bold push for the summit. The slopes just below the top on either side of most passes are easy going. On the summit, if the pass is a well-marked one bounded by steep walls, the force of wind is felt at its worst, but there is an extraordinary difference a few feet below. Of course much depends on the direction of the wind.

It appears to be generally agreed that in the Alps the most suitable months for mountaineering are July and August. No doubt the fact that these are the principal holiday months for our countrymen has largely determined the choice, but in many years it is not altogether a wise one. As a general rule, the weather is more unsettled during July and August than during any other months of the year. On the other hand, the days are still fairly long, and the winter snow is

THE PRINCIPLES OF MOUNTAINEERING 125

probably reduced to a minimum. Snow passes and snow mountains are consequently less laborious. Sometimes the steepest rock peaks are only accessible during August, but in a great many years rock peaks will be found as free of snow and ice in September as in August, and the weather is, as a rule, much more settled during September. In the central parts of Switzerland it is a very common thing for the weather to break altogether for a while during the latter half of August, and often very heavy snowfalls take place. When this occurs September is nearly always a fine month. The shortness of the days is less of an objection than it used to be, now that the routes over the lower parts are well known, while what may be called the hotel line is much higher. In short, the mountaineer may start at as early an hour in September for an expedition as in July. If he loves a crowd, if he likes to go to a place where luxuries are provided and comfort is studied, at a time of year when he can neither obtain the luxuries nor enjoy the comfort, let him select July and August. He must be prepared often to return discontented. The valleys are then intolerably hot and dusty. Fine days are few and far between. The sub-Alpine scenery is seen at its worst, and the scenery above the snow-line is very frequently veiled by clouds. In June and September sub-Alpine scenery is seen almost always at its best, and many of the higher mountains and the principal passes are perfectly practicable.

The truth is that people have somewhat servilely followed the example of those who first explored the high Alps. Most of the pioneers achieved their expeditions during July and August. It is still the case that for any new or difficult expedition that may be devised these are perhaps the best months, though it is often necessary to wait long, and then to snatch a fine day. But now that roads are well known and that facilities for travelling are enormously greater, not only to the base of the mountains but up them, the period over which mountaineering is possible has been really extended far more than people seem to imagine. The Swiss themselves are not less keen about the mountains than our own countrymen. Finding themselves crowded out of their own land during July and August, partly by the rush and partly by the raising of the prices during the brief tourist season, they have to visit the mountains at other times—in May or June, in September or October. During these months they are able with perfect comfort and safety to achieve much. The aspect of mountains varies greatly, and he who sees the Alps only during the same period of the year knows them only in one of their phases.

It is possible, as many have shown, to climb the high mountains in winter. Mont Blanc, the Schreckhorn, the Jungfrau, and many others, have been ascended safely enough in winter. Rock peaks should be avoided. All who have practised winter mountaineering speak most enthusiastically of the charm. A little of the attraction may be due to the fact that the novelty has not yet worn off. But there is much more to recommend the pursuit. In the Alps during the months, say, of January and February, long spells of absolutely perfect weather are the rule. Day after day is cloudless, still, and bright. During the hours of sunshine the warmth is such that the temperature indicated by the thermometer seems incredible. The clearness of the air is phenomenal, and the delicacy of the tints on the snowfields a revelation to those who know the high regions only under the fierce glare of a midsummer sun, when the eye is dazzled and confused by the intense illumination. To sit on the summit of a peak, basking in the rays of the sun at a time of year associated with fogs and slush in the minds of so many of our countrymen, is a sensation that is well worth realising. Travel is cheap, hotels are moderate, and the hotelkeepers are able to do themselves justice. On the other hand, the days are short, and high mountain expeditions occupy more time. A full moon is a great assistance.

The expense is greater than in summer. It is generally necessary to sleep out more than one night. Extra firewood and provisions have to be taken, and the bill for porterage quickly runs up. The huts will usually be found in good order inside, but much work is often needed in clearing away snow before access can be obtained. Extra rugs or blankets should be taken. At times, owing to the culpable carelessness of late summer visitors, the huts are found full of snow. The most laborious part of a high ascent is often the walk to the hut, and there is some risk of snow avalanches. Much of the baggage, however, can be conveyed in sledges for a great part of the way, and the last part of the descent by tobogganing, when practicable, is by no means the least enjoyable.

The ice-falls and more rugged parts of the ice scenery are seen to better advantage in summer, and the higher snowfields, save in the matter of colour, show no difference. To the writer's mind the greatest beauties in winter are to be found in the sub-Alpine regions. The completeness of the transformation is a never-ending attraction in itself. It must not be assumed that in winter the lower mountains—hills with mule paths or broad tracks—are necessarily easy or safe. No hills or mountains where the track lies in part over rock faces should be attempted in winter for some days after a 'Föhnwind' has been blowing. There is then the greatest risk from the form of avalanche known as the 'Schild-Lawine' in the Oberland. The cold at night is intense, and the sudden lowering of temperature directly the sun goes down very striking. Unless due precautions are taken there is risk of frostbite. To guard against this, keep warm, but, above all, keep dry. Winter expeditions to the Alps are becoming year by year more popular. Almost as many now go for pleasure as for health. Still, probably on account of the drawbacks set forth above, comparatively few high expeditions are made. Perhaps the seductions of skating, tobogganing, and the like, restrain some ; perhaps the attractions to be found in the sub-Alpine regions suffice for others, who must be hard, indeed, to please if they cannot satisfy themselves with their infinite variety and beauty. M. Loppé, whose experience of winter in the Alps as a traveller is unrivalled, considers that the latter half of January and the month of February is the best time in most years for settled weather. In March the weather begins to break. Snowfalls alternate with warm winds and rain. The playground is in disorder while preparations are made, as it were, for the tourist season.

Pickel'd herren

CHAPTER V

RECONNOITRING

BY C. T. DENT

Übermorgen's Arbeit

F all the departments of mountaineering, reconnoitring, to good purpose, is at once the most difficult and the most neglected. It is one that requires long and thoughtful experience; one in which any man with some natural aptitude may excel if he will take the trouble, and one in which it is very rare to find an expert. Were it not for the fact that this book is chiefly composed of unasked-for advice, the writer would hardly dare to offer the first item of counsel to be given to a party desirous of attacking a mountain unknown to them. That advice simply is, to look at the peak before they try to climb it. Yet this apparently obvious suggestion is not so unnecessary. Many an expedition has failed because the climbers neglected to see beforehand where they were going to. Comparatively few have failed simply because a wrong line was chosen from a suitable reconnoit-

K

ring spot. In the Alps, unfortunately, these preliminary 're-connoissances' are almost things of the past. Every step of the stock expeditions is either known to the guides, or minutely laid down in descriptions of previous ascents. The broad general view of a given mountain is ignored so far as the traveller, at any rate, is concerned. The utmost that a climbing party in the Alps now commonly does is to make sure beforehand of certain points of detail :—whether, e.g. a partially concealed slope is of ice or of snow, whether some great crevasse is easily passable or not, and so on. Too often, in order to obtain the information desired, the party proceeds on entirely wrong principles.

A mountaineer, even of some experience, is frequently at a loss if he endeavours to trace out from a distance a route which he has followed perhaps only the day before. He will sweep all over the mountain with his telescope, in the hope of lighting on tracks or ice-steps. Failing to recognise any, he will evolve a line of ascent, sometimes a very surprising one; for remarkably brilliant climbs can be made, in imagination, through a field-glass. Satisfied with the reflection, often insisted on, that appearances are deceptive through the telescope, he will lay it by and remain in sublime and contented ignorance of what he has done, and of what he has been looking at. The lesson to be learned is writ large all over the field of view, but too few take the trouble to learn even the alphabet of the language in which it is set forth.

Mountaineering is, beyond all other sports, one that it is imperative to take up, to a certain extent, seriously. If a man intends to climb high, it is his bounden duty to acquire all the proficiency he can develop in himself, for the sake of his companions.

Now, the broad principles of mountaineering must be first learned if proficiency is, as it should be, aimed at; and a cardinal principle is, that a mountain should be, for climbing purposes, considered first as a whole and subsequently in detail, the detail being necessarily best studied on the spot.

Such a method is, however, exactly the reverse of the ordinary practice. It is of no use to work out minutely a passage through some crevasses which leads in a wrong direction, or to discover a practicable climb up some rocks that are crowned by masses of hanging glacier depending below an inaccessible ridge. Errors such as these are very elementary. Into mistakes, however, almost as obvious, mountaineers and guides alike are prone to fall in strange countries or unknown districts. A good judge of a mountain is at least as rare as a good judge of a horse, but it is as easy to get an opinion on the one subject as on the other. There is no actual need in known districts to reconnoitre every peak or pass that the mountaineer intends to ascend or cross; it will be possible to get up the mountain or over the pass without any preliminary survey. But the practice may be turned to good account some day, and the man who delights in learning all that can be learnt on the spot will never fail to make the best of his opportunities for comparing a previous estimate with the actual experience.

Following routine, or acting on the persuasion of the guides, mountaineers usually start from the valley late in the afternoon and make straight for their sleeping-place, thus taking the most laborious part of their walk at the hottest time of the day, and seeing nothing of their route for the morrow. If the party left their hotel early enough to reach some suitable vantage point for survey, the time would never prove to be ill-spent. It is really less fatiguing to be walking or climbing at a high level than making preparations to do so at a lower one. If the camp is far distant, and the walk to it in itself constitutes a good day's work, the mountaineer who has started in good time, though unable to survey the peak on which he is bent, has ample opportunities of studying many other points. He can hardly employ his leisure better than in planning lines of ascent to every mountain and pass that he can see around him, and especially in tracing out routes to the same places from varying points of view. Such observations can be readily checked by inquiries of the guides, or by reference to written descriptions.

The earliest lesson to be learnt by the traveller is that he is liable to fall into egregious errors. He will find it difficult at first, even with the map spread before him, to understand the broad features of the scene he is gazing at, or to recognise the principal landmarks from a photograph taken from a slightly different point of view. It is not enough to identify landmarks such as an ice-fall or a huge buttress of rock. Their position and relation to other landmarks have to be grasped, so that any point may be recognised when looked at from an altered distant point of view, or still more difficult, recognised when actually reached. Their broad relations and comparative levels form the best guides. Curiosities of shape or character vary inconceivably when gazed at from afar and viewed near. A fantastic tower of rock, for instance, appears a colossal and characteristic landmark to the climber on the mountain, but when looked at from a mile or two away, merges into and is lost in the face. The curiously-shaped patch of snow seen from a distance and noted carefully appears to melt away when approached. Yet the mountaineer may feel absolutely certain of the position of the rock-tower and the snow-patch if he has some broad indication of the level at which they are to be found. It is excellent practice for the traveller to carry a photograph of the mountain or pass on which he is climbing, and endeavour to recognise the places in detail. The better known the mountain the more profitable will the exercise be, as the corrections can be made with more ease and certainty. Take, for instance, a photograph of the Matterhorn as seen from the Riffel. The great white streak like a high-road, shown in the photograph to run right across the eastern face, will not readily be recognised by a man on the peak. Yet it looks a strongly marked guide. Innumerable towers of weird and wonderful shape will unfold on the Zermatt ridge (the easiest line of ascent) of which the photograph shows nothing. The Matterhorn is taken only as a familiar instance of a mountain known to many. The same principle applies with equal force to the smallest hills.

Constant practice alone gives the mountaineer the ability to appreciate minutely the scale of the surrounding scenery, a matter of no slight difficulty, but which, when once acquired, will always stand him in good stead. Even when he has acquired some proficiency, he must bear in mind it is probably in but one district, say the Alps. Good and delicate judgment is necessary, if he finds himself in other mountain regions, to make a proper allowance ; and an acute sense of proportion in this respect is indeed a rare gift.

A person standing on Mont Brévent and looking across the valley of Chamonix sees the rocks of the Grands Mulets at a distance of about four and a half miles (see fig. 11. p. 386). These rocks are some 300 feet in height. Yet persons wholly unaccustomed to the scale of the scenery have been known in all seriousness to mistake them for a party ascending Mont Blanc, and tourists interested in high mountaineering sometimes watch the buttress diligently for long periods of time, and express surprise at its slow progress. The writer well remembers standing on the commencement of the Crib Goch ridge of Snowdon one bright winter's day, with Melchior Anderegg, a Swiss guide of almost matchless experience and of great intelligence. The little peak was covered with snow from top to bottom, and looked like a miniature Weisshorn. As the party looked up, the guide was asked how long he thought it would take to reach the summit. It was then one or two o'clock in the afternoon, and he answered that he thought we should take some three hours, and could scarcely credit our assurance that we should be there in an hour. 'In my country,' he remarked, 'I should have said that we were already too late to have any chance of completing the expedition.' As it happened we reached the top in almost exactly sixty minutes. It is more serious, of course, when the mistake is made in the opposite direction, and when height and distance are under-estimated.

In the Caucasus, the Himalaya, and suchlike mountain-ranges, Alpine guides will almost uniformly under-estimate heights and distances. So, too, will the traveller, unless he proceeds on some principle. It is in the detail of climbing that the guide shines pre-eminent, and his 'reconnoissance' of a mountain from a distant view-point is often a very haphazard affair, in which he depends more on his memory than his judgment. In unknown districts, on the other hand, the amateur, who can make some use of map and compass, who is able to deduce information of practical value from the writings of others, and who will take the trouble to use his reasoning powers as much on the mountain-side as in the ordinary affairs of life, is capable of forming judgments which will prove of signal value, and he should not neglect an essential branch of the sport, a branch, too, in which he is fitted naturally to excel.

The first point in reconnoitring is to select a spot that gives a comprehensive view of the mountain. If the peak juts up from a ridge bounding one side of the valley, the proper view-point will be found on the opposite side of the valley. When the mountain has been studied fully, make, if time allows, a level traverse, and observe the face again. The detail of the modelling will be rendered clearer by comparative observations, especially on rock mountains. Probably all the necessary observations can be made by ascending to a height of about half that of the objective peak. In the diagrammatic 'Beispielspitz' shown in the illustration, the observer is assumed to be standing on such a point. It may be again insisted on that if a mountain is, say on the east side of a valley, it is better to ascend for 1,000 feet or so on the west side, than to go up 6,000 feet on any part of the mass of the mountain itself. If, however, the mountain closes in the head of a valley, the traveller has the choice of the two sides, and assuming both to be equally easy of access, had better select the slopes which are most free from trees, and most detached from the main peak. If the mountain is a culminating or central point of some high stretch of snowfields, the final peak will almost certainly be easy of access, and the best route to the upper snow-fields would then form the principal point for consideration. From any such view-point as has been described the upper part of the mountain must necessarily be foreshortened, and due allowance must be made for this. Compare, as an illustration, the apparent slope of the roof of a house as viewed from the street and from the second-floor window of the house opposite.

With a view to reducing theory to practice, a reconnoissance of a typical mountain such as the 'Beispielspitz' may now be made. Certain broad features at once strike the eye. The mountain is peaked, culminating in a sharp top. Long rock ridges run down, enclosing corries at the base of the faces. The faces are all steep, and resemble each other in character. In structure the peak is assuredly composed of a hard schist that has been worn and weathered to its present pointed shape. The agencies of wind and weather, heat and cold, that have determined its present shape will still be found at work; crumbling and disintegration will still be going on as for ages past. The general dip is from east to west.

We may assume that the mountain is on the south side of a valley, and is 14,000 feet high, and that the observer stands at a height of about 8,000 feet on the north side. Two distinct ridges are seen, running broadly east and west, while a more distant edge, having a south-westerly direction, is visible over the right-hand shoulder. The apparent distance of the peak is very deceptive, but of course can be gauged accurately if the district has been mapped. The glacier flowing down from the flanks will appear steeper than it really is; the complexity and extent of the ice-falls will afford a better test of its real inclination and difficulty than the appearance, while from the shadows some idea can be formed as to the extent and angle of inclination of the more level stretches. Such shadows are hardly shown in the diagram. The extent of the upper snow-basins, which are pure in tone and almost unbroken in line, cannot be judged by reason of their foreshortening, but this is a matter of no great importance. The inclination of the rocks bounding the snow-fields, and the shadows thrown on their surface, give the best measure. The run of the ridges visible, and the character of the face, which is composed partly of rock and partly of steep snow slopes, render it probable that the mountain is more or less bayonet-shaped, and that the side towards the observer is at least as practicable as the south, hidden from view, is likely to be. We have then to devise a route up this north side.

The highest part of the peak may be considered first. There is here less choice of route, and the details of the line to be adopted over the glaciers and snowfields below depend on the point at which the principal attack on the actual peak is to be made. Moreover, clouds are apt to settle on the higher parts after mid-day, and cling to them obstinately. Every advantage must be taken of a clear view.

The final part of most mountains, for reasons explained elsewhere (p. 222), consists of two or more rocky walls which may be looked upon as coalescing at the summit. Between these walls are enclosed the faces of the mountain. The top edge of the gable is termed the *arête*, or ridge. It is along the very crest of this ridge that the proper route lies in all cases, and the main object of the mountaineer is to get on to the ridge at as high a level as possible, and when once on the crest never to quit it if he can avoid doing so. The face of the peak is to be used only to give access to the ridge. Only in the rarest instances is it right to attempt to climb straight up to the summit by the face, without touching the ridges at all. It is of vital importance to remember that, in dealing with a mountain, the descent must be considered as well as the ascent. Every passage must be looked at from two points of view, therefore, and the possible alteration of conditions of snow and rock between morning and evening must be borne throughout in mind. The route up obviously, then, lies either by the north-east or north-west ridge ; the next point is to consider which of the two is better. The north-west is more foreshortened than the north east, and it is therefore more difficult

THE BEISPIELSPITZ

to judge of, but, on the other hand, the steepness may be less than it appears.

The chief troubles the mountaineer experiences on the ridges are due, as a rule, to cold and wind. North winds prevail in fine weather, and these are often bitterly cold. There is an extraordinary difference between the force and coldness of a wind on an actual ridge and on a mountain face. The climber while on the face, even on the windward aspect, is but little hindered by cold wind. It is sometimes judicious to keep a little longer than usual on the face, when possible, if strong winds are blowing. Wind is more trying to a party on snow than on rock ; most serious of all on an icy ridge where a long series of steps has to be cut. The north-east ridge, being on the whole less exposed to wind, has an advantage over the north-west.

The last part of the north-west ridge, curving up to the summit in an unbroken line, is clearly possible. If the crest, however, is of ice, step-cutting will be found necessary for the whole length. Now an ice slope shines in the sun very much more than one composed of snow, and in brilliant sunlight glistens with a metallic lustre ; in shadow it has a wet, blue-grey appearance. The icy surface is the result of the daily melting of snow above. Rocks in the neighbourhood of ice slopes are apt to be coated with a thin layer of ice formed in the same way. The water trickles gently in the thinnest possible sheet over the hard face, and is caught and frozen at night, or when the face is thrown into shade. These are the 'glazed' rocks, always to be avoided. The rock patches on the west ridge might be in this condition. 'Glazed' rocks shine when the sun is on them and their surface is wet.

If the ridge is mainly of snow it is feasible, and has an additional advantage in that the sun will be on it earlier. Further down, the north-west ridge is interrupted by some patches of rock. Their direction reveals the important point, borne out by a comparison with other parts of the mountain, that the general direction of the dip is downwards from west to east. It would be easier, therefore, to climb these rocks, or any others on any part of the mountain, from west to east, as explained in the chapter on Rock Climbing (p. 258). This is a point in favour of the north-west ridge. Looking further down the mountain for a moment, it is tolerably clear that access can be obtained to this north-west ridge, and so far everything seems in its favour. A hasty survey, in fact, might not improbably have led the party to select it. Close examination of the rocks, however, shows that these are cut very straight down on the west side. Such faces appear very small from a distance, but it must be remembered that a vertical slab of rock twelve feet or so in height will be quite sufficient to stop a party in ascending. Elsewhere, when we look to the rocks facing west, the same character is found to obtain. This feature alone must give the mountaineer pause, for on the north-west ridge these rocky interruptions are repeated more than once. If these rocky obstacles prove insurmountable, the climber will be forced to one of two alternatives : (1) He will have to desert the crest of the north-west ridge at first, and make his way up the north face, so as to strike the crest above the last patch of rocks ; or (2) He may keep on the snow-crest between the rock interruptions, and circumvent these by crossing more or less horizontally on to the north face : then turning straight up he can make for the ridge again. The alternative routes are indicated in dotted lines in the illustration.

The first alternative would lead him under a hanging glacier dependent on the north slope, the treacherous nature of which can be estimated by the avalanche fragments seen at the base of the slope at A. In the very early morning, and before the sun is on the glacier, it might be possible to cross over this *débris* without much risk. Ice avalanches, the result of breaking off from the end of the hanging glacier, fall principally in such places in the heat of the sun ; when covered up early in the year by much snow they occur less frequently. But at such a time there is much fresh snow also on the slopes above, such as those of this north face. Masses are constantly sliding down over the smooth surface, and falls are of perpetual occurrence throughout the day. One form of avalanche forbids the ridge as peremptorily as the other. There is a further objection to the north face. A great crevasse (the 'Bergschrund') is seen bounding the head of the corrie. Opposite the middle line of the north face it is interrupted, and it is here that it would be easiest to cross. But this is the very line of the avalanches that fall from the face, all of which are directed, by the inclination of the slopes above, towards the centre. To cross this crevasse, therefore, the climber would have to go deliberately to the most dangerous place, and then ascend in the line of the avalanche tracks.

If the small rock face (B) proves too steep, the climber will be forced on to the face, making what is technically known as a 'traverse.' Now a traverse should be avoided whenever it is possible to do so, for reasons given elsewhere (p. 175). It is almost certain that the traverse in this instance would involve a great amount of very difficult step-cutting, first horizontally or downwards, and then again directly upwards. The amount of time thus consumed might prove fatal to the success of the expedition. The southern side may be better than the face in view. The mountaineer had better assume that it is worse, as indeed it is likely to be, for reasons to be presently given. For the moment it suffices to urge that the invisible had better be regarded as the inaccessible. Direct evidence is better than circumstantial. This north-west ridge, tempting though it appears, must be abandoned as a bad and a dangerous line.

We have, therefore, to turn our attention to the north-east ridge. This, as so often is the case, is 'mixed,' that is to say, partly of snow and partly of rock. The final snow portion is plainly feasible. At the very summit of the mountain may be noticed, overhanging to the south, an cave or cornice (C). Far away to the left (at D) this appearance is repeated. The mountain is probably, therefore, corniced on the south aspect, and as the whole of the snow crest may have this feature, care will have to be taken not to keep too close to the edge. The proper track will lie a little on the north side of the snow ridge. The same corniced formation might readily obtain on the southern side of the north-west ridge, constituting an additional objection to that ridge, for it would preclude a traverse in safety on the south side. Information on such details as presence, direction and extent of cornice, all of which are of the highest importance, can be got more effectively by a preliminary survey from a distant point of view than by any other method.

The next point is to consider the route by which the north-east ridge can be reached high up. The main part of the snow-covered north face, swept by avalanches, as the longitudinal marks show, is out of the question. The great rock-rib running right down to K, and dividing off the central from the east glacier, looks tempting. Owing to the dip of the strata it would be easiest to attack it from the west, but above (at E) a gap occurs where a great rock cleft has been formed, bounded on its upper side by one of the steep cut-away faces with which we are already familiar on the north-west ridge. Now, of the depth and detailed character of this cleft it is impossible to judge from a distance. It is highly probable that the gap may prove far less formidable than it appears, but this must remain uncertain. Careful note should be taken of its level, which is nearly on a line with the terminations of the hanging glaciers on the north face and below the saddle (H); on a level, too, with the lowest rock patch on the north-west ridge. If the gap (E) proves feasible, success is pretty well assured. Undoubtedly, by means of this rock-rib, the climber could ascend to a great height, and such buttresses generally form an excellent line of ascent. By keeping along their crest there is perfect safety from avalanches, however frequent these may be. Falling snow or ice must of necessity pass down on either side of the projecting rocks; falling stones too, if any, would be deflected to the right or left. This rock buttress may be kept in reserve as a likely line in case no better can be found.

Further east, running up the rock wall, two snow gullies (F

and G) are noticed. Such gullies are spoken of as 'couloirs,' another of the terms which, with numerous alien companions, has emigrated into the mountaineer's vocabulary, settled down, and become partially naturalised, to the discredit of our language. Gullies, filled with snow, form an admirable and a constantly employed line of access to the ridges above them. But gullies are not invariably lined with snow for the convenience of the mountaineer. They may be full of ice, and if shallow and exposed to the sun they often are. They may be full of ice covered over with a lightly adherent layer of snow, a treacherous condition which cannot be accurately judged of from a distance. Some idea may be formed by noticing the extent to which the great crevasse or fosse that defends the base is filled up by snow that has slid down. If the lining of the gully is glistening and wet-looking at one place and white snow at another, there is a probability of its being treacherous. If it can be made out that water is coursing down the gully in little runnels, avoid it.

Snow gullies may terminate in blind ends above, or branch off and become lost on the rock face. Where they terminate abruptly above, short of the crest, it is certain that the rocks crowning them will be steep. Where they branch off and are gradually lost, the rocks above them will probably be smooth and in slabs. All these forms are shown in the drawing. The east hand branch of the gully (G), however, runs right to the crest at G' and forms the most promising track. So large a gully will certainly continue the whole way down the rock face, though below it is partly hidden from our point of view. Fewer stones, perhaps, will fall down this gully, for there is less rock above for them to start from. There is not much, however, to choose between F and G in this respect. The gully G is more sheltered from sun; the more likely, therefore, to contain snow and not ice. If during the day-time, when the sun is on this part of the mountain, the surface of the snow is seen to glisten from the wet, the gully will certainly be found full of ice ; and if this is the case, the ascent will have to be made by the rock immediately to its right or left. The ascent further east on to the ridge to the saddle marked H is forbidden by the hanging glacier above, and also by the rocky towers [1] on the ridge. Such towers look very inoffensive from a distance, but a jagged, splintered crest of rock often presents very great difficulties in detail, and a single uncompromising pinnacle, though small, may bar all progress. Towers on a ridge, being exposed to extremes of heat and cold, are commonly weathered, and worn into fantastic shapes that are most admired by the mountaineer when he is on the right side of the obstacles.

It remains now to consider the best line of access to the gully, and the best place to get over the great crevasse that will be found at its base, supposing that ultimately this line has to be adopted instead of the rock buttress. Such crevasses are usually best crossed at the foot of a gully, for the reasons already pointed out. The same route over the snowfields will lead to the rock buttress, so that the two lines of attack can be considered together.

From this point of the survey we may descend to the lower levels and work upwards, noticing only before we quit the higher snowfields, that the east glacier between the ice-falls lies at a lower level than the central one.

The whole of the rocks on the extreme east side of the glacier bed, that is to say on its right bank,[2] are steep, and in accordance with the general character of the peak will probably be cut off towards the glacier. On the left bank the trees extend up higher. Above the trees will be found scrub, and at the upper level of this scrub will certainly be a good place for a camp, as at J. No firewood need be carried up; water will be abundant, for if there are no springs, the snow will still be found lying in patches, as the general aspect of the slope faces

---

[1] A tower on a ridge is sometimes called, in Alpine slang, a 'gendarme.' There is absolutely nothing to recommend the adoption of so senseless a term.

[2] As in the case of a river, in speaking of the sides of a glacier, the observer is assumed to be standing on the ice and looking in the direction of the flow.

north. If the lower slopes (below the illustration) are very steep, and it is necessary to get on to the glacier, the lowest ice-fall shown will be best circumvented on the right side of the glacier. On this, being the concave side, the marginal crevasses will be less numerous. The glacier can then be crossed above its lowest ice-fall to the camping place, J. The terminal moraine will probably give a good line of ascent.

Starting up from the camp the track will lead along the trough between the left lateral moraine and the rock slope. Here the winter snow is still lying, and this offers always an admirable route. Moreover, lateral moraines at so high a level offer bad going, and the ice itself will be irregular and troublesome. At the end of the snow-patch it will be necessary to take to the glacier, keeping well to its left side round the buttress of rock forming its left bank, for here again, as the glacier is more compressed at its turning point, the crevasses of the ice-fall will be less troublesome. A lateral glacier, too, joins the central ice stream at this point, and further compresses the ice and obliterates crevasses. Once on the *névé* level, all is plain sailing for a time, and a straight course can be made to the right side of the central glacier and the termination of the rock buttress. In crossing these snow-fields, however, it must be remembered that the party will be proceeding, generally, parallel to the direction of the crevasses, (which, though they may be hidden, are none the less sure to be present), and not crossing them at a right angle. A long rope will be necessary. No attempt should be made to cross to the east glacier, for there will be no passage to it above the patch of rocks seen in the middle of the ice-fall. Such a patch occurring in the middle of a glacier is known as a 'hot plate.' Portions of the ice mass, as the glacier moves on, break off at the summit of this cliff, and the base will be strewn with the fragments of ice avalanches. These fallen masses choke up the crevasses conveniently, but it would be folly to pass within the line of fire, and if the glacier is crossed at a lower point still, trouble will be found in working through the marginal crevasses. In addition, this east glacier has a narrow neck above, at the level of the termination of the great rock buttress, and then expands below this point. Where such lateral expansion takes place longitudinal crevasses form, intersecting the transverse cracks, and complicated ice-work must be expected. The rocks on the right bank of the east glacier are ice-worn and terraced—that is to say, ground smooth by the ice when the level of the glacier was higher. Such rocks are always bad, and are best avoided. It is of cardinal importance to find the easiest and quickest track to the base of the final wall in all cases whenever a difficult mountain is being attacked.

An hour spent in disentangling a route through a trifling ice-fall, necessitating perhaps retracing the steps two or three times, or the exertion consequent on climbing early in the morning difficult rocks out of the true line of ascent, may just ruin the success of the expedition. There will be sport enough in the gully and on the ridge; and by lantern-light or in the very early hours of the morning men are not at their best, and difficulties always appear exaggerated. A tired man will give up as hopeless obstacles which he is perfectly capable of surmounting with ease and safety. Nothing is more disheartening than to find the sun well up, and some of the best hours of daylight gone, before the most serious part of the expedition is even in sight. When men are a little fatigued, and before the more interesting part of the climbing comes to brace them up, the mental barometer is very sensitive, and very trifling causes bring about a great depression. Work out, therefore, with the utmost care, the route over the middle part of a contemplated ascent.

Crossing the snowfields of the central plateau, the rock buttress would be touched at K. While crossing the snowfields the rock buttress itself should have been carefully scanned. A fair view of the gap would have been obtained from the point where the lateral glacier joins the central ice-stream in the left, and if the interruption appeared promising on a nearer view, the conclusion to try the rock-rib would probably by this time have been adopted. If it fails above, the gully route will still be open and accessible. If the lower rocks prove very troublesome and smooth (which is unlikely in a schistose mountain), the gully can be made for at once. The very crest of the rock buttress is the right line ; it is safer, and gives the best view of the gully and the mountain generally. Moreover, the sun will be on the climbers, and progress is infinitely faster when the rocks are warm.

If it should prove necessary to adopt the gully G throughout as the line of ascent, the lowest rocks may be skirted as far as the narrow part of the east glacier, or the rock-rib crossed to the base of the gully G, the commencement of which is just perceptible.

Before leaving the point of survey, attention should be turned to the route up to the camp from the valley. The situation of a bridge, if there is one, is to be noted, or the best place to cross the stream. A very small path can be recognised when the observer is looking down on it. In unknown districts the route up from the valley and through the forest requires almost as much attention as any other portion of the contemplated expedition. The heavy loads have to be carried to the camp, and all care is needed to save undue labour and waste of energy.

If the climber can make out even one-fourth of the information about his expedition which has been here indicated, he will have been more than rewarded by his survey. He may find it difficult, when actually on the mountain, to recognise the points that seemed so clear and simple the day before, even if the rough outlines have been jotted down on paper, a precaution it is always prudent to take. But he will be working on a principle, and the broad lines he should follow will be present in his mind throughout the day and cannot fail to prove of advantage. Moreover, he will, unconsciously perhaps, have acquired information which at any moment in the expedition may prove of inestimable value as to the best

L

line of retreat. Fortified by this knowledge, he may push on much further in doubtful weather. He will have done everything, in short, to deserve success. But if the details, when confronted, are puzzling and difficult to reconcile with the distant survey, how much more must they be when encountered without any knowledge of what there is to follow! Too often climbers in a country new to them, after working their way laboriously over a long stretch of ground, look down at the end of it only to discover that all the complications might have been avoided by simply keeping a few yards to the right or left. Imagine a party crossing for the first time the pass marked H, taking it from the south side. Very little detail, indeed, can be seen on looking down from such a point as H, and what is visible, such as broad stretching basins of snow or the flat ice-stream below, will be easy. The steep portions are foreshortened. But it is just on these that the difficulties will be met with. In all probability the party would endeavour to work down the east glacier, which presents almost every objectionable feature that a glacier can compress into so small an extent of surface. The central glacier would be cut off from view, and it is hardly likely, without any previous knowledge, that the party would skirt the rocks on the left bank of the *névé*, and then cross over the rock-rib on to the great central plateau. Yet this would, as has been shown, be the right route.

Finally, it is not sufficient to merely reconnoitre and note down, or even (the best plan) to sketch what is seen. Unless a man keeps his wits well about him, his mind will but stagger feebly over the information he has himself collected, and fumble among his amassed facts. It requires a cool head and judgment to utilise knowledge and experience. People collect facts, but do not always know how to pack them so that they can be readily got at when wanted. And often they leave the key of their intelligence at home when they go mountaineering. Captain Bunsby's classical remark may well be applied to reconnoitring: 'The bearings of this observation lays in the application on it.'

To sum up :—

The best place for the camp is at J, on the left bank at the upper level of the scrub. A sufficiently high level can be reached. Water and firewood are plentiful.

If the route up from the valley leads to the right bank, the lowest ice-fall can be ascended on the extreme right, and the glacier crossed to the bivouac.

Above the camp, the line at first is along the avalanche snow.

The second ice-fall is to be crossed on the left side, for it will be here compressed and less crevassed.

No attempt should be made to touch the east glacier, though the route looks shorter, because : (1) it lies at a lower level than the central, and would be difficult to reach ; (2) the vicinity of 'hot plates' is to be always avoided ; (3) above the level of the hot plate the glacier expands, and will be crevassed longitudinally and transversely ; (4) the rocks on the right bank are cut away.

The rock buttress K is a good line of ascent. If chosen, keep on its ridge. The gap at E may stop the climbers : if so, a traverse will have to be made, and possibly the steps retraced in order to make it. The gully G is also a possible route, but less safe than the rocks K. It is a better line than the gully F because : (1) it runs to the actual ridge ; (2) it strikes the north-east ridge higher up than the gully F, and the latter ends blindly ; (3) it lands the climber beyond most of the rock towers on the ridge.

The remarks apply to the east branch of gully G. The west branch ends blindly, and is probably crowned by steep rocks. If feasible, the west branch will be a good line, and the ascent there would be made by a combination of the gully (G) and rock-rib (K) routes.

The north-east ridge is the best : because (1) it is more readily and more safely accessible than the north-west ; (2) although the climber might find the dip of the strata against him, there is but little rock to climb in a westerly direction ; (3) it is more sheltered from wind than the north-west ridge;

(4) it is not more exposed to sun than the north-west ridge, and not, therefore, more likely to be icy.

The snow-crest may be corniced on the south side.

The north face proper is impracticable. The lines showing snow avalanches above, the *débris* at the base and the hanging glacier on the face all alike forbid it.

The north-west ridge should be avoided : because (1) the rock slabs are cut away, and traverses may be necessary; (2) the ridge cannot be reached at as high a level as the east; (3) it is more exposed.

The survey is ended, and the mountaineer may descend to the valley with the pleasant assurance that his time has been well spent, and that even if he fails from unavoidable causes to climb his peak, he will at least have learned something that may be turned to profitable account.

A Chamoniard

CHAPTER VI

SNOWCRAFT

By C. T. Dent

Incidence and reflection

CIENTIFIC questions are presumably beyond the province of a book which deals with mountaineering as a sport. Certain phenomena, however, connected with the glaciers and the mountains must be borne in mind by those who traverse or climb them, and certain physical features of the snow world can be turned at every moment to practical account even by the least scientifically minded of climbers.

We may consider broadly that the main mountain masses were the result of gradual upheaval at some period of the world's history. The stratified layers of the earth's crust, no matter how they were deposited, were, in the process of upheaval (due probably in great part to lateral squeeze), bent up and contorted; their continuity was interrupted often by huge rents as the harder igneous rocks, the crystalline granitic rocks, and such like were forced up under, and often through, the overlying crust. No sooner were the mountains formed than the process of levelling again began which has been going on ever since and is still progressing. The softer stratified rocks worn by the weather, by the action of water, frost and heat, gradually disintegrated and, as they crumbled, fell, finding temporary resting places on the slopes or in the valleys. The harder crystalline rocks preserved their forms and general character more sturdily, thus accounting in great measure for the divers shapes and for the two great classes of peaked mountains and round-topped mountains.

Wherever water found a channel it ran, and as it ran cut into and eroded the crust. Valleys, if due at all to subsidences, were at any rate deepened by erosion. Great masses of snow accumulated in the hollows and corries at a period when the fall was in excess of that removed from them by melting. Partly by melting and refreezing, and partly by pressure which led to regelation, the snow became pressed down and squeezed into ice, and glaciers were formed, which streamed down the valleys. Then came the period in which, whatever one may hope to the contrary, it is obvious we are still living, of the retrocession of the glaciers: a period when the melting was in excess of the deposit.

With the disintegration of the rock peaks and their practical results to the mountaineer we have little to do here. It is sufficient to point out that the moraines give immediate evidence that the ice streams move downwards as they stretch out into long thin lines the fragments that fall on their surface. This movement of the glaciers is perhaps the most striking physical phenomenon of the snow world. We may assume that the movement of a glacier is due in part to pressure from above and in part to gravitation. At the higher levels where the snow is deposited in any great bulk it is merely snow in the form of crystalline flakes containing much air. At the heads, therefore, of the great snow basins the snow will be soft. In the summer the surface snow attacked by the sun's heat melts to a certain extent and the air is liberated. The great piles of soft snow therefore sink a little and the surface becomes a little harder, sufficient to bear the weight when the temperature is low, but softening in the daytime, so that the foot sinks in until a firm step is gained at a greater or less depth by the process of regelation. Lower down the course of the glacier the snow is compressed into a mixture that is not quite ice, though it has ceased to be typical snow. This is called the *névé* region. By this time the glacier is sufficiently compact to crack and split when subjected to tension. Still lower down the masses of *névé* subjected to further pressure are formed into clear ice containing little air save in the form of bubbles. This is the beginning of the glacier proper. Lower down still, a level is reached at which the annual melting is balanced by the annual snowfall. This is the snow-line, which must of course vary with the time of year and the direction in which the slope faces with respect to the sun. The average height of the line is usually computed at 8,000 feet in the Alps. Lower down still the winter snow entirely disappears in summer, and the clear ice beneath is revealed. This lower portion is spoken of as the dry glacier; the extremity is called the snout. The term dry merely signifies that the surface is free from snow during the warm months : for the glacier is scored over with a thousand little runnels which collect together, and here and there plunge down vertical shafts (*moulins*) to the bed.

So far, then, two great principles have been mentioned; first, that the glacier moves, and secondly, that the process of regelation plays a most important part in glacier phenomena. The average rate of movement of a large glacier may be computed at from 10 to 20 inches in the twenty-four hours, a little slower in winter than in summer, but still constant. The hanging glaciers—that is to say, the smaller masses of ice clinging to the slopes and faces of mountains—have not, so far as the writer knows, had their rate of motion estimated. Certain points in the movement of a glacier must be constantly kept in mind. The mass of ice behaves like a river. The surface ice moves faster than that lying on the bed, which is retarded by friction. The central portion moves more rapidly than the sides, which again are retarded by friction. When a curve occurs in the course of a glacier, the line of quickest movement is thrown out towards the concave side of the valley. There is consequently more friction on the convex side of the curving glacier and greater tension. The ice splits at right angles to the line of tension, and thus rents are formed at the margin of the glacier which are called marginal crevasses. They point upwards always at an angle of about 45°. The practical point for the mountaineer is, that on the convex side of the curve he will have more difficulty in getting off the ice on to the lateral moraine, and the way down near the bank will be more intricate though the ice will be less steep. (See fig. 1, p. 154.)

Suppose that the glacier flows in a straight line, and that its bed curves downwards towards the valley : it is obvious that the surface ice will be subjected to more strain than the deeper layers. The ice then is torn across, and the tension being now longitudinal, the lines of fissure must be transverse. In this way transverse crevasses are formed, and seeing that a glacier always flows down hill, these must necessarily be abundant. The direction of the transverse crevasses where first formed will be tolerably vertical. As the surface ice moves faster than the deeper layers, the inclination of the crevasse will tend to become more oblique as the rent becomes wider. It follows that the practical depth of transverse crevasses becomes greater as the part is approached where they are first formed. For example, in descending a *névé* field the transverse crevasses first met with above the ice-fall will be the deepest. They will also be more dangerous, because more concealed. On the more level stretches of the upper snowfields the very superficial layers subjected to less pressure do not fissure under the strain. Bridges are thus not formed, but left over the crevasses. The fissure widens as the glacier moves on : the snow bridge arching over the hollow sinks a little, and the slight groove due to subsidence may alone mark the site of the crevasse (fig. 2, p. 155). Wind, too, drifts the loose snow into the hollows, and still further obscures their situation. Isolated crevasses, again, will be deep : such are met with chiefly in the more level parts.

Where the curve of the glacier is sharp the way must be intricate. If the bed of the glacier is convex from side to side, the crevasses will be formed of course longitudinally. Where the trough containing the glacier widens out the mass of ice tends to spread laterally, and again longitudinal crevasses will be formed. Longitudinal crevasses high up will be deep, though not to the same extent as the transverse, and where the glacier expands at the snout they are often little more than slashes in the ice. The two sets of crevasses may be formed on the same portion of a glacier, as perpetually happens in ice-falls. The fissures then intersect, and square-topped towers of ice are formed. These are the *séracs*, indicated in fig. 1. Where the ice is thus split up the sun has access to a much larger surface, and disintegration is more rapid. It is well, therefore, in attacking an ice-fall about which nothing is known in detail, to try for a route on the more shaded side. Here, too, there is more chance of finding the crevasses choked up by winter snow. The bed of a glacier very commonly consists of a series of huge, irregular steps or terraces. The ice which was bent and strained, and consequently fissured as it passed over the curve, becomes compressed again as it emerges on to the level ; the cracks are closed up by the process of regelation, and these level stretches will therefore prove easy walking. In the same way, where lateral glaciers join the main mass, the additional pressure closes up the crevasses, a point worth remembering in traversing an unknown glacier (p. 143). There is another kind of crevasse, usually spoken of as a 'Bergschrund,' which is uniformly found where the very head of the snowfield abuts against the face of the mountain. Bergschrunds often form a serious impediment to the mountaineer, and are dealt with in detail later on. All the foregoing points are indicated in fig. 1, which gives a bird's-eye view of an imaginary glacier.

The longitudinal section (fig. 2) illustrates the formation of crevasses, and the manner in which the bridges are left. The dotted line on the right shows the amount of overlying snow.

It is worth noting that the dry glacier surface represents a series of more or less parallel, transversely arranged ice-waves.

FIG. 1.—Map-diagram of a glacier

The slope of the waves will be more gentle always on the side chiefly exposed to the sun. This fact might be turned to account if the way were lost in the mist on a very large dry glacier. The glacier tables (as the large rocks perched on ice stalks are called), too, dip in the direction in which they are most exposed to the sun. In fact, the compass bearings can be determined roughly by the inclination of a table. But the general direction of a glacier can of course always be known by that of the streams on its surface.

Small detached bits of grit absorb the sun's heat and sink a little, and the whole glacier is further disintegrated superficially by the rays of the sun, so that the surface is rough and almost invites the mountaineer to set foot on it. During the whole of the daytime the nails bite in, and no easier walking can be found. There is so much variety of step that the way can never be monotonous save to the very prosaic or the extremely gymnastic person. A jump is necessary now and then across a crevasse, but there is absolutely neither difficulty nor danger. At certain times, however, the surface may become much more slippery. In the early morning the little glacier pools

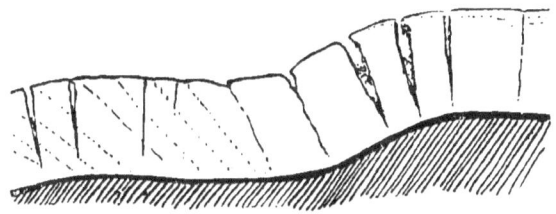

FIG. 2.—Diagram of crevasses; section along line A—B in fig. 1.

are frozen over, and a careless walker or one unable to see well in a dim light may easily put his foot through the ice. Wet feet are always to be avoided, and not merely on the score of discomfort, for frost-bite is unlikely ever to occur so long as the feet are dry. The little 'sand cones,' consisting of conical masses of ice covered over and protected by adherent grit, are extremely slippery, and may easily trip up the unwary. In descending even the simplest glacier after dusk, when the party is at all tired, it is generally wise to keep on the rope. In descending a glacier at night, too, avoid following the course of a stream too closely, for the neighbouring ice is often more slippery, being glazed by the spray.

If the bed of the glacier is at all irregular and crevasses consequently frequent, the best route will lie along one of the lateral moraines. The medial moraines are not so convenient, for they consist of little more than a line of irregularly placed blocks of stone adherent to a raised rib of ice beneath, which they have sheltered from the action of the sun. The warmth absorbed by the stones leads to melting on the sloping surfaces and, further, the rocks protect the ice from the disintegrating action of the sun so that the ice surface in their neighbourhood will be smooth and slippery. The lateral moraines, consisting of heaped-up débris, though often very loose and crumbly, will give a continuous route. It is best to keep along the very crest of their ridges, and to note that the large blocks of stone that occur are often very loosely attached, especially after rain has washed away the finer material round their bases. As long as there is any vegetation on the moraine it will prove good walking. In the trough between the moraine and the proper bank of the glacier long stretches of firm winter snow will often be found which are excellent to walk over.

Progress on *névé* requires more attention. The crevasses are no longer plainly visible as on the dry glacier, but they are none the less certain to exist wherever the bed is at all irregular. They are constantly half or entirely concealed. They are always wider than they really appear. The moment the dry glacier is left the rope should be put on. Do not wait till somebody has put his leg through a bridge into an unsuspected crack. As long as the snow is hard, progress on *névé* is really faster than on dry ice, for the surface is less irregular and a straighter line can be taken.

When the snow becomes soft a different gait has to be adopted. Directly the whole weight is thrown on to one leg the foot sinks in. By a movement of shuffling rather than walking the weight is more distributed, and the gradual pressure allows the principle of regelation to come into play more effectively. The swing must still be kept up. If the foot sinks, think more of bringing the disengaged leg forwards than of trying to arrest the downward progress of the errant member. The man who stops dead after each step will not go far. A fourteen-stone man who understands exactly how to walk in deep snow will sink in far less than a tyro who is half his weight, and the labour of walking, especially of walking badly, on deep snow is extremely great. An even more trying condition than unmitigated soft snow is crusted snow. Here the surface has been rather quickly melted and then rapidly frozen. A few hours of 'saft' weather, followed by a clear bright sky and sharp frost at night, produce this condition. A brilliant sun does not soften the snow so rapidly as might be imagined, as so many rays are reflected. Muggy, misty weather turns firm snowfields into a sort of quicksand in a very short time. Deep and soft snow is usually associated more with clouds and mist than with bright sun. When the snow is crusted the surface is sufficiently hard to bear part of the weight, but directly the body leans forward to take a step, the foot plunges through with a jerk and sinks in up to the knee or further. There is no condition so laborious as this. By planting the foot down as flat as possible, and by shuffling still more than in deep snow walking, the crust can often be induced to break in large triangular pieces and a few steps can be obtained in comfort, and this is about the best that can usually be done. The most aggravating conceivable condition of crusted snow is when the crust is so thick that the mountaineer slips upon it at one step and plunges through at the next, and he may even find one leg grasped by the hard edges of the crust and firmly encased in the snow, while with the other limb he describes a flourish in the air and performs an involuntary extension movement. If to his discomfort he has the added consciousness that some critic is seated comfortably in front of the hotel watching his flounderings through a telescope, his cup will be filled to the brim. As a rule, however, this crusted condition of snow will not be found to extend very far. It is worse on the level than on the slopes. By keeping well to the side of the valley it may be avoided. If snow is likely to be soft, the earlier the mountaineer gets on to it the better. The extra half-hour in bed may mean two or three hours extra plunging and wading and execrating in the afternoon.

Hummocky snow, in which the surface presents an appearance like that of sand when 'ripple-marked,' is tolerably constant to some level snowfields, and is supposed to be due to the action of wind. The explanation is not very satisfactory.

In crossing high snowfields it is often a good plan to follow chamois tracks, provided these lead in the right direction and are fairly recent. These animals show great intelligence in avoiding crevasses, and make long détours in order to find safe snow bridges, returning then to their chosen line of march. Old chamois hunters have often noticed too that they are extremely quick to detect, and careful to avoid, ice slopes covered with a loosely adherent layer of snow, a dangerous condition that will be described later on.

On mountains that run to a much greater height than any in the Alps, and in the Alps themselves in winter, especially after a high wind has drifted all the surface snow into ravines and on sheltered ledges, the snow is found in a fine granular or powdery state more like that which would be produced by the collection of extremely minute hailstones. The particles contain little air. In walking on ordinary deep snow, the temperature is not low enough to prevent the flakes from easily undergoing regelation on pressure, but on powdery snow this effect is not so easily produced. Walking then becomes a matter of enormous labour. The traveller may sink up to his neck in the loose stuff, and can get no foothold at all or fulcrum. Mr. Whymper found the snow on Chimborazo so yielding that at a height of 20,000 feet he had to flog it down and go on all fours. Often the climber is reduced to floundering along as best he may. Exertion at a very great height, and after travelling long distances up hill, is necessarily so great that this drawback will probably be found to prevent, except under very special conditions of weather, the ascent of the higher mountains of the world. Still days do occur, as Mr. Graham found in the Himalayas, when the snow is in fairly good condition at a height much over 20,000 feet, and Mr. Whymper was able to walk upright over the last part of the same ascent of Chimborazo. On mountains such as Elbruz (18,520 feet) powdery snow might easily prevent a successful ascent. In winter expeditions in the Alps the condition may be met with low down, at a height of 6,000 or 7,000 feet, and progress can only be made by a series of fatiguing gymnastics which more resemble swimming than walking.

Above the *névé* district proper on a snow mountain the ascent will have to be made by snow slopes or by the snow gullies, and often on rock peaks the route by the gullies is preferable. On these upper slopes few crevasses will be met with. The walking is easy enough when the snow is in just the right order, allowing the foot to sink in a few inches. Constantly, however, these high slopes do not consist essentially of snow or *névé*, but of ice on which a more or less thick layer of snow reposes. This condition may be exceedingly treacherous. If the surface layer is firmly adherent to the ice underneath, all will be well, and the gully is practically an easy snow slope; but if the layer merely reposes on the smooth icy surface, it will often happen that the slightest touch will send the whole of the snow layer sliding off with a gentle hissing, which is to the mountaineer as unpleasant as the hissing of a snake to the tropical traveller. Granular snow especially behaves in this manner. In summer expeditions in the Alps this condition of snow is only occasionally met with. But in winter it is of very frequent occurrence. The powdery or granular snow has no chance of becoming adherent to the firmer layers. The wind sweeps it off the more exposed rock faces, and whisks the drifts into sheltered corners. Naturally the mountaineer is tempted to avoid the bare glazed slopes, and seeks to make a way where the snow lies. In the Caucasus this condition of snow is extremely common, and in fact constitutes one of the special difficulties of the Caucasian snow peaks.

A slope, as constantly happens in the Alps and with yet greater frequency in the Caucasus, may be pure, unmitigated ice. Such slopes can be recognised by their glistening wet appearance, and by their grey tone. Rapid thaw and quick refreezing in the shade are essential to the formation of ice slopes. They are very common in gullies where the warmth of the rocks has led to some melting, and the ice has rapidly formed as the gully was cast into shadow. Ice slopes are also gradually formed by the water that trickles down the natural drain of the gully as the patches of snow above melt. Slopes facing north are especially apt to be ice-coated. On those facing south, the influence of the sun being felt for a longer time, the snow originally filling the gully becomes softened to a greater depth. Rocks covered with a very thin layer of ice or 'verglas' are termed 'glazed.' They are discussed in the chapter on Rock Climbing.

Deep snow is less fatiguing on a steep slope than on the level. The climber slides down a few inches at each step, and so does not gain the full height at which he first plants his foot, but he does not sink in deeply as on the level, and the back of the leg or the heel being nearly or quite uncovered, he can easily raise his foot out of the step. Here again, as in walking on scree or elsewhere, the climber must accept the slip that he makes. To make a violent effort to strike off from a foot that is sinking, is only to make matters worse. In one respect attitude must be studied. The more a man stands upright in his step as he is ascending or descending a steep snow slope, the more will regelation come into play and the sooner he will be brought up with a firm foothold. In traversing such slopes the vertical position is more important than in either ascending or descending.

In descending deep snow inclined at a steep angle the heels should be well pressed down and the toes kept up. Little more hold is required than the heel of the boot can provide. A close row of nails on the front edge of the heel is a great help. The climber has to be tolerably quick on his feet,

SNOWCRAFT

even though going slowly, for as in deep snow the front foot slides down, there may be difficulty in inducing the hinder leg to follow. On such slopes some degree of spring is needed, but it should be very slight, and scarcely anything more than a quick picking up of the back leg.

A plunger

Above the slopes will come the final ridge. Here, of course, there will be no crevasses at all. The ridge may be fairly broad, and perfectly simple. If of ice or very hard snow, it is still safe enough, but will take time. Many ridges are corniced—that is to say, they have great overhanging eaves of snow, which project on one side or the other, and are due to the action of the wind. On any given mountain the existence of cornices should be most carefully looked out for on all parts, and not only in the line of route selected. The compass bearings of any cornice which can be seen should be at once noted, and their existence assumed on all ridges having the same aspect. For example, on a peak such as the diagrammatic 'Beispielspitz' (p. 135), a cornice is visible far away to the east, projecting over the south side of the mountain. The ridge of the peak has to be reached from the north side. It is proper then to assume that the whole length will be heavily corniced on the south side. If a ridge has any curve in it, a viewpoint will probably soon be found which will make it clear whether any cornice really exists or not. Very often it is wise, on gaining a ridge, to descend a little way on the opposite side, so as to look at the ridge from below, and ascertain its condition as regards cornices. Guide-book, or even local information, is often untrustworthy. On some mountains there may be no cornice one year, and a very large one the next. If there is no cornice, all is straightforward, but if a cornice exists, it must be avoided by every manner of means.

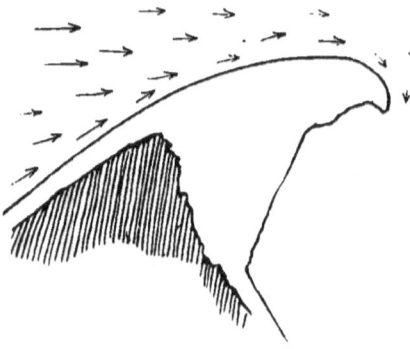

FIG. 3.—A snow cornice

The climber is often forced to avoid the true crest of a ridge. He may keep a few feet below the crest on either side; the corniced side is invariably the worst to select. The walking is very much harder, and occupies much more time when a way cannot be made on the actual crest. Moreover, the party are practically making a long traverse on the face, and this is to be avoided always if possible, but when there is a cornice no other course is open. It is tempting, no doubt, when the summit of the peak is still far off, and the day slipping away, to walk along the crest, which offers perhaps easy going. To keep deliberately a few feet away from the easiest track may just prevent the party from gaining the top. These are the occasions when the beginner, thinking more of his record than of his safety, sticks to the crest. Often, far too often, he merely follows his guide (if an inferior man) in so doing. The mountaineer of experience will keep well below the crest, even at the risk of having to turn back with the top of his mountain well in view and close at hand. Cornices are the most treacherous things. Enormous masses of overhanging snow will break away at any hour of the day, almost without rhyme or reason. It must be remembered that the cornice has to sustain not the weight of one man, but the weight of the whole party who are on it at once. The cornice will be weakened, too, either by blows from the axe, or by the footsteps. Even very experienced guides have before now underestimated the dimensions of a cornice. Twenty feet is no uncommon width, but the extent may vary very considerably at different parts of a ridge. The upper surface of a cornice is generally more or less flat. The moment the traveller is distinctly on the slope, away from the cornice, he is safe as regards the danger of the snow breaking away. If he is able to look straight down the slope on both sides while standing on the crest of a ridge he is perfectly safe, and may keep right on the crest. Snow slopes that are extremely steep on both sides are very unlikely to be corniced at the top. Ridges that are mixed—that is, partly snow and partly rock—will often take in the inexperienced. The little snow saddles that interrupt the rock ridge here and there are very frequently corniced. Very often, too, the actual summit of the mountain may be corniced when no part of the ridge has been found to be so. It has happened, on more than one occasion, that the danger from a cornice has deceived the most careful of guides, and that the whole mass has broken away under the party. On more than one occasion, too, it has happened that a member of the party has had sufficient presence of mind to recognise instantaneously what had happened, and to throw himself out of his steps, and down the slope on the opposite side to the breaking cornice. The rope stretched across the crest, and the fall was checked. Although in this book for the most part the mention of names has been avoided, an exception may be here made. During an ascent of the Gabelhorn, a cornice some way below the summit broke away in the most unexpected manner. One of the guides, to his infinite credit, at once threw himself back, and was able to bring up the party. The name of the guide was Ulrich Almer of Grindelwald, and it was an instance of presence of mind and good mountaineering worthy of a man who bore such a name.[1] Good mountaineering, because the accident was unavoidable under the particular conditions, because there was only one way to meet it, and because that way was promptly adopted.

Cornices sometimes project so far, and are so stayed up by huge icicles and ice buttresses, that it is possible to creep under them, when no road offers on the opposite side of the slope. In descending from the summit of a pass where a cornice overhangs, it is generally desirable to break with the axe a gap right through it, as was done by Michel Croz on the first traverse of the Moming Pass.

So far we have only dealt with the very elements of snowcraft, and have not called into play that indispensable article of equipment, the ice-axe. When the snow is so hard that the foot does not sink in readily, it is often possible by stamping or kicking to make a step more quickly than it can be made with an axe. All that is necessary is to put a little additional weight on to each step, bringing the foot down with force enough to make a sufficient step. In ascending very steep snow slopes, sufficient foothold can be obtained by kicking with the toe of the boot, and in descending by kicking

---

[1] Ulrich is a son of Christian Almer, one of the finest guides who ever put on a rope. A full notice of Christian, with portrait, will be found in Cunningham and Abney's *Pioneers of the Alps*.

with the heels. But directly the snow becomes too hard for this plan the axe has to be called into play. In the early days of mountaineering in the Alps it was customary for the party to equip themselves almost entirely with alpenstocks, often of prodigious length, while one or more of the guides was provided with a little hatchet with which he used to cut steps. On many expeditions—the first passage of the Schwarz Thor, for example—the party, consisting of two only, were provided merely with alpenstocks. With the spikes of their poles the explorers were wont, when occasion arose, to make a series of jobs at the ice, and so punch out a little ledge on which their feet might rest with more or less security. In many other expeditions it was thought quite sufficient if one of the guides carried an axe. The art of mountaineering is supposed to have greatly improved during the last thirty or forty years, but it must remain a moot point whether there are many climbers now who could ascend a difficult mountain equipped only with an alpenstock. This much is certain: that the safety of the party is most materially increased when every member is armed with an ice-axe, and that on a vast number of snow mountains the risk is reduced to almost a minimum, other conditions being favourable, when every member of the party can use his axe efficiently, and understands also the use of the rope.

On moderately hard snow it is sufficient to scrape out a step with the broad part of the blade. A single stroke will be found sufficient with a little practice, but this practice is required. The action of scraping is perfectly different from that of cutting steps, and unless the mountaineer takes pains to acquire the knack, he will find that he soon gets tired. Little swing can be employed, and the step is scraped out more by dragging the adze or broad part of the blade through the snow than by a direct blow. The steps still should be made on a definite plan, at regular intervals suitable to the slope, and should slant inwards. In a simple place a little scratch is often quite sufficient to give the requisite foothold. Perfectly easy snow slopes often require this scraping early in the morning, and with a good leader progress can be as fast as on level hard snow. The axe need only be held in one hand and slightly turned, so that the angle of the cutting blade first strikes the snow. If the edge is driven in it is apt to stick. The adze-shaped blade has the disadvantage that, as it is turned, an edge practically meets the snow surface first. It is desirable to make the steps always in zigzag, and as a rule no thought need be given to any necessity for descending by the same steps.

Step-cutting with the pick is quite another matter. The use of the axe is just as difficult to acquire as of any other implement, bat, racquet, or oar, and just the same principles must be observed with regard to learning it. The right style has to be acquired; for the right style in this, as in every other form of sport, means the method which involves the least expenditure of muscular energy and waste of power. The right method can only be properly learnt from a good teacher. The best teachers are the guides, but most of them are clumsy at imparting information and slow to perceive and correct faults. The method has to be acquired more by imitation than by following instructions. A theory of step-cutting is not of much use : practice is everything. It must not be imagined that every guide who holds a certificate is skilful in the use of the axe, and even good step-cutters will vary most distinctly in what may be called their form from one day to another. Many guides are execrable step-cutters, and, never having really learnt a good method, are incapable of improvement.

A step is well made when it is made with the fewest possible strokes, and of exactly the right size and shape and in the right place. It should allow the man who stands in it to stand upright. The floor of the step should have a very slight slant inwards, and the length of it requires to be no greater than will accommodate one foot. The action of cutting differs materially in the various conditions of snow and ice. The blow is different according as the traveller is ascending, descending, or traversing: whether he is dealing with hard *névé*, clear ice, or the disintegrated ice found on a dry glacier. The main principles are much the same in all cases. A good step-cutter can work away for a couple of hours with ease on hard stuff, while a bad step-cutter, though perhaps a much more powerful man, will begin to tire in a few minutes. The

Step-cutting (A)

difference in the rate of progress between a party led by a good and one led by a bad iceman is immense. Yet the leader who is clumsy with his axe works much harder than the adept. It follows that a greater part of the exertion must have been wasted. The less the blow is delivered from the arms the better. A certain amount of swing from the shoulders, and of the body from the hips, will always be found an economy of force. It is just this swing that amateurs find so hard to learn.

Without swing it is impossible to cut a long series of steps. The centre of the head of the thighbone is the fixed point, as the vertical position is that which should be adopted. The trunk, arms, and ice-axe form a jointed radius. Little movement of the trunk is necessary, therefore, to enable the pick of the axe, which is at a great distance from the centre, to describe a very large circle. The extent of the circle is further increased by the play of the shoulder-blades and movement at the shoulder-joint. Meanwhile the grasp on the axe-handle must be firm. The swing is in itself, therefore, a complicated combination of movements, rendered much more difficult by the fact that the step-cutter has to preserve his balance, often poised on one foot. For if he gives much conscious thought to the maintenance of his balance he misses his stroke, and, *per contra*, if he thinks solely of his stroke he disturbs his balance.

A common fault is to use the muscles of the fore-arm too exclusively, and to cut from the elbows without swing of the shoulders. Stiffness of the muscles below the elbow after step-cutting shows that the cutter has not used his weapon properly.

Little force should be put into the first two or three blows. Begin by light strokes: when the step is half finished more power may be used. Sketch out the step first. If the pick is driven too far into the ice it will stick; if too little, it will merely scratch the surface and glance off, perhaps to expend its energy on the shins of the step-cutter. A well-delivered blow fetches out a great block of ice cleanly. As in the use of many other implements, one of the hands rather guides the axe while the other is mainly concerned with the swing. The hand nearest to the head of the axe is the guiding hand. It thus becomes obvious that the mountaineer must learn to cut steps off both shoulders, and must, in fact, become ambidextrous. The forefinger of the upper hand may be a little advanced in making the grasp (A), and this forefinger is chiefly concerned in giving a very slight turn to the axe when the blow is half completed and the pick is in the ice. The action is much the same as that which has to be learnt in splitting wood blocks. The guides are often perfectly unconscious that they do give this slight turn, which is, however, of importance. Some first-rate step-cutters—Melchior Anderegg, for example—do not advance the forefinger save when using the axe with one hand only. The grip will vary slightly according to the conformation of the hand. It is sometimes said that the blow should be delivered at the ice as if the pick were to be brought out a foot's length in front of the point of impingement. This is only expressing the same idea in other words. For those wishing to learn the proper swing it is essential to remember that some thought must be put into the stroke, and it is as necessary to keep the eye on the place where the pick should strike as it is to keep the eye on a golf ball. If the eye is 'lifted' during a stroke, the blow will be delivered foul, and so much energy is wasted. And in these days, when the language of golf is understood of all mankind, it may be pointed out that to 'press' with the ice-axe is as bad a fault as to 'press' the golf ball.

Step-cutting (B)

The number of strokes required to make an efficient step varies enormously according to the state of the snow or ice and according also to the character of the place. Where the ice cuts well, as on a dry glacier when the sun is full on it, half a dozen strokes may suffice where a single step is required. On a hard ice slope where the leader is cutting his way up, and can deliver the blow with but little swing, the number will be enormously increased, and a good guide has been known to take seventy strokes to fashion a step. The greatest number is required in cutting steps for a traverse of a very steep ice-filled gully. Not only has a ledge to be provided for the foot, but the ice far above the step must also be cut away to give room for the leg. The step then assumes the shape of the recesses provided for statues on buildings, without the overarching top.

The easiest step-cutting of all is on the dry glacier, where the surface is disintegrated by the action of the sun, and it is here that the beginner will do best to practise at the outset, selecting places where a slip will do no harm. Let him bear in mind that at first it is more important to pay attention to swing and style than to the actual result as manifested on the ice. It is far easier to cut steps down than up, though it is more difficult to descend than to mount an ice staircase. The beginner will do wisely if he confines his practice for a few days almost entirely to cutting down hill. The common tendency in cutting down is to make the steps too far apart. They should be cut only at such a distance that the balance can be preserved without effort as the feet cross. As the upper foot is brought down there is a great temptation, which must be rigidly withstood, to lean towards the slope. Good balance is one of the most difficult things to learn in mountaineering, and there is no better way of acquiring it than by

Step-cutting (C). A beginner

practising descending. By the time that a man is able to cut a staircase, of, say, a hundred steps without fatigue, he will have improved his balance in a manner that will stand him in good

A dry glacier

stead throughout his mountaineering career. As his power of balancing developes, so will proficiency in step-cutting. Practically the poise is on one leg while the step below is being cut, and as the foot is brought down the balance has to be maintained with the knee bent, at no time a very easy thing to do (D). Often a better balance can be secured by resting the disengaged knee lightly on the slope, by the side of the step (F).

No two steps should be cut vertically one below the other, and the series, whether ascending or descending, is best made in slight zigzag ; six or eight steps in one direction and the same number in the reverse. Not only are the steps easier to fashion, but they are more convenient to descend by. An additional advantage is that the beginner is forced to learn to cut his steps off either shoulder, and the falling chips do not fill up the steps below. The turning step may be made a little longer than the others, sufficient to accommodate both feet. A dangerous practice, but one that timid mountaineers constantly employ in descending ice steps, is that of going down as a child goes down a staircase, putting always the same leg foremost.

Step-cutting (D)

This necessitates standing with both feet in one step. Either the step has to be cut unnecessarily large, or else the foothold is insecure and there is room only for the two heels side by side. This slovenly method is sure to lead to a slip sooner or later. In cutting descending steps, the axe can be held near the end, and a full swing obtained as soon as the mountaineer is so secure of his balance that he can neglect its anxious consideration. As a rule, in descending, the disengaged leg may be passed in front of the other. But it very often answers better to turn sideways when the course of the staircase alters, and to slide the disengaged leg behind the other. As the skill increases, and the balance improves, steeper and steeper slopes can be selected, and when a man can cut down a slope that appears nearly vertical from above without once touching the ice with his hands, he may begin other forms of step-cutting. These he will now find himself able quickly to master.

In ascending a slope, much less power can be put into the stroke, but, on the other hand, the foothold is firmer ; it is much easier to preserve the balance, and the beginner will stand much more readily upright in his steps. As the slope becomes steeper the arms have to be bent more. The power of swing is diminished. The grasp has to be higher up the axe-handle (E). More strokes are consequently required even to fashion the floor of a step, and it is difficult to make it slant properly inwards. The hand nearest to the axe-head is now largely used as the guiding hand. Some advise that the guiding hand should slide up and down the handle, after the manner in which a blacksmith uses his hammer. Guides seldom use the axe in this way, and most of them view the method with disfavour. 'Always grasp the handle tightly with both hands' is Melchior Anderegg's counsel. A very common fault is to cut away the floor when making the vertical strokes to enlarge the step. In making these strokes the pick should be directed well inwards. The general tendency in cutting up hill is to make the steps too far apart. The climber can easily ascend a staircase in which the floor of the steps are twenty inches or so above each other, but there would be great insecurity in descending. The climber might easily find himself in an awkward position if a sudden retreat were necessary down such a staircase, for in many conditions of snow or ice, any attempt to make intermediate notches on the descent might damage the original steps. In ascending, therefore, steps must always be cut the right distance apart for descending, say twelve inches. There is no real saving of labour or time effected by making steps further apart, for they will have to be made larger.

To make 100 steps in an hour in hard stuff is not bad work. Assuming that a party are ascending a slope and propose to return by the same route, the vertical height between the floor of two consecutive steps can only be about 12 inches. This will give a height of 100 feet per hour. The result seems insignificant, yet on a dangerous ice-slope much less than this is often the best that can be accomplished. Supposing, now, that a second guide had taken the lead, and that he averaged, say, but ten seconds longer in making each step, the party would have taken a quarter of

Step-cutting (E)

an hour longer. In other words, the better man would in four hours go up an ice slope that the other would take five to accomplish. An astonishing amount of time will be wasted on a long slope where a man pauses to look about him after every three or four steps.

Much time is saved by changing the leader from time to time when a long staircase has to be made. When climbing a gully the order is best altered when the party find themselves near the side rocks.

Steps are more easily made in the centre of a gully, but this is the course of falling stones and there is less shelter. Very frequently in gullies, and often on snow slopes, it becomes necessary to traverse, that is to say, to make the steps on a horizontal level. Here another stroke is required.

Step-cutting (F)

To cut traversing steps is harder than to cut steps down hill, but much more power and swing can be put into the blow. Most strokes require to fall vertically, but two or three may be delivered in a horizontal direction. Here again the difficulty is to avoid cutting away the floor of the step in trying to deepen it. When a party is traversing, the rope gives far less security in the event of a slip. If a man loses his foothold when the party are all on the same horizontal level, the strain falls sideways on the man in front of or behind him ; and if the rope is at all slack the strain comes with a jerk. The utmost care is needed, therefore, to guard against accident when the step is completed. The pick should be dug into the slope directly above the step, on a level above the waist, before the foot is moved. It is customary in making a traverse to cut the steps a good deal larger than on other occasions. The additional security is more imaginary than real. The all-important point is the attitude. There is no time at which it is of more importance to stand properly in the steps than when a man immediately in front or behind has slipped badly and fallen clean out of his step on a steep ice-slope. If in attempting to check a man who has fallen, those who are still in their steps follow the natural inclination and try to get a handhold by leaning towards the slope, the feet are almost sure to be jerked out by the pull, whereas, if the knees are extended while the legs are kept rigidly vertical, and the feet well planted in a step that slightly slants inwards, an enormous weight can be sustained. The body may be bent forwards or backwards, but must not lean towards the slope. At the same time the emergency is always a very serious one, and when a gully is less wide than the full extent of the rope which the party have in use, one or another of the end men should be firmly anchored on the rocks at the side. The moment that there is a sufficient length of rope, a bight should be thrown over any projection or rock that lies above. In case of a slip, then, the drag on the body will be in an upward direction, and with a good rope there is perfect security.

Many gullies will be met with in which a more or less thick layer of snow rests on the ice. These are very frequent in the Caucasus. Often enough the snow in the morning is just of the right consistency to give good steps, and there is a great temptation to hurry on when this favourable condition is met with. Whenever a lining of ice at a little depth is suspected or revealed, the only safe and proper proceeding is to scrape away all the overlying snow and make the steps in the hard ice beneath. It is very trying when descending late in the evening to meet with a gully filled in with this mixture of snow and ice. Two or three minutes might suffice to run across it. Perhaps the gully shows marks suggestive of falling stones, and a plausible excuse for avoiding delay by neglecting

BACKING-UP

proper precautions is easily found. Nevertheless the climbers must take their chance of the stones if the bad state of the snow is certain. If the place is one where a fall will be likely to entail any serious consequences at all, the only proper method is to cut right through the soft snow and into the ice. The layer may be so deep that it is impossible to do this, but in such cases the snow will usually be safe. In a wide gully, if the spike of the axe driven in through the surface layer strikes on hard ice, and especially if it slips readily through the snow and rings as it strikes on the harder material underneath, it is absolutely essential to cut into the ice no matter what the labour may be or what time may be occupied. But if the gully be narrow and time a great object, the party may rope singly at the edge and pass over one by one to the rocks. Very often the condition of a gully may be efficiently decided by throwing a big stone into it and watching the behaviour of the snow. On one occasion in the writer's experience in the Caucasus, it became necessary to descend a long, steep snow slope, on the face of a mountain late in the evening. Suspecting that the snow, which had been in good order on the ascent in the early morning, and allowed the steps to be kicked or scraped, might be treacherous, as there had been a good deal of warm cloud on the mountain face during the whole of the day, it was thought prudent to test the slope by throwing stones on to it. The second or third stone, which was of no great size, carried away a huge layer of snow, 12 to 13 inches in thickness, which slid down and revealed an ice slope almost as clear as glass. Undoubtedly if the party had trusted to the layer of snow, all must have fallen.

The greatest variety of step-cutting is to be found in working a way through an ice-fall. In threading complicated crevasses a step or two may have to be cut on the far side of a fissure, which can only just be reached. Such steps are extremely difficult to fashion : they have to be well and truly made, for occasionally the climber has to jump into them. A step large enough to admit both feet side by side is made on the near edge, and the step-cutter leans well forward while those behind support his weight with the rope kept taut.

Occasionally on difficult places in an ice-fall small handholds are great help. Two or three vertical blows with the pick dig out a little notch, sufficiently large to admit the fingers. If the ice is at all hard, it is safer to make such handholds than to trust to using the point of the pick as an anchor.

The axe can be turned to many uses besides its principal one of step-cutting. In descending, or in traversing fairly firm *névé*, the safest plan is to thrust the stick vertically into the snow, grasping the shaft as low down as possible. But if the *névé* is too hard to admit of the stick being buried to half its length, the pick is used as an anchor, and in traversing every member of the party should so use his pick, except the leader who is actually cutting the steps. On hard ice, a hole may be cut to admit the point of the pick at the level of the waist. The pick is generally better than the broad end to use as an anchor.

Again, the degree of consistency of snow or the thinness and stability of an ice-bridge can be very accurately gauged by grasping the head like a walking-stick and using the stick as a probe. The axe thus used gives invaluable information, especially required in crossing high *névé* fields, where there are many concealed crevasses. It requires, however, a very educated sense of touch to estimate accurately the bearing capabilities of snow by a single thrust of the axe, and this department the amateur cannot too sedulously practise. Once acquired, once a man learns confidently to interpret the sensation, he will never forget it. It is a mistake in the writer's opinion to have the point of the axe extremely sharp, as it is commonly made in England. An abrupt shoulder just above the point where the spike merges into the collar will often catch in even quite soft or treacherous snow, and give a false sensation of firmness. The ledge or ring that is sometimes placed at the lower part of the axe-handle is also apt to deceive when the axe is used as a probe. A traveller who goes with guides, and the vast majority of course must do so, may find comparatively few opportunities of practising step-cutting, but in the course of a short time, by diligently persevering in snow-probing, he may acquire a sense of touch as good as that of any guide, and this will stand him in good stead if he seeks eventually to mountaineer without guides. In crossing high *névé* the leader must perpetually have his axe in front of him, and should test every step before he places his foot on it if the presence of crevasses is in the least degree likely.

Bearing in mind the conformation of ice and *névé* that leads to the production of crevasses, the mountaineer will have no difficulty, if he uses his intelligence, in anticipating where these obstructions are likely to occur. In great hollow basins or gently sloping stretches of snow, crevasses will be infrequent, yet they must be perpetually looked out for. To look only straight ahead is not enough. Immediately above icefalls there are sure to be a few large transverse crevasses, very probably concealed. A glance to the right or to the left of the line of route that it is proposed to take will often reveal a slight crack, or it may be a large fissure. If a corresponding appearance is noticed on the same level on the other side of the line of route, there is a strong probability that the crevasse runs really right across it, but that a certain extent of the crevasse is bridged over. It is always wise to assume that such is the case. As the party approach the line of fissure the snow must be carefully probed. Occasionally a very slight depression on the surface, which is not easily noticed when the glare is strong, may give reason to suspect a crevasse. The tone of the surface may give the clue. Amateurs are constantly in the habit of recounting tales of their guide's extraordinary sagacity in detecting crevasses where the slope seems smooth and unbroken. In a majority of instances the clairvoyance of the guide amounts to nothing more than assuming the probability of a transverse crevasse running right across the slope when he sees it gaping on the right and on the left of his line of march. A careful study of a good photograph of high crevassed snowfields is most instructive. The slight differences of tone are admirably rendered by photography. The sensitive film is not dazzled like the eye, and, in a photograph, a whole system of 'concealed' crevasses can often be made out clearly. The recognition in nature will become much more easy if the mountaineer is familiar with the appearance in photographs.

One cardinal rule must never be lost sight of in crossing crevassed regions, especially when the crevasses are concealed, and that is to cross them always at a right angle to their main direction. When transverse crevasses succeed one another with great frequency and many of them are open, or the snow bridging them is insecure, it becomes necessary to pursue a very winding course. The party has in consequence to walk for long distances parallel to the line of the crevasse. The utmost care must be taken not to walk really over the crevasse itself. As there must always be an interval between crevasses, the safest place is close to the fissure. It is better to keep near the lower lip of the crevasse above, rather than on the upper edge of the crevasse below. Throughout, if the *névé* is at all complicated, diligent probing must be practised, for longitudinal intersections cannot always be detected by the eye.

However careful the party may be, and however well led, it may happen from time to time that a bridge may give way and one of the party go in. The more each member of the party keeps this possibility in mind, the less likelihood is there of his falling to any depth. As a matter of fact, if the rope is kept taut and the crevasse crossed at a right angle, and the possibility of making a false step constantly borne in mind, a man will very rarely sink in much above his knee, especially if he knows how to save himself. The great point is to throw all the weight possible off the foot that has given way. If a man directly he feels his leg go through, clutches his axe, throws himself forward on all fours and thus distributes his weight as much as possible, he will commonly be able to extricate himself, scarcely delaying his companions a moment in doing so. It is seldom, if ever, safe to jump *névé* crevasses.

Albert Smith (1851) describes a little incident illustrating an ingenious method of obviating the consequences of a foolish act. 'Tairraz, who preceded me, had jumped over a crevice, and

'The right angle'

upon the other side alighted on a mere bracket of snow, which directly gave way beneath him. With the squirrel-like rapid activity of the Chamonix guides, he whirled his baton round so as to cross the crevice, which was not very broad, but of unknown depth, transversely. This saved him.' A modern guide who jumped on to a mere bracket of snow would be compared to an animal other than a squirrel. The word 'crevasse' is more applicable here than 'crevice.'

Beginners and those who are clumsy walkers will sometimes disappear into a crevasse with an abrupt completeness that is altogether surprising. Frequently, this is the result of attempting to go fast over the insecure ground. If a man is crossing a treacherous snow-bridge awkwardly, throwing the whole of his weight rather suddenly first on to one leg and then on the other, he is very likely to bore a neat little round hole through the crust and disappear totally. It then becomes necessary to get him out, and this is not such a very easy matter. With the rope taut he is safe enough, though not in an enviable or dignified position if dangling with absolutely no support for his feet. The process of extricating the temporarily absent friend must be conducted in a methodical manner. He will not be got out by hauling vigorously at him from both sides, though he may be partly suffocated by this proceeding. First slacken the rope a very little. The submerged man will often then be enabled to work his way up again through the hole by which he entered, or can break down the edges and widen the gap, and so make a way out. Enormous power would be required to actually pull a man out of a crevasse even though he only weighed eight or nine stone, for the pull can rarely be exerted in a vertical direction, or even at a convenient angle. The rope, in fact, cannot do much more than give him material assistance, and he will have to do the major part of the work for himself. Guides do not always show much method on these occasions. They exert themselves with a hearty good will, it is true, but rather on the principle on which surgeons in olden times set to work to 'reduce' dislocations, by exerting most powerful traction. A variety of interesting results followed this treatment, but the restoration of the bone to its proper place was not always one of them. Too often guides pull until they can pull no more, and until the submerged individual is reduced by compression to a state in which he is scarcely capable of doing anything to assist himself. It is better, and less painful, to negotiate a man back to the surface than to drag him.

In the majority of cases, on fairly level places, the man who goes through a snow-bridge into a crevasse will tend to fall forwards. The best way out being in the reverse direction to that by which he entered the hole, it follows that the man in front can do less good by hauling on the rope than the man behind. If the traveller has completely disappeared beneath the snow, the angle which the rope takes will be a guide to the depth to which he has penetrated. Supposing that he is able to reach the wall of the crevasse, he will probably be able to get out without much difficulty. If he is suspended in mid air, or finds only soft loose snow to support his feet below, the difficulty of getting out will be very considerable. It will then be necessary to break away more of the snow-bridge. The submerged unit can generally do this best for himself while supported by the rope, kept taut but not strained on each side : if he is helpless, it might be necessary to break away the bridge from the sides, which must be done carefully. If the traveller in the crevasse can get his head and shoulders out, a steady pull on the rope from in front or behind, as is found best, will often enable him to spread himself out a little on the surface, and once he can do this, and distribute his weight, the work of deliverance is easily accomplished.

It is not only, however, the concealed crevasses that travellers visit at times unintentionally. An unexpected descent into one of the crevasses met with in ice-falls is a common enough mishap, the result of a downright slip, the breaking away of an ice-step, or the giving way of the foothold, just as the traveller attempts a jump. The more visible the crevasse, the less difficulty will there be in extricating a person from its depths ; on the other hand, as such crevasses are often lined by hard irregular masses of ice, the result may be much more serious than a fall through a rotten snow-bridge on high *névé*. In a steep ice-fall it may be safe to approach the edge and haul a man nearly straight up. The walls of the crevasse, in shadow, may be firm enough in consistence, but still the thin hard rope that is let down will cut deeply into the lip of the crevasse, and it will not be prudent to stand at the very edge. By laying an axe or two in the snow at the margin, parallel to the direction of the fissure, and passing the rope over the stick, the pull can be exerted at much greater advantage. The rope plays over the hard smooth wood as if it were a pulley, and does not cut into the snow. All the other members of the party should stand well back, and make a continuous chain, as if preparing for a tug of war. This is the method adopted by cliff-climbers. An axe can be lowered if the fallen traveller has lost his own, and is able to assist himself by cutting steps. If at all hurt by his fall, it will be necessary for one of the party to descend to him. Any overhanging snow had best be broken away from the edge of the crevasse before descending, and large steps cut on the way down. If the fallen traveller is too much hurt to assist himself, a good plan of attaching him before hauling up will be with a bowline on a bight. The loops can be passed under the knees and armpits. It is by no means pleasant to be dragged up by a single loop of fine hard cord constricting the chest. It might even be necessary to get a ladder from the nearest available place. One rule is absolute, that when a man is in a position of peril, he must not be left altogether alone. At least one of the party must stay, if additional help has to be summoned.

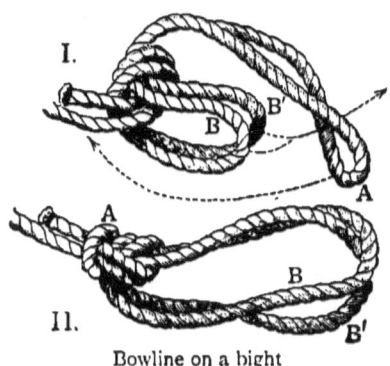

Bowline on a bight

In a crevasse on an ice-fall, the direction is rarely vertical, (see fig. 2, p. 155), and owing to the more rapid movement of the surface ice, the lower wall of the crevasse, i.e. that nearest to the end of the glacier, will always be less steep than the other. It will consequently be best to endeavour to get a man up on the lower side, and those who have crossed the crevasse must find a way back to its lower edge. The same remark applies, though with more limitation, to deep crevasses on level snowfields. The difficulty at all times of hauling a man out of a crevasse, even when he is able to assist himself a good deal, is very much greater than most people suppose, for the force has to be exerted in a most inconvenient direction. It would require superhuman strength for one man to pull another out of a *névé* crevasse of any depth. Even for two men favourably placed, the exertion of dragging up a third is very great, and this fact alone shows the necessity of never going in less number than three on snow mountains. The most minute topographical knowledge may be at fault. On some mountains the situation of certain large isolated crevasses is very constant from year to year: but at any time new cracks may form, snowbridges alter in stability, or old crevasses widen. Do not assume because recent footsteps are seen that a bridge is therefore safe. A few minutes' sunshine may render a bridge rotten; a few minutes' shadow may render it sound again. It may be necessary at times to wait till the snow-field is in shadow before trusting to a doubtful bridge. However carefully a man may probe the snow with his axe, however skilful he may be in using his legs, or in distributing his weight, there is at least the possibility of an accident happening, and people have no right to go on mountains without allowing a proper margin to ensure safety in case of misadventure. The stake always at risk is too heavy to be treated in a gambling spirit, and the mountaineer is bound to have all the odds he can in his favour.

On a dry glacier only extreme carelessness, such as walking in boots without nails, could lead to any accident of this description, and it will not be found very difficult to extricate a man directly from any fissure into which he may have chanced to fall; nor will he descend to any great depth. At the same time he is far more likely to be injured by falling into a dry crevasse, and will very often sustain severe cuts or grazes.

(*To be continued.*)

A Professor on a T-table

THE most formidable variety of crevasse is known by the special name of a 'Bergschrund;' a more correct term, if we must borrow from the German, would be 'Randkluft.' These are great chasms constantly found at the head of the snow-fields, and marking off their separation from the mountain wall. They are continuous fissures following the contour of the mountain; very frequently they are double, a second fosse lying a few yards above the first. Here and there the line of the fissure may be difficult to trace; sometimes, in snowy seasons, the snow on the slopes of the wall appears to be continuous with the head of the *névé* basin, but it is always best to assume the existence of these moats. Below gullies filled with snow and inclined at a steep angle, or walls thickly coated with snow and ice, one of these fosses will always be found, where the angle of inclination changes suddenly as the snow-field commences. Immediately below a gully or surfaces acting as snow-shoots the fissure is very often invisible, the explanation being that in the spring the accumulation of winter snow slides off the harder layers beneath and chokes up the chasm; a point of much significance to the mountaineer.

Bergschrunds are often twenty and thirty feet wide, and are at their worst and widest after a hot summer. Usually they have heavy overhanging eaves on the upper lip. The avalanche snow impinges on the lower lip, and falling back into the cavity blocks it up to a great extent, and masks the real dimensions. Some of the fallen snow adheres and forms an eave also to the lower lip, smaller than on the upper. As these fissures are situated at the base of slopes and their direction is more or less oblique, they are but little exposed to the sun's rays. The cold in the interior being intense, icicles, often of an enormous size, form in them, for there is sufficient warmth from the neighbouring rocks to cause melting and dripping. Regular ice-bridges too will often be found extending over the fissure if the direction of the crack is such that the sun's heat strikes forcibly on it for a short period only. A thoroughly satisfactory explanation of the formation of these fissures is still to be desired. By some it is supposed that the snow adhering to the steep face of the rock, being retarded by friction, is practically stationary, and that the movement of the glacier beneath as it tears away the upper border of the snow-fields from the wall accounts for these gigantic rents. Probably the warmth of the subjacent rock, most potent to exercise its softening influence where the depth of the snowfield is comparatively slight, is an important factor in the production of these formidable obstacles.

Generally speaking, these chasms are more easily crossed in descending than in ascending, and it is to the latter that attention must first be directed. As already explained, the apparent width is much less than the actual extent, especially early in the year. In the month of June, a slight depression, interrupted here and there by an irregular hole revealing some blue depths below, may alone indicate the presence of a Bergschrund. The situation, however, can always be approximately determined, and care must be taken in approaching their vicinity. If under a hanging glacier, the blocks of ice that have been detached and fallen into the chasm, choking it up more or less, will be found hard and welded together; but loose snow that has slid off the slopes into the shade of the fissure will remain soft throughout the year. Bearing these facts in mind, it is necessary when nearing a Bergschrund to test the consistence of the snow very carefully by probing. The rope must be kept taut, and the Bergschrund crossed at a right angle to its direction. A party on nearing a half-choked-up Bergschrund are very apt, while searching about for the best place to cross, to find themselves actually standing on the projecting cornice of its lower lip. If the rope is kept tight, the leader can creep in safety well into the crevasse. He is thus enabled to judge to the best advantage of the extent, the strength, and the nature of the bridges as he views them from below. A mistake often made by an inexperienced person is to suppose that the bridge with the smoothest surface is necessarily the best. The angle of the crevasse being directed backwards and away from a party ascending, the bridge with the steepest inclination is often practically the best to choose, for as a man would fall vertically, he would descend to the least depth where the upper edge projects most over the lower. In addition, if the bridge is insecure and thin, the steeper the angle of its inclination the more will the lower lip support the weight, in the same way that a man might climb up a thin

SNOWCRAFT

pole placed upright that would not for an instant support his weight if laid as a horizontal bridge. The soft snow should be well patted and stamped down so that the particles of snow may be compressed into ice by regelation, but it is much better to work with the hands and knees and feet; if beaten with the axe the bridge will be broken through. The bridge

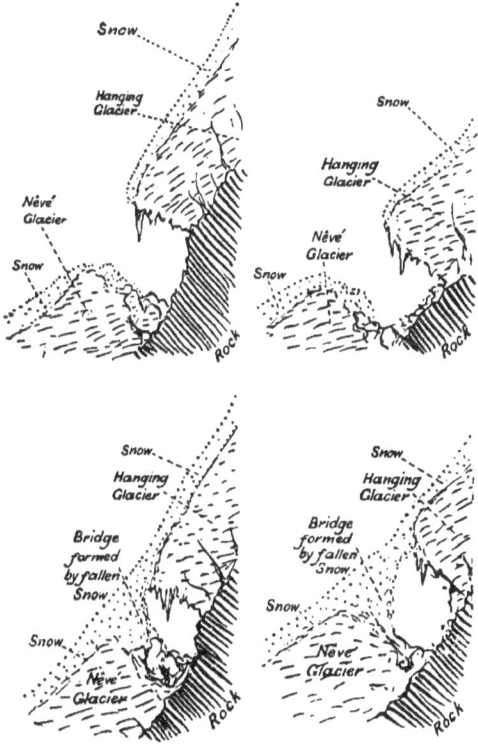

Diagrams of Bergschrunds

should be crossed on all fours in order to distribute the weight as much as possible.

Suppose that the length of the sole of the boot is 11 inches and the average width 3 inches. When standing on a bridge on both feet the weight is supported on 22 × 3 inches, and as one leg is moved in front of the other the weight is thrown entirely on the employed foot. We may assume the width of the leg to be much the same as that of the sole of the foot. Now, it will be found by measurement that if a man is crawling on all fours, the length of support from the knee to the extended foot is about double, or 44 × 3 inches on the two legs. Consequently the supporting surface is twice as great, while the whole weight of the body no longer presses vertically down. In addition, in the 'all fours' position, when the hip is bent and the knee brought forward to make the step, the support of the moving limb is not altogether lost, for it has to be advanced by a shuffling movement. If the elbows are bent and the hands extended, the front pair of limbs give an additional support, of say 32 inches, averaging 2 or 3 inches in width. A man therefore when on 'all fours' spreads his weight over an area of say 80 by 3 inches : while the man who walks erect throws his whole weight on 22 by 3, and as he moves the disengaged limb forward on a surface of 11 by 3 inches. The more he extends himself, the greater the distribution of weight and the greater the safety. It is a little startling when in this attitude to feel the hand plunge through a thin bridge into space, perhaps giving a view into the blue icicle-garnished jaws of the chasm. The more slowly the limbs are moved the less likelihood is there of realising the sensation. When the mishap occurs the climber should not be in too much of a hurry to extricate himself, or he may go deeper and fare worse. By slowly dragging himself along, and by slowly pushing with what in the attitude adopted may properly be termed his hind limbs, he will easily cross inconceivably rotten and treacherous looking bridges and snow and ice crusts.

It is often better to wriggle over very rotten bridges by movements like those of swimming. The only man to be feared is the member of the party who is provided with a hand camera. The process of crossing a bad bridge is not graceful when viewed from below, nor is it agreeable, as the climber finds the soft snow working into his pockets, up his sleeves, and into the space between his apparel and his skin, but it is the

'UP YOU COME!'

safest method, and the best for those that have to follow, though it is often difficult to persuade a nervous man of the fact.

What is spoken of as 'good form' may be disregarded, and security and the comfort of others considered first. Judge of a man's 'form' rather by the condition in which he leaves a bridge he has just traversed than by the statuesque variety of the attitudes he adopted in crossing it. His poses may have been the perfection of grace, but if he broke the bridge in one place and punched half a dozen holes in it at another part his 'form' was bad. The only 'form' worth anything is that which comes unconsciously as the result of years of experience. If the expression merely signifies doing things in the right way, it is misleading. The objection to any such terms in mountaineering is that novices take them to imply that showiness and brilliancy are more valuable qualities than safety, carefulness and finish.

Each man as he crosses a soft bridge makes it thinner, but makes it stronger if he treats it fairly; but if the bridge is once badly broken there may be great difficulties for the last members of the party. As the climber lies prone on the bridge he should reach out as far as possible with one hand, bury his axe well in nearly up to its head, and then drag himself up by this support. Inexperienced people commonly desire to cross such bridges on their feet, and it is of no use to point out to them that if they fall through it makes very little difference what attitude they are in at the moment. Moreover, it is no easy matter to extricate the leg when thrust through a bridge. Purchase must be got by throwing the weight on the secure limb, and this is too apt to go through also. Then indeed the climber is stuck, like a pitchfork in a bundle of hay. The end of the bridge must be treated as gingerly as the middle. Novices frequently, when very nearly across, make a vigorous effort to get over the last two or three feet and damage the bridge irretrievably. When the leader is well across he should ascend a little way on the slope, stamp out a good seat in the snow and a place for his feet, bury his axe well in and take a turn with the rope round it, drawing in the rope as the succeeding members of the party pass up to him. The rope should be kept tight, but should not be hauled upon. No attempt should be made to move on till all are over the bridge, but all who have crossed can assist. It is well to avoid assembling too close together.

Ice-bridges can be crossed in much the same way. They often look more alarming, but are in reality safer. Frequently, however, there is no real bridge. The steepest part is still the best at which to cross. It will be necessary to descend into the crevasse, working the way up on the opposite side in the manner described when dealing with soft steep snow slopes. It is essential to proceed slowly, to probe carefully and thoroughly, and to stamp out the footsteps energetically. Sometimes, though rarely, Bergschrunds are impassable without the aid of a ladder. In the first passage of the much crevassed Jungfraujoch a ladder was found necessary, the party having had to turn back the day before owing to a huge fissure. Mr. Auldjo (1827) describes a plan by which he crossed a bad snow-bridge. 'Our "batons" were placed on it, and in so doing the centre gave way, and fell into the gulf. However, enough remained on each side to form supports for the ends of these poles, and nine of them made a narrow bridge, requiring great steadiness and precaution to traverse.'

In the descent less difficulty is, as a rule, experienced. Occasionally it is possible and safe to 'shoot' the crevasse. The climbers unrope, seat themselves on the slope some little way above, say forty or fifty feet, so as to gain momentum, and then, as they glissade down, spring off the slope just as they approach the upper edge of the fissure. This is not an easy thing to do well at the best of times, and one that an inexperienced person had better not attempt. The sensation of flying through the air is very delightful so long as it lasts. The method is only to be attempted when the snow on the slope above is sufficiently hard to allow the mountaineer to slide down rapidly, unless the Bergschrund is single, with well-defined non-corniced edges, and unless there is certainty of landing on soft snow and of being able to arrest the progress at once when over. Considerable knack is required to make the leap or 'take off,' as athletes term it, and the first time a man tries the exercise he will probably spring too soon and roll into the schrund in a very inglorious fashion. There is little chance, however, of his falling to any depth. It is a good deal easier to follow than to lead on these occasions. Axes are awkward companions, and care must be taken to keep them well out of the way of the landing-place. If the descent is effected by the snow- or ice-bridge, it is better to sprawl across it, and it is easier and safer to go head first.

Sometimes the only way is to jump the crevasse. The baggage must then be taken off the shoulders and lowered down after the first man has crossed. To jump straight down though, with a certainty of landing on soft snow from a height of 8 or 10 feet, is more nervous work than it appears when such matters are discussed quietly by the fireside. Mr. Whymper on the Col de la Pilatte was met by a crevasse of this nature in which a downward jump of 15 or 16 feet and a forward leap of 7 or 8 had to be made at the same time. Such formidable obstacles are not often met with. Care must be taken not to stand too near the edge of the upper lip in 'taking off,' for if the foothold gives way at all the spring will be lost. It is safer to lower the first man down by means of the rope. He can see what the landing is like, and can arrest the progress of anyone who jumps too far—a mistake, however, that is very seldom made. See before you jump that you have ample rope and that it is all clear, for if a coil catches in an axe or under somebody's feet the jerk is terrible. The writer has known ribs broken by such an accident. If it is safe to jump at all, it is safe to jump unroped or roped only to the man in front, the coil being thrown back after each one has crossed. The best plan of all, though by far the slowest, is to lower the party one at a time, paying out the rope round the head of the axe buried well into the snow. It is best to lower a man rather quickly, for the

o

sensation of being suspended by a thin hard rope round the waist is not a comfortable one. In any case the last man has to jump, and must be careful just before he takes his spring not to bury his axe in the snow lest he be forced to let go and leave it above out of reach.

When the snow is fairly hard and inclined at a favourable angle, it is frequently possible to descend it by a very delightful mode of progress known as glissading. Slopes that have cost hours of hard work and step-cutting in the morning can be descended in a few minutes. But the delights of a glissade must not be indulged in without careful reference to two things : first, the condition of the slope, and next, the state of the ground below it. It is never safe to glissade down any slope where the progress cannot be very quickly arrested. Where cracks are seen on the surface or where rocks jut up through the snow, a glissade is not safe, and on an icy slope is mere foolhardiness ; for here the traveller can neither control his pace nor arrest his descent. Snow slopes, too, on which patches of ice intervene, are unfit for glissading ; ice slopes, covered over by a layer of snow, may only be descended by step-cutting. There is no surface better for glissading

A moot point

than the remains of winter snow lying at the side of the glacier or on steep slopes. When there has been a heavy winter snowfall, the most unlikely and largely crevassed places may offer magnificent glissades, and the writer can recollect on one occasion descending in this way, in two or three separate shoots, almost the whole ice-fall leading up to the Mönchjoch. As a rule, however, it is not wise to glissade in the neighbourhood of an ice-fall, however choked up with snow, unless the detail of the route is well known to the guides, or has been carefully noted during the ascent. Glissading down a gully, however good the condition of the snow, is seldom prudent ; ice patches may occur, and there is sure to be a big crevasse at the bottom.

A good glissader can go fast and stop quickly. The balance is not very easily acquired, and no man can get up a great pace with safety without a good deal of practice and plenty of confidence. The main point is to keep the knees straight. At the outset it will be found easier to lean rather back on the axe, which is held in both hands nearly horizontally, the spike pressed into the snow. The toes are pointed downwards so as to allow the flat of the foot to skim through or over the snow. The least turn of the foot deflects the course right or left. By throwing the weight well on to the axe through the arms, which are kept rigidly straight at the elbows, a most powerful break can be applied, and progress almost instantly stopped, if the toes are raised and the weight thrown on to the heels at the same time. The more upright a man keeps the faster will he go if he points his toes well downwards, for if he leans back at all to the slope he must throw weight on to his axe and thus put on brake. The best attitude or 'stance,' as a golfer would say, has always been a matter of some dispute. While learning, at any rate, the beginner cannot do better than keep both knees perfectly straight and the feet side by side. The feet should not travel down in the same trough, though they are only separated an inch or two. Another school holds that the proper attitude is with one leg perfectly straight and the other a very little bent at the knee, so that the toe of one foot is on a level with the heel of the other. This is a little easier, and if the glissader trips or loses his balance he is better able to recover himself. It is universally agreed that to glissade with both knees bent is an ungraceful and indefensible proceeding. A man who glissades with bent knees soon tires on a long slope, and is

Glück

certain sooner or later to adopt an attitude—somewhat abruptly, too, at times—characteristic of the sitting glissade. This latter method is adopted only by the very timid or the very bold. A really good glissader finds it wholly unnecessary to employ his axe at all, and will hold it carelessly in front of him;

or if of an ostentatious disposition, will carry it above his head. The attitude is dramatic, but the result of a slip or the least loss of balance leads to surprising calamities, and the elegant glissader is sometimes observed, without any apparent incentive thereto, to suddenly depart from a statuesque attitude, and perform a gyration in mid air, known to the London street boys as a cart-wheel. There is no doubt that a mountaineer who can glissade without leaning on his axe will go very much faster than the more cautious person who likes to put on a little brake : but the tortoise sometimes reaches the bottom of the slope before the hare. When the party are glissading roped together, all its members should press their axes in the snow so as to prevent any jerking of the rope, and the distances must be carefully preserved. Very little suffices to throw a man off his balance. A climber who is quick on his legs will often seat himself for a moment and right himself immediately, but his legs must be pointing straight down the slope as he does so, or otherwise he will convert a slip into the most inglorious and complicated tumble. On hard snow slopes experts can come down seated on a flat stone adapted to use as a toboggan, or they may omit the stone. Crusted snow is never fit for glissading. Either the mountaineer skims too fast over the surface for safety, or he may sink in suddenly and sustain a very violent wrench. Crusted snow, however, is very rarely met with on slopes of any steepness.

It has not been thought necessary in this work to draw up any detailed list of the dangers of the mountains. Many of these are real enough, and bulky volumes have been devoted to little more than their mere enumeration. The art of mountaineering consists in the avoidance of dangers where they can be avoided, or in reducing the risk run to a minimum where a certain degree is inevitable. Avalanches constitute a source of danger which cannot be wholly guarded against. Yet, for the most part, the risk is extremely small if the mountaineer will take the trouble to use his intelligence and see where and under what conditions they are likely to occur.

With the most destructive form—that is, the snow avalanche—
the summer mountaineer will have little acquaintance. Prodigious masses of snow lodge on the slopes of the Alpine valleys
in winter, and being but loosely coherent are prone to fall when
the spring melting begins. Enormous areas may be denuded
in this way, and masses of snow large enough to overwhelm
a village sweep down in a few minutes. The forests are the
only defence against these visitations, and the action of the
peasants formerly in some Alpine districts in cutting down
trees recklessly was little short of suicidal. Within the last few
years saner counsels have prevailed.

Ice avalanches, on the contrary, are principally met with
in the summer. Great masses break off the terminations of
the hanging glaciers as a result of the motion of the glacier;
or they are forced off by the freezing of water that has
trickled into the interstices. Ice avalanches may fall at any
time of day or night, though they are far more frequent from
the ends of hanging glaciers in the daytime. The broken-up
masses that have fallen are piled up, where arrested by the
curve of the slope, into huge conical heaps whiter than the
surrounding ice; sometimes great blue blocks of ice may roll
far down the slopes. The neighbourhood of such places should
be avoided, and, whenever possible, the most outlying block
should be given a very wide berth. Guides often assume too
readily that there is little danger from ice avalanches if the sun
is not out, and go far too near these places in the early
morning. The neighbourhood of a 'hot plate' is always
extremely dangerous. From time to time, as a glacier moves,
the séracs in the ice-falls topple over. There is still an idea
prevalent among some of the peasants that in the neighbourhood
of séracs it is not desirable to speak above a whisper, lest
the vibrations prove the last straw to the tottering tower of ice
and bring about its downfall. Many of the early accounts of
ascents of Mont Blanc describe places where the guides en-
joined the strictest silence, while at the same time they were
probably cutting away with their hatchets to make steps, a

proceeding far more likely to bring about this form of ice avalanche than a yell in unison from the whole party. Silence will certainly do no harm, but the leader of the party need not be afraid of using his voice to urge rapidity when under a sérac inclined at a doubtful angle with the rays of the sun well on it.

Early in the season, after a heavy winter snowfall, a small form of snow avalanche may often be met with in attacking the mountain faces. These avalanches are small only by comparison with those that fall into the deep inhabited valleys, for they will be found quite large enough to convey a large party quietly down a slope and cover them up for good and all. Such avalanches usually start from smooth rocks high up on the mountain-side, and in the Alps they are not likely to occur in July and August. At the beginning of the summer they constitute a formidable risk, and places where they are likely to occur must be carefully looked out for. Their tracks are easily enough seen : the masses of snow come down in small quantities, always following much the same line. Their natural tendency will be to slide down the gullies, which are often impracticable in consequence. If the traveller should be unfortunate enough, owing to an error of judgment, to be caught in such an avalanche, there is little that he can do beyond striving to the utmost, as he is carried down in the midst of the seething mass of loose snow, to keep on his face and keep his head up the slope. There is usually but little chance of using the axe in any way as an anchor. It has happened more than once that a party has been able by running to escape out of the line of a descending avalanche. If there is no chance of doing this, each man should anchor himself to the best of his power, stretching himself to his full length with head up the slope. On fairly soft snow, with axe driven well in, he may be able to withstand a small avalanche that has not come from far above him. If on ice he has next to no chance, and must trust to being carried to a safe place. This winter snow contains so much air that a party, even though overwhelmed by enormous masses, are able to breathe in perfect comfort, and can usually extricate themselves unless carried into a crevasse. As so little can be done when the mishap occurs, the utmost precaution must be taken by careful inspection beforehand to prevent its happening. Even in perfectly fine weather and with snow in good order, it may be wise early in the year to turn back from a mountain face if any large masses of snow are seen collected on smooth rocks at its upper part. Such masses can hardly be mistaken for true hanging glaciers of which the lower edge is generally sharply cut away. A more formidable avalanche that the mountaineer meets with has already been mentioned, that is, the form of snow avalanche which consists of a stratum of snow loosely adherent to hard *névé* or clear ice inclined at a steep angle. Here the avalanche is started by the party themselves, and no weight of falling snow has to be withstood. The precarious foothold, and the impossibility of anchoring, constitute the difficulty. Little or nothing can be done to arrest the progress; the whole mass of snow moves with the traveller, and the hard stuff underneath gives no chance. There is always a risk of being carried into a crevasse at the base of the slope, but often enough parties have had sufficient way on to be carried safely over this. Nevertheless no man who has ever become acquainted with this form of descent will care much to repeat the experience. Attempts to glissade in improper places would lead readily to the starting of such avalanches.

A curious variety of snow avalanche is one in which the falling masses are broken up into countless numbers of round snowballs, which descend rolling over one another with a very peculiar and characteristic sound. Seen sometimes in the Alps, it has also been met with in the Himalaya.

Ice avalanches, more formidable in themselves, are more easily avoided. It is very easy to recognise where they are likely to occur. Masses of ice may, of course, break off from the snout of a glacier in the same way as at the level of the hot plates. The falling stones that are so liable to slide off the smooth surfaces of this part of a glacier should alone sufficiently warn people not to go into the neighbourhood of such places. The majority of avalanches already mentioned are formidable chiefly on account of their bulk, and the ease with which they will envelop and sweep away a party. A very small fall of ice or hard snow blocks will carry away a party quite as readily, especially if standing in ice-steps, and will pound them as well. A falling mass of ice and snow, due to the breaking away for instance of a cornice, may look very insignificant from a distance, but it is a formidable danger.

The simple rule with regard to all forms of avalanche is to avoid their track, and all that is necessary in the majority of instances is to recognise the marks on the snow surfaces that denote their course, and to steer clear of them. These marks are really so plain as to be unmistakable once they have been pointed out. Any photograph of high snow-fields and hanging glaciers will show them plainly enough. Sometimes the marks consist of long straight lines, due to the troughs ploughed down the sides of the snowbeds. Sometimes detached blocks of ice lying in the snow, blue and glistening, indicate the dangerous localities. Often, and especially when the ice avalanches have fallen a great distance, or where their course has lain in part on rocks, the pulverised masses are piled up in great conical heaps whiter than the surrounding ice. It is best to assume that these are due to avalanches, and not to inspect them too closely to make sure of the point. In the height of the summer avalanches constitute but a slight danger, and are principally or almost entirely limited to blocks or masses falling from hanging glaciers, or the giving way of cornices. If it becomes absolutely necessary to pass across the track of any possible avalanche, it should be done with all rapidity, and always when the most suspicious part is in shadow. The best time, therefore, generally is early in the morning. In the High Alps the effect of sunshine in starting avalanches or falling stones is astonishingly rapid. The writer well remembers descending a long mountain face, partly of rock but with extensive masses of snow and ice plastered on it, late in the afternoon on a changeful day. As the clouds drifted from time to time, the direct rays of the sun were cut off and an almost complete stillness seemed to reign, but the moment there was a break in the clouds, and the sun shone, the whole face seemed instantaneously alive with falling stones and occasionally with small fragments of ice. A route up a mountain or over a pass that forces the travellers under a cutaway cliff of overhanging glaciers is never a safe one. Yet sometimes no other way offers. It is safer, as a rule, to keep close under the cliff, which will act as a shelter from any avalanche that may start at a higher level.

The final ridges of the mountains, or 'arêtes' as they are commonly called, in many cases form to the climber the most enjoyable part of his whole expedition. On some snow mountains there is a savour of monotony in a long succession of snow slopes, on which the footing may be bad or the snow deep or crusted. It may, however, be remarked in passing, that the vigilance must never be relaxed, however dull the actual walking-surface may prove to be. It is on the ridge very often that the first opportunity occurs of exercising snow-craft. Ridges that are corniced have already been dealt with. Occasionally the crest presents but a steep gable of ice with some thousands of feet fall on either side. This becomes a mere matter of step-cutting and steadiness, and very laborious the progress is. The closer the party keeps to the actual crest of the ridge, if free from cornice, the better their position: better hold is got with the axe, which can be anchored on the opposite side of the slope. The great point is to cut the steps at the right distance apart, and to make them sufficiently deep to be utilised with a little extra work on the return. Such ridges, however long, are safe enough for a steady party in fine weather, but they are in the highest degree formidable when a high wind is blowing. Mountain sickness is a malady now almost out of date in the Alps, but it is well on a high peak to keep a very sharp look-out on any member of the party who is weak or not in good condition. At the beginning of a season, or when men are altogether new to the mountains, the novelty, the keen enjoyment and the excitement will carry them a long way, but the spirits evaporate very quickly when a man is out of sorts, and nothing tends more to weary and render him careless than some hours of step-cutting on an ice-ridge, when a bitter wind is blowing and the top of the mountain seems never to approach any nearer. A night in a hut or a bivouac does not give quite as much rest as a night in bed, and fatigue may first show itself just at the moment when the greatest care is needed. Although a man may not suffer from mountain sickness, he not infrequently finds himself seized with an extraordinary desire for sleep, and people have been known to give way to this desire even while standing in ice-steps. The expression 'falling asleep' is in some danger of coming rather literally true. Whoever is captaining the party should watch closely each man, if the ridge is long and icy and steps have to be cut at a slow rate, and not scruple to call attention sharply to any relaxation. Do not neglect to keep an eye on the porters, if any are with the party. Porters are often chosen rather at haphazard, and without any knowledge of their powers. These men work desperately hard in the summer, and have to manage with very little sleep.

More commonly the summit ridges are 'mixed,' that is to say, partly of rock and partly of snow, and offer the most splendid climbing. Hardly two steps are alike, as the course lies now on the very crest of the ridge now up a steep rock face with a moment's shelter from the wind and warmth to the hands, now turning a rock tower by a traverse, or descending a few feet into a great cleft. It is on these ridges that the elements spend their fury, and their towers and pinnacles give infinite variety. Rocks on the ridges are often very loose, being exposed so much to the action of alternations of frost and heat. It is best, as far as possible, to keep on the sunny side of the rocks, where warm handholds can be obtained. Passages may be found here and there where the safest mode of progression is *à cheval*, but these are few and far between. It is said however of one peak in the Alps, that though it has been ascended several times, the foot of man has never yet been set on it. A good deal of balance is required on these sharp ridges, especially on the actual crest, for often the axe can give no help, and there is little chance of handhold. A gusty wind adds very greatly to the difficulty, and perfect stillness on a high ridge is the exception. It is often necessary to lean well over towards the wind when it is blowing with any force. The sudden cessation of a gust must always be anticipated. The body should be bent at the hips, and the legs kept upright in the steps. The more level parts of these ridges are more difficult, for the simple reason that there is then no handhold. A feeling of giddiness in looking down a slope is very rare indeed, even in a beginner. People will look unconcernedly down a slope at 70° or 80° if they can only see 20 or 30 feet of snow or rock stretching away immediately beneath their feet. The majority of people will find it much more trying for the head to stand on a brick wall 20 feet high than to look down a steep ice slope of a couple of thousand feet, on which they would not have the remotest chance of stopping themselves if they once fell. The grasp of the axe gives a feeling of security, even where the stick is of no help. It is by no means easy to stand well in steps on a steep place without an axe or stick in the hands. In such places the moral influence of the rope is very great indeed, but the actual security it affords is comparatively slight. Still, the rule must be remembered in all such places, that if a given passage is too bad to be crossed when a party is roped, lest the slip of one should drag down all, then it is too dangerous to be crossed at all.

On many mountains in the Alps various routes have been devised which enable the mountaineer to enjoy the pleasures of ridge-climbing for the greater part of his ascent. By following the very crest of the rocky ribs and buttresses jutting out from the face, the climber can find scope for gymnastics during the greater part of the day; and these routes are often

ON THE MESSER GRAT

safe enough. No way up a mountain, whether it be of snow or of rock, is reasonable, however simple the climbing may be, if it involves danger from falling stones or ice. Far too many of these 'variation' routes—that is to say, variations from the accepted track and shortest line—have been introduced of late years. The art of mountaineering will be better developed and better maintained by following as well as possible the old routes, than by attempting at any cost to fashion out new ones. The craze is rather a seductive one, and the desire for novelty in the Alps, where almost everything reasonable and unreasonable has been accomplished, has often led men to do things of which their better judgment was really ashamed. In any case, let the beginner not strain after novelty, if that takes the form of following some line of ascent on well-known mountain which good judges have pronounced unwise or unsafe. There are peaks and faces innumerable in the Alps where much rational 'variation' may be made, though for the most part they lie away from the more hackneyed districts, and where the fleshpots are not so well filled.

A fourposter

A mountain arab

CLIMBERS in the High Alps are not uniformly favoured with clear weather enabling them to steer their course by reference to prominent landmarks. It is constantly necessary to find the way in mist and fog, or even in driving rain or snow. There is something rather alarming at first, when on high levels, in finding the mists sweep up over the snowfields and shroud all the surrounding scenery. The beginner at mountain-craft will feel astonishingly helpless under such conditions. The way, which a few minutes before may have seemed so clear, seems now almost hopeless to find. One part of the snowfield looks precisely like another. Small objects around lose all proportion when viewed through the mist, and a sense of desolation and solitude, that at times becomes almost overpowering, impresses itself on those who are new to the mountains. But to the man of some little experience, possessed of confidence in his own reasoning powers, a little bad weather in the Alps serves only to heighten the pleasure. The necessity for using the wits, and the consciousness of the difficulties of finding the way, intensify the keenness of the pleasure. No one has really seen the mountains who has not seen them under all their aspects, in storm and wind, as well as in stillness and sunshine. It is very easy indeed to lose the way in bad weather, but it is also very easy to keep the true line if the proper rules be observed. Under ordinary circumstances, in fine weather the tracks made in the snow will always guide to a line of retreat, but with snow, or rain, or wind, these become effaced with marvellous rapidity. It is often, however, possible to follow old tracks, especially when they have been made in deep snow. A step that has been made in soft snow leaves some little trace, which can be detected by careful scrutiny a long time afterwards, even after a moderate snowstorm, or when a high wind has partly filled up the holes. Here and there, by scooping away a little of the surface-snow, the deep imprint of a suspected footstep, crusted over and protected, can be revealed, Frequently, after a wind, steps appear to be raised above the surrounding level in the form of little conical snow-heaps. When crossing familiar passes, it is a good plan to practise the recognition of these slight traces, and to note how tracks made many days before, practically on the same line of march, leave some little evidence on the snow-surface, even after bad weather. Swiss guides, who have a good deal of familiarity in the winter months with this department of snowcraft, have a very keen eye for traces that the ordinary observer ignores. The amateur chiefly fails to rival the guides in any degree in this respect from his neglect to make any effort to train his observation in the same manner.

At the best, however, an attempt in stormy weather to follow an old track is a lengthy process. Often there is no track, old or recent, to follow. Rough weather will soon efface the small footsteps made on hard snow in the early morning. Other means, therefore, must be ready to hand. The ordinary rules that obtain when a traveller has lost his way apply as forcibly on a high snowfield as in the middle of a sandy desert. The first principle when you lose your way is to admit that you have lost it, and to stop at once to consider when and where you first deviated from the proper track. The way is never really lost until, by wandering in an aimless manner, all chance of finding it again is lost. If on a snowfield, mark the place distinctly where you first recognised that you were off the track, by freely scratching and digging at the snow, so that you may recognise the spot again. Then sit down and endeavour to work the matter out. Do not guess; try to reason. Tracks in the snow can be seen better from above than below. Ascend, therefore, till you are certain that you are above the track, and scan the snow carefully. Through a momentary break in the mist footprints can be distinguished a long way off with the naked eye. If the weather renders it useless to search for any marks in the snow, let the party ask themselves the following questions : What is the general compass-direction in which the proper route lies? When we last left the proper track, did we turn to the right or to the left? Now the traveller on a snowfield has one advantage over a traveller in a sandy desert, and it consists in this fact—that the snowfields are not level. One general direction therefore goes up, and the other downhill, and it is absolutely certain that the latter will conduct him ultimately to the valley. Whether he should go on or turn back is a matter that should be decided by the conditions of the weather, rather than by other considerations. A practicable, rather than a short, route has to be aimed at. On high snowfields, to take a direct line usually requires clear weather, when the detail can be seen at some distance. In thick weather, the great point is to avoid all obstacles, crevasses, and such-like, as much as possible. On wide snowfields, the track will be simpler at the sides than in the middle, though more circuitous. If the party can once hit on a slope bounding the side of a snowfield, they will be able, in the thickest mist or the worst of weather, to follow the course of the glacier up or down by simply watching closely the dip of the slope towards the middle. On a still foggy day, the position can be determined fairly accurately by shouting, and listening for the echoes. If the sounds are reflected back from the two sides simultaneously, the party is in the middle. A still further assistance to the snow-traveller lies in the direction of the crevasses ; in fact, these alone are quite sufficient to enable him to find his way down. On fairly level snowfields the crevasses must all be transverse, except at the very margins, and all the traveller has therefore to do is to cross them at a right angle to their main direction. On fields where there are no visible crevasses, he must remember that the great tendency is to walk in a circle. By lengthening the rope to its utmost extent and constantly looking back at the line of tracks, this tendency can be counteracted to a slight extent. It is said that travellers on moorlands and on deserts when lost generally wander to the left and describe a circle. The Scotch shepherds are stated to be so conscious of this, that when caught in thick mist they systematically step a little longer with the left foot than with the right. A compass becomes, of course, invaluable, but it must be perpetually referred to. In a minute or two a party who have started off walking, say due north, will find themselves unconsciously, on next referring to their compass, going as much as fifteen or thirty degrees to the east or west. Two compasses at least are wanted, and the leader should not carry one. The second and fourth men are the right people. The leader has quite enough to do to work out the details of the route, and it is for the others to keep him straight. The compass-bearer watches the leader as the helmsman watches the bow of his ship, and keeps him incessantly informed of the direction. If a détour has to be made to turn or to cross a crevasse, it is well to count the number of steps taken to the right or left of the true course, and to take a similar number after the crevasse is crossed, in order to come back to the proper line. Otherwise, the party is apt to keep edging away along one side in order to keep a true line. For some inscrutable reason, people, even with a compass pointing to them the right direction in which they should go, prefer at times to follow what is little more than the inclination of the moment. On snowfields it is better to keep a straight line than to endeavour to work out the easiest route. When the leader can only see a few feet or yards ahead of him, he can at best only choose a line which is suitable for the moment, while the general direction is the one that it is all-important to keep in mind. A break in the mist is sure to occur from time to time, and the fullest advantage must be taken of it to glean all the information possible. If any landmark becomes visible, its compass-bearings had best be at once noted and written down, together with the time at which it was noticed. This will often prove a valuable assistance on consulting the watch again. An hour or two later on the travellers may be able to reason out their probable position pretty accurately. It is generally safe to allow, in mist, that they have taken about half as long again as they would have in clear, fine weather.

As long as the daylight continues matters are fairly easy. But a time may come when the leader has to determine whether further progress is proper. The decision is a very difficult and important one to make. If it is decided that it is safer to stop, the best must be made of the circumstances. Recollect that if the party is not composed of experienced mountaineers, the moral effect of being apparently lost on the mountains is a very great one, and the experienced person will tire far less soon than those who are physically more strong, but newer to the work. A tired member of a party is always a source of danger, and the strength of a party can only be measured by the strength of its weakest unit. The party may be benighted actually on the upper glacier, though every effort should be made to get off the snowfields, and at least to obtain such shelter as the nearest rocks afford. It may be desirable to accept the situation, and to admit that the party is really benighted when they have still plenty of vitality left in them. Do not persevere until exhaustion sets in. It requires more energy to resist cold than to go on walking. On no account wander about to keep warm. A party doing so might find themselves all together on a bad snow-bridge, without being aware of the situation till they go through. In windy weather there is more shelter and more warmth inside a crevasse than on the open snowfield, though it does not sound very inviting; but the great point is to get shelter from the wind. If the shelter of a suitable shallow crevasse has to be sought, great care must be exercised in the selection and in probing the snow before descending into it. The rope must never be taken off through the whole night.

If there is any food left, some of it should be taken directly the halt is called, and some, however little, kept back for the night; the portioning out even of a crust of bread helps to pass the time. If there is any brandy or spirit among the stores, it had better be kept. The idea that swallowing spirits 'keeps the cold out' does not work accurately in practice, and the brandy would be of much more use for rubbing the hands and feet as an external application than used internally. The cold due to evaporation is much more to be dreaded than a low temperature. If any man, then, has anything that is dry, let him put it on. The great point is to check the evaporation. A Scotch shepherd, if his plaid is partly wet through and he is caught at night, turns his plaid with the wet side in. It is often desirable to take the boots off if they are wet, and also the stockings. The boots can be tucked beneath the coat, under the arms, to prevent them from freezing; and if there is a rücksack, the feet of two of the party can be put in that and will keep fairly warm, especially if they are wrapped in paper; a sheet of newspaper is almost as warm as a rug. If in any dangerous part of a glacier, the travellers had better not be in too great a hurry to start the next morning as soon as daylight breaks, for they will be benumbed and stiff with cold, and not very secure in the matter of balance. In the High Alps, at least when led by competent guides, a traveller who is not afraid to turn back when he finds the occasion demands it, is not likely ever to have the most unpleasant experience of being benighted.

Once low down on the glacier, the course can be determined by following the moraines. On many dry glaciers it is difficult to find the right way off the glacier. But here, with the crevasses visible, there is no danger to a party benighted in still weather.

Apart from any question of losing the way or being benighted, and more serious than fog or mist on a glacier or a high snowfield, is a driving snowstorm accompanied by a high wind. The Föhnwind in the Alps is very formidable. The beating of the hail or snow, and the rush of the storm, seem to paralyse the wits, and to knock all energy out of even the sturdiest. 'Many can brook the weather that love not the wind.' Under such circumstances the compass becomes of extreme value. The rope must be kept out at its full length, and thoroughly taut between each of the travellers. It is wise, whenever the smallest opportunity for shelter offers, to take advantage of it, and make a brief standing halt. A really bad storm on the mountains is at all times a serious thing. Guides and porters are very often much more affected by bad weather than the travellers. Bad weather, indeed, finds out the weak points of a guide very quickly. The second-class man, of whom, unfortunately, there are only too many now-a-days in the Alps, becomes worse than useless. Not only is he unable to exert his mountaineering powers, but he loses his head entirely; and if the travellers are not accompanied by guides whom they really know and can trust, they should not hesitate to turn back if the weather appears very threatening. Here is a matter in which the travellers must use their own judgment. The bulk of the Swiss guides are drawn really from a rather low-class peasantry. Like most residents in mountainous districts, they are astonishingly superstitious, and have a curious dread of thunderstorms. There is nothing very terrible, though it is very impressive and dramatic, in a thunderstorm in the high regions. It does not last long at the time, and will be followed by a spell of better weather. A Föhnwind in the Alps, or a really bad break with a low barometer, is far more serious than any summer thunderstorm in the high mountains; so, too, are heavy snowstorms, which in the Alps are not infrequent in August and early September. In really bad weather a mountaineer is far safer on snow-fields than he is on rock-mountains. There is nothing on snow-mountains quite equivalent to the rapid glazing of rocks that occurs under stormy conditions.

Snowcraft is undoubtedly the department in mountaineering in which there is most to be learnt. In the Alps, a magnificent field is offered for acquiring the knowledge. Snowcraft possesses the great charm that the head must be used as well as the hands and feet. Guides, though they vary much in this respect, will be found to possess in the majority of instances a good deal of knowledge of snowcraft, though the really first-rate icemen are very few and far between. Men who begin their climbing, then, in the Alps start in the best of training-grounds, and they can always associate themselves with people who, at the least, are capable of teaching them a good deal. The information has rather to be dragged out; for the instructors are men who have never learned how to teach. A single day over a difficult pass in doubtful weather with a first-class guide is a grand experience for the beginner. He will see at once how varied and intricate a matter snowcraft really is. Till he realises this, he has not learned the alphabet of mountaineering.

In countries such as the Caucasus, and the mountains of New Zealand and the like, the people of the place will be found to have absolutely no knowledge of snowcraft. Swiss guides will never exert their powers as efficiently in other countries as in their own. Their national home-sickness appears almost uniformly to depress them and render them far less useful when away from their own districts. On great mountain ranges, which have hitherto only been partially explored, snowcraft is the all-important requisite. A pure rock climber will find comparatively few opportunities of exercising his powers. The peak will be won, the passes crossed, and the districts above the snow-line explored and revealed by the iceman, not by the rock climber. The man who can drag himself up a vertical rock face when he can just get the finger-tips on to one little ledge, will be of far less use in an exploring party than the man who can cut a series of steps truly and well, who can judge quickly of the state of snow, who can estimate correctly what effect the weather of a given day will be likely to have on the snowfields; who, following always a definite principle, can thread his way with ease and certainty through an ice-fall; who can recognise with a single thrust of the axe the bearing-power of snow; who can choose his line over a snowfield, and follow it with a confidence born of his own tried judgment and experience, and who becomes cooler and more full of resource when bad weather sets in. Rock climbing is largely a matter of detail : in snowcraft much broader principles have to be followed, though no detail may be neglected. Yet snowcraft, unhappily, year by year appears to be becoming a matter in which men take less and less interest ; for the fashion of the day, which people follow as much in Alpine as in other matters, all tends to the glorification of the greased pole.

A 'Schnee-zug'

CHAPTER VII

ROCK CLIMBING

BY C. T. DENT

Over it goes!

OF the various branches of mountaineering, rock climbing is undoubtedly at present the most popular. It may safely be asserted that, if it were not for the rocky peaks in the Alps, the sport would not there have obtained the favour that it now enjoys. Nor is the fact to be wondered at. So long as the minute topography of the higher Alpine regions was but partially worked out, the tracks led, naturally, over the snowfields. Men sought to make new passes, to connect old routes, or they searched for expeditions which enabled them to traverse much ground. Difficult climbing was rather an incident in an expedition than the main object and attraction. Now it is thought that the exploration of the Alps is practically as complete as it needs to be. Consequently, the lines followed by the early pioneers are comparatively deserted in favour of what is deemed to be a more exciting branch of mountaineering. Far be it from the writer's intention to depreciate the new departure, or to join with those who speak of mountaineers as 'mere gymnasts' because they find in rock climbing their chief pleasure. The nature of the attraction is not far to seek. There is more apparent variety to be got out of rock mountains, the exercise is much less monotonous, and more muscles are brought into play. The climber can modify his route in detail to an almost indefinite extent, and is able to feel more dependent on himself and much less on his guides. A certain degree of proficiency can be quickly attained by any man, on the right side of forty, who is accustomed to athletic exercise. A chapter indeed on rock climbing may seem almost superfluous, and the best success that the writer can hope for is to convince the reader that there is really something to be learnt.

Rockcraft bears to snowcraft much the same relation that batting does to bowling. There are more good bats than good bowlers, and yet the bowlers win the match. Beginners will recognise early that snowcraft is a difficult matter; one that requires judgment, and one in which the head must guide the limbs. It is, unfortunately, on these very accounts in some cases that they neglect to qualify themselves as icemen; or else they take the trouble, if wise, to learn something of snowcraft while they look upon rock climbing more as an amusement. Such it may be, but it is an amusement which, for pleasure and safety alike, must be taken seriously. In the writer's judgment a weak member of a party is far less a source of danger on the most difficult *névé* than he is on even a simple rock mountain. On snow and ice the responsibility of each man to the party as a whole becomes obvious directly the rope is put on. But on rocks, where each climber has to shift more for himself, his relation to the party is less constantly in mind, and even when fully recognised, cannot so uniformly or readily influence the course of action. The vast majority of dangers on snow and ice are reduced to insignificance by the rope, when properly used; on rocks the rope, though still indispensable, is of far less value. It matters little if a clumsy man slides out of a snow-step, goes head foremost down a glissade, or breaks through a snow-bridge; the risk is commonly foreseen, and the party are forearmed. A slip on rock is a very different matter. It is very apt to take the party unawares, and, generally speaking, entails worse consequences. The qualities of a good leader are more valuable on snow than on rocks. An unskilful person can be got over the very worst passages of snow or ice by the assistance of good guides. He may blunder, but if all precautions are taken, no harm is likely to come. On rocks, on the other hand, the inexperienced climber will meet with places which he is utterly unable to climb at all, even with all the aid that it is possible to afford him. Nor is it only on the snowfields that concealed risks are met with. Dangers that are not recognised are hidden. A loose mass of rock that is imagined to be firm is quite as much a hidden danger as a bad ice-bridge that is not suspected, or an unseen mass of snow that may fall as an avalanche at any moment on the party.

Probably few would admit that rock climbing is the more popular in that it is rather the more showy branch of mountaineering. Snowcraft consists largely in the avoidance of difficulties and dangers, and in doing things in the simplest way when the untutored person would run into difficulties. The rock climber, on the other hand, is for ever being confronted with passages which appear more difficult than they really are. If he overcomes the obstacle, he is proportionately elated. It appears much more prosaic, profiting by the experience and judgment of a guide, to avoid a concealed crevasse or pass across a weak snow-bridge without any mishap, than to climb up a steep place whose difficulties are plainly perceptible and have the merit of looking worse when they have been surmounted. The question in the latter case is : How is it to be done? in the former : How is it to be done best? The snow problem really makes more demands on the finer qualities of the mountaineer than the other. Even though rock climbing be inferior as an art to snowcraft, it must still be practised properly. Let not the seductive charms of rock climbing occupy too large a place in the mind of the young mountaineer to the exclusion of snowcraft, lest he be but preparing for himself in matters athletic a sad old age.

Elsewhere it has been urged that scientific knowledge can

always be turned to practical account, and this holds good of rock climbing. The mountaineer may dispense with any theory on waterspouts if he understands certain laws which commonly hold good with regard to the weather. So, too, he may climb well without any abstract knowledge of pure geological science; still, the more information he possesses about the chief features of the physical geography of the mountains, the more security has he against making any fundamental mistake in his mountaineering. Geological knowledge alone will not give the key to the right route up a rock peak, nor will it replace mountaineering skill and experience, but it forms an invaluable supplement. Keen and thorough mountaineers find in rock climbing something better than merely overcoming intrinsic difficulties of detail. An acquaintance with the chief formations of rock mountains and their individual peculiarities is as useful as a knowledge of the physical phenomena of glaciers and snowfields. The general character of the climbing, the method of attack, the risks to be avoided or guarded against, vary with each class of mountain. Much information is furnished by the outline and modelling of the range or peak. The diagrams that follow may serve to assist in the recognition of the main characteristics. The principal types can be well studied in the Alps.[1]

FIG. 1.—Limestone

Limestone mountains are very common. The lower hills, covered with grass and forest, illustrate the formation distinctly enough. In winter the architecture is still more plainly visible, as the walls and banks are mapped out by white lines of snow. Viewed, so to speak, in profile, these mountains are often seen to have one side gradually sloping to the top and one side

---

[1] Admirable drawings of the bolder types of mountain forms will be found in Elijah Walton's *Peaks in Pen and Pencil for Students of Alpine Scenery*, Edited by T. G. Bonney, folio, London, 1872.

abruptly cut off. Long lines running parallel to the crest of the mountain reveal the stratification. The sides are broken up into huge steps, the tops of which more commonly present sloping surfaces. Such slopes are due to interbedding with layers of softer rock which have worn away faster than the limestone. Vegetation may flourish on the tops of the steps or on the slopes. One typical shape of a hard limestone peak has been most happily described by Mr. Leslie Stephen as the 'writing-desk' formation (fig. 2). When a mountain presents this shape, its structure and character, though snow-covered, can with certainty be determined at a distance. Good instances are the Wildstrubel and the Titlis in the Alps. On such mountains there will obviously be one easy route up, while many sides will be difficult, or altogether impossible, to climb. The more

FIG. 2.—' Writing-desk'

or less vertical rises between the steps will be smooth. The great ledges may offer easy enough going, but will not lead up to the summit. An observer might be so placed while surveying the mountain that he sees only the gently sloping side, but if he recognises the character of the rock, he may feel pretty confident that on the further aspect the mountain will be cut away to a cliff-like shape. It must not, of course, be assumed that there is no variety in limestone mountains, and that this single type is characteristic of all. The limestone may be of a hard or soft description, much or little worn by water, interbedded with layers of softer rock, or comparatively homogeneous.

Dolomite is a hard magnesian limestone which has its own strongly-marked characteristics. Exceptionally hard limestone peaks may take the same forms. The aspect of these mountains has been most aptly compared to ruined masonry, and it is difficult often to believe that the summits of dolomite peaks and ridges are not crowned by crumbling towers, castles, and walls built by man. A *square* columnar formation, as shown in fig. 3, is characteristic. This is due, of course, to joint planes. A wall-like architecture is the rule, and one side of the mountain is likely to be as steep as the other. The whole of the face shows transverse and vertical markings (fig. 3), the transverse lines running more or less continuously across the face. The colour of dolomite is exceedingly varied and beautiful, cream colour and grey predominating; the tints of these bare rocks in the glow of a late September sunset are marvellous. The jagged outline of the crests forms a principal feature for their recognition. The outline is usually 'embattled,' to borrow an expression from heraldry. As this is due to weathering, and as the process of decay goes on continually, the climber must expect to meet with many loose stones, piled up at the bases of the cliffs or wherever they can rest, with insecure hand- and footholds, and with extensive surfaces of smooth rock. On the crests of the ridges the rocks will be broken up and practicable, but once on the crest there will be little chance of traversing the faces.

FIG. 3.—Dolomite

The Aiguilles, which form so prominent a feature of the Mont Blanc mass, might, on a hasty inspection, be taken to resemble dolomite peaks. The outline of the crests is serrated, but more in the pointed style of architecture. To borrow again heraldic terms, it is 'indented,' or 'dancetté' (fig. 4). The square columnar form is absent. The pinnacles sometimes terminate in actual spikes. The apparently sharp points are often hatchet-shaped in reality. Thus, the needle on the right of the main mass in fig. 4 might owe its look to the fact that it is seen end on. The Aiguille de Charmoz, which from many points of view seems a sharp needle of rock, is in reality a tolerably long narrow crest, with numerous sharp towers jutting up. Large blocks are often perched on the very summit of the pinnacles (fig. 5). The aiguilles have more the appearance of 'Cyclopean,' as contrasted with ordinary masonry, still solid and stable though weathered (fig. 5), in contrast to the ruined look of the dolomite crags. Horizontal and vertical lines there are as in the case of the dolomite rocks, but arranged on a different plan. The architecture is that of great undressed blocks of 'compact crystalline,' fitted together upright, irrespective of size and shape, and destitute altogether of the precise formality of the human architect. So the transverse lines are not continuous, and the squares and oblongs that reveal the jointing together vary greatly in size. Here we have to deal with the hardest materials of which mountains are built. They weather and wear like the rest, it is true, but their softer coverings have been stripped away, and only the hard and solid interior remains. Long cracks may often be noticed running irregularly up as in a church tower that has 'settled' a little. The fissures are, conceivably, in some instances, due to the same cause, viz. settlement. Constantly these cracks offer the only practicable route. In these schistose granite aiguilles they will be true fissures, and not due to the erosion of a streak of softer rock bedded into the

FIG. 4.—Aiguilles (mass)

harder material. Great slabs and faces of rough granitic rock, small ledges of no greater extent than the width of the block, and not continuous in one line over the face of the mountain, firm hand- and footholds, few loose or falling stones, will characterise chiefly the climbing in the aiguilles. Little deviation is permitted from a line once chosen. For the most part the line of attack lies up the face of the aiguilles, and they are consequently difficult.

Granitic mountains, in one of their typical forms (fig. 6) bear some resemblance in outline to the dolomite walls, but are far less jagged and fantastic in form. As on the aiguilles, the rocks will be solid, occurring in large slabs or blocks, little broken up and rough in texture. These mountains are usually

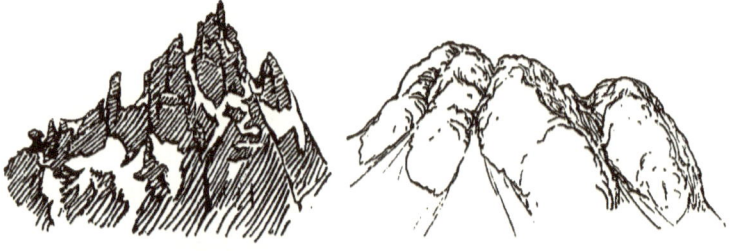

FIG. 5.—Aiguilles (detail)   FIG. 6.—Granite

flat-topped. Occasionally they present the 'writing-desk' form, but not in the Alps. Granite is a far less durable rock than is commonly imagined. The mass breaks away in large blocks and boulders, which again often crumble away by the rotting of one of the constituents. The mountains are not bold in form, but massive and low, with rounded outlines to their ridges. Bosses project from the masses. Climbing is often very easy on such hills; the rocks are firm, the slopes gentle, and steep 'pitches' and cut-off places few and far between.

The peaked mountains (p. 135) owe their shape to the comparatively friable nature of the rocks of which they are composed. The schists, of which they mainly consist, disintegrate rather freely under the influences of the natural forces that ceaselessly attack and erode every exposed surface. The higher portions of the mountain, more exposed and unprotected by snow, strew the flanks and corries with their débris as they waste. Mountains, truly pyramidal or conical in shape, are not so numerous as they appear to be, and the term peak is rather loosely applied in Alpine literature. A climber's classification would include, roughly, in this group all mountains with long broken ridges and peaks more or less well developed. In structure these mountains consist of hard schist. The precise shape depends on the 'bedding' of the main material, the readiness with which the different portions wear away, the dip of the strata and other causes. On such mountains the rocks will be broken up; very smooth faces and slabs will occur rarely. The unequal wear of different parts will give rise to hollows and gullies filled with ice or snow. The climbing varies much according as the line of attack is with or against the dip of the strata, or of the dominant structure planes. The crests of ridges and rocky ribs form the best line of attack. Stones are certain to fall on some parts of the mountain. Monte Viso, the Weisshorn, and Matterhorn are often mentioned as examples of peaked mountains. The Bietschhorn is a very good instance of symmetry due to rapid erosion. The rocks of this peak are notoriously rotten.

There are certain first principles that apply generally to all rock climbing. In practice the more a man reverts to the quadrumanous type the better. At no time, if it can be avoided, should he depend on one limb. When he is shifting a foot, he should have hold with both hands; when he is shifting a handhold both feet should be firmly planted. Three out of his four limbs indeed should be constantly used as anchors. Handhold is more valuable than foothold, and with a good grip of the fingers alone a man can stop as much weight falling as if he stood in an ice-step with his axe properly planted, the rope being equally taut in both cases. Every step chosen, and every handhold taken, should be tested for security before any reliance is placed upon it. The eye can help the foot more than the hand. A glance may be sufficient to assure the climber that a foothold is trustworthy, but as a rule every handhold should be carefully tested before it is used. It is best to assume that every rock touched with hand or foot may be loose, and for each man to judge this for himself, however many may have employed the same hand- or foothold previously.

When inspecting a piece of rock, trace the footsteps and the handholds as far as possible, and consider before starting which is the best leg to put foremost. The neglect of this very obvious rule will frequently convert an easy rock passage into a very difficult one. A and B, both equally gifted, start one after the other to climb up say twenty or thirty feet of rock. A finds infinite difficulty in transporting himself to the top, and takes a quarter of an hour in doing so. B follows in two minutes with perfect ease. The explanation of the difference lies constantly in the fact that one started with the right leg for his first step, and the other with the wrong one.

When a man has got hopelessly entangled on rock, he can often set himself straight by simply taking hold with his left hand in the same place that he had anchored himself with his right, or by changing his foot in the step. Frequently, however, it is a very difficult matter to change when the climber has once started up. The best climbers, if carefully watched, will seldom be found to change, save deliberately, and when the conditions require that they should do so. The hand- and footholds seem to fall naturally into the right places for them from the beginning to the end. If you have a good climber for leader and are following him, note carefully the foot with which he takes his first step and do likewise.

Men stand upright more readily in snow-steps than on rock, and on the latter it would seem that greater confidence is really required. A man is really just as safe when standing on a firm ledge of rock, say three inches wide, as he is when on a snow-step of double that width, the inclination of the slope being the

ROCK CLIMBING

same in both instances; but very few indeed can be found to lean away from a rock slope sufficiently to approach in any degree the perpendicular On such a slope, inclined, say, at an angle of 45°, the climber has the choice of two attitudes:

A very worm

either he must stand upright and neglect handholds, or else he must incline his body to nearly the same angle as the slope in order to grasp with his hands. The latter position, though less graceful, is generally the safer; and the broad rule may be laid down, that whenever there is any handhold obtainable it should be utilised. The difference in pace is very great between the two methods. On a moderately smooth sloping rock face, where a man can, if he keeps upright, walk in safety, he will pass over it in a third of the time required by the climber who leans to the slope and uses handholds. Guides sometimes trot over rock passages with the greatest unconcern, on which their employers have spent much time and energy and fancied that they have achieved rather a neat piece of scrambling. But safety must be considered before 'form.'

When climbing straight up a rock face inclined at a steep angle, the hands do the greater part of the work. See that you start with the proper foot, having a good hold with one hand at about the level of the chin; reach well up with the other for a further hold, and test it carefully. Now, if it can be done, while both hands tightly grip the holds, lean back to see how the feet are to follow, but do not move a foot until you have satisfied yourself where you are going to place it next, and unless you feel certain that you can replace it if necessary in the place whence it has been moved. In climbing up rocks a man who has perfectly good hand- and foothold for all four limbs will at times, after a single ill-judged movement, find himself perfectly helpless—unable to go up and equally unable to regain his former position. There is nothing quite comparable to this in snow and ice work. The climber may slip out of a step or go through a snow-bridge, and if the rope is properly used, finds that the slip has no serious consequence. It is far less easy on bad rock for a man to recover himself or for his companions to assist him. Climbers even of some experience will lose their heads with extreme suddenness when they find they have moved a foot and failed wholly to discover a place in which to put it. The handhold, which a moment previously seemed so good, suddenly deteriorates. The grip, which served to steady, proves wholly inadequate to stand a strain. A struggle or two follows, then a brief descent of extreme rapidity, and the climber finds himself again in a position to inspect the best line of ascent. There

'Kommen Sie nur!'

are probably very few mountaineers who, in scrambling about the hills of our own country, have not had this experience. Such incidents do not, as a rule, find their way into print.

In rock climbing more than in any other department of mountaineering, it should be made a rule absolute never to go up a place that you do not feel certain that you can come down again, or down a place that you could not, if the occasion required, re-ascend. Further, no single step up or down should be taken without observing the same precautions.

The rope is far less valuable as a safeguard against a slip on rock than on snow. It delays progress, for the coils are perpetually catching in cracks, hooking round sharp projections, or sweeping down loose stones. Yet it is as essential to use the rope on rocks that are at all difficult as it is on high snowfields. Every member of the party, too, whether ascending or descending, must attend to the length of rope in front of him as well as to that behind. It is more difficult, or, rather, requires much more attention, to keep the rope properly stretched between the several members of the party when climbing rocks. The pace being irregular, the intervals are much more frequently varied. But when rocks are at all stiff every man must throughout be on the alert. Indeed, rock climbing, when at all difficult, may be almost summed up in this: That from the first moment to the last, the utmost attention must be paid to every detail of every movement, while additional care must be given to the rope.

In ascending rocks the axe is of comparatively little use; frequently, indeed, it is an encumbrance. Slings, consisting of a strap or loop of string, secured round the axe-head are convenient. The arm is pushed through the loop up to the shoulder, and the axe is thus slung, leaving the hands free. Axes are sometimes made with large notches cut on the under-surface or edge of the pick; by means of these the climber is supposed to anchor the pick above, and so draw himself up. It may safely be asserted that there is no place on any rock mountain in any country where such a method ought at any time to be employed. Even on steep grass mountains the method is seldom a good one, while on rocks the axe had better be left behind altogether than employed in this way. The notches in the pick are undesirable, if only for the reason that they might suggest to the beginner to adopt an unsafe method : they are useless for any other purpose.

Young climbers are prone to treat rocks as if they were a mere staircase or ladder ; they endeavour to ascend them too much in one straight line. It is far easier to see a good foot- or handhold on the right or left of the immediate line of ascent, and, by making a series of very small zig-zags, progress is much safer and quicker as well as less laborious. In a narrow rock chimney, if the disengaged foot is raised in a straight line, directly in front of the body, say for 18 inches vertical, the exertion of straightening the bent legs, so as to raise the body to that height, is a very great one, for the leverage is enormously against the muscles on the front of the thigh concerned in the action. But if the disengaged leg be raised to the same height and carried 18 or 20 inches to the right or left as the case may be, an immense increase of power will be obtained. Suppose that on a given rock the knee is bent at a right angle when raised 18 inches in a straight line. Now raise the limb to the same height, but carry the foot 18 inches to the right or left of the middle line. The knee will be bent at a much larger angle, say 120°. Much less exertion will in the latter case be required to straighten the limb, and further, a point of great importance, a powerful lateral thrust can be obtained. Moreover, the body can be bent forwards much further when the limb is raised laterally. A person stands more securely when his feet are separated at some little distance than when he stands at attention, as may be noticed any day on board ship. When the step up is a long one, the advantage becomes still more obvious. Imagine that a man is standing in front of a smooth, nearly vertical rock slope, both feet being planted together on a ledge at the base. The first sound handhold is 7 feet above the level on which he stands. He has the choice of two footholds ; one is directly in front of him at a height of, say, 36 inches ; the other 2 feet to the right or left, and some inches higher. The latter will be found much the easier step to take, merely because it lies to one side. On many rock passages a man can easily get up, when he carries the whole of the lower limb out at almost a right angle from the body. With a suitable rock face to push against, he can dispense with any ledge or resting place for his foot; finally, he has the great advantage of being able to see where and how his foot is placed. Often, however, the position is so cramped, that it is not possible even by leaning back to raise the foot to the required spot. If the knee is well bent, even though it rests on a smooth surface of rock, very good security can be obtained, and a step up of 36 inches or so made with ease. As the body is raised the climber comes gradually into an ordinary

Giving a hand

kneeling position, and will find no difficulty in raising himself from this until he stands again on his feet. The resistance from friction is very great where a large surface of limb with its rough covering is brought into contact with the rock. On irregular rock a very small portion of the sole of the boot really touches the surface. The weight of the body is thus transmitted through a small area of contact. There is direct pressure, but very little friction. The method of climbing with the bent knee is, in suitable places (which are of very frequent occurrence), a most admirable one. The principle can be carried still further, and in narrow rock gullies, if the back is pressed against one side and the feet against the other, a climber can work his way up with great security and ease after the manner of the chimney-sweeps of a bygone age. The hands are fairly free, and an immense weight can be supported in this position. As the body is moved up, it should not be scraped too much against the wall, or the clothes will suffer ; anything carried over the shoulders must be slung in front before starting. The weight must be slowly shifted from one point to another. A slow drag, not a series of convulsive movements, is the true method.

Steep walls of rock will often confront the mountaineer, on which for a considerable height there is no foothold at all. But it does not follow that there is no handhold. Small ledges, cracks or projections that will not accommodate the foot, will often suffice to give a good handhold. If a suitable place can be found at a height of 6 feet or 7 feet up, the climber is generally able to surmount the obstacle even when the wall is described, nominally, as perpendicular or overhanging. The power of the arms alone is hardly sufficient. Few, except very practised gymnasts, could draw up their whole weight from a handhold alone, say 7 feet above the level at which they stand ; an artificial foothold has to be supplied. If there is a good standing place below, it is simple enough to climb on to a companion's shoulder, and often in this way the place can be surmounted. It is not, however, very pleasant for the companion while the nailed boot rests on his shoulder, still less as the step up is made. The top climber should throw all the strain he possibly can on his arm as he takes his step up. A better plan, when circumstances allow, is to grasp the leader's foot well in both hands and raise him up as in assisting a lady to mount her horse. The axe-head furnishes a small and not very trustworthy step if planted against the rock. By pressing the point of the axe well into a cranny and raising the stick above a horizontal level, a better hold can be given.

Even these expedients are insufficient on some steep walls, and rather hazardous methods have been resorted to in the endeavour to erase the epithet inaccessible from the Alpine dictionary. There is one well-known passage on the Italian side of the Matterhorn, which to those who first attempted the ascent proved an insurmountable difficulty. The place was turned at last by the traverse of a narrow ledge offering little or no handhold, and still known as 'Carrel's Ledge.' It has only once been traversed since and ought never to be traversed again, for the cliff is now made easy by fixed ropes and ladders. Once in a way climbers have been enabled to make their way up such difficult passages by throwing up a rope after the fashion of a lasso, and endeavouring to make the loop catch in some projection above. The performance requires the skill of the Mexican cowboy, and is a most dangerous expedient. Climbers, more ingenious than prudent, have endeavoured, by means of rocket apparatus, to fire up a rope armed with a grapnel. This plan was tried unsuccessfully in an early attempt on the Aiguille du Géant. Such expedients savour too much of the romance of mountaineering to be discussed in this work. Mr. Whymper devised an arrangement, figured in his 'Scrambles amongst the Alps,' which consisted of a steel claw, to which a rope was attached. Mr. Whymper says,[1] 'The claw could be stuck on the end of the alpenstock and dropped into such places (cracks or ledges above the arm's length), or on extreme occasions flung up until it attaches itself to something.' On the inventor's own showing the device was rather a hazardous one; the 'something' might have been anything that was incapable of standing a strain. If the leader of the party can be got up a particularly steep piece of rock, there is no objection to his fixing a

---

[1] *Scrambles amongst the Alps*, p. 110, 2nd edit.

CRACK CLIMBERS

spare rope to assist those who follow. Indeed, it may be impossible for the last man to ascend at all without this aid. The uncertainty as to the real attachment of a rope flung up renders the plan insecure. It is quite another matter if a suitable place for fastening a rope can be deliberately chosen. The use of fixed ropes in descending will be discussed later on in this chapter.

Taking the oath and his seat

ESCENDING rocks is popularly supposed to be more difficult than ascending. A partial explanation of this commonly received view is to be found in the effect on the imagination of looking down any place. When standing on a steep rock face but little of the route to be followed can be seen, and a slope inclined at even a moderate angle appears worse than it is in reality. The frequent use of adjectives, such as sheer, vertical, and overhanging, in Alpine literature affords evidence that even practised climbers are prone to be deceived. A climber of limited experience will undoubtedly ascend rocks better than he will descend them, but as soon as confidence is gained most of the difficulties vanish. To estimate correctly the difficulty of rocks when viewed from above is not an easy matter. So much, indeed, does the appearance of rock alter when viewed from above or below, that the majority of mountaineers find it hard to recognise the line of descent, supposing that they are going back by the same route. A beginner will sometimes fail entirely to follow the identical line if he endeavours to go down twenty or thirty feet of rock by which he has a few moments previously ascended. This inability to recollect a route which is not marked in any way as a steep snow slope would be by steps, arises almost wholly from the fact that most people neglect as they ascend to look continually backwards and note mentally the detailed appearance of the rocks viewed from above. If the climber begins early in his mountaineering career to do this, he will insensibly acquire the power; not only will he learn the detail of a particular route, but he will accustom himself to understand when looking down any rocks where the best passage exists. Such a power of judging is of course of the highest importance. Amateurs are too apt to assume that their guides possess this faculty as an instinct. There is no instinct in the matter at all, if the term means some quality which is inherited or inborn, and incapable of being learnt. The 'instinct' is an acquired quality, the outcome of experience. What the guide can acquire the amateur is also capable of learning. People often fail, though they have given time and thought to the matter, because they habituate themselves too much to depend on minutiæ. They look for a stone of an odd shape noted in ascending, a curious ledge, or a nest of crystals. No feature is worth attending to as a guide to a line of descent that is not likely to be seen clearly from above. The route down a particular place cannot be noted to good purpose while actually climbing up it, but just after it has been surmounted. Mental notes of cross bearings in the detail of a line of ascent are as valuable as written notes of the same nature in reconnoitring.

The most difficult place to fill properly in rock climbing is that of leader in the ascent and last man in the descent. On the whole, it is true that in descending difficult rocks the last man on the rope has a more onerous task than the leader during the ascent. But this is in great measure due to the fact that less assistance can be given during the descent to the last man than to the leader during the ascent. If a man were absolutely alone he ought to find no greater difficulty in going down a particular passage than in going up it, and unquestionably he could get down many places which he could not ascend. The descent of rocks, as of snow, involves more strain on certain groups of muscles, but is, as would naturally be expected, less trying as a matter of general exertion; in athletic phrase, it is more a question of strength than of wind. Good balance is of more value in descending than in ascending; yet even here, good balance is of less consequence than in snowcraft generally, and in this respect proficiency in rock climbing is on the whole less difficult to acquire.

Two distinct methods may be adopted in descending rocks. The climber may turn his back to the rocks, or he may face them as he would in going down a ladder. The latter method is the slower, but on steep rocks the more secure, and it is the best to follow when the angle is such that the climber, leaning back from his handhold with arms extended, is able to look down and see where he has to place his hands and feet. Descending with the back to the rocks is the proper method to employ when the face is only moderately steep, and when rapid progress is necessary. In ascending rocks, the greater part of the work falls on the legs and comparatively little strain is thrown on the arms, which are more used as anchors than as means of progress. In descending, on the contrary, the arms have the more important duty to perform : not only has the weight of the body to be lowered by the arms in making each downward step, but the hold must, in addition, be such as to be capable of withstanding a pull, such as might be brought about by the slip of any other member of the party.

It is chiefly, however, on very steep places that the difference of strain thrown on the arms and legs respectively is felt. On slopes of moderate steepness, where the climber has his back to the rocks, he is usually able to secure a strong hold, and to support others below him in exactly the same way as he would on snow-steps. On the rocks he has the advantage of a firmer foothold, and can generally add to his security by using his arms.

The axe becomes of great use in descending moderately steep slopes of rock. Frequently the climber is able to gain an additional point of support by the proper use of the axe, which gives also a delicate and quick test of the firmness of any step that he desires to select. The head of the axe is grasped like the handle of a walking-stick, and the chosen foothold tested with the point. By thrusting the point of the stick strongly forwards against any projecting ledge or into any crack, the climber is able to use it as a sort of crutch, or can lean on it as a lame man would use the balusters in walking downstairs; but this form of support, while excellent to assist the climber's own downward progress, is of little or no value in enabling him to withstand any sudden downward pull. On loose rocks with many small stones, it is an invaluable method, for it distributes the weight as much as possible, and a slight slip is neutralised at once.

On smooth rocks the resistance occasioned by friction has to be utilised to the fullest extent. When hand- and foothold are precarious, additional brake power can be put on if the climber spreads himself well on the rock, and lowers his weight slowly. Care must be taken that the clothes do not catch on any projection, or the climber will find himself hung up in positions which are more than humiliating. The apparel oft proclaims the man. It may be a matter of considerable difficulty to detach any portion of the clothes securely hooked on a projection. Tailors are but mortal men, and the climber's repentance will too often be accompanied by rending of the raiment. Yet another precaution must be observed in descending. It is quite as necessary to make sure that the foot can leave any particular hold as that the step can be made on to it. More than once the writer has seen a climber get his foot into a little crack just the width of the sole of the boot, then take a step downwards with the other foot and find himself wholly unable to move the limb now far above him, which had become securely wedged in. The resulting attitude would

Hookey Walker

have been considered highly commendable in a *première danseuse*, but less decorous in a rock climber, who was as incapable of drawing back his lower foot to the level of the upper one as of moving the latter in any way whatever.

On one occasion, within the writer's recollection, a member of the party found himself in this predicament in an awkward place on a very difficult mountain, and it cost nearly a quarter of an hour to release the imprisoned limb. With a good deal of trouble he was pushed back to his former position, but even then it was found impossible to extricate the wedged-in foot. A guide attempted with extreme vigour to pull the limb backwards, but without avail; some strong language proved equally ineffectual. Finally, the boot had to be unlaced, which was a work of some difficulty, for the party was in a very narrow chimney, and then by a vigorous haul the climber was pulled out of his boot. The mishap really might have occurred in an ordinary ramble, and all who witnessed the situation doubted whether our friend, if he had been alone, could have freed himself at all.

It is not easy with heavily shod feet to feel accurately for a good hold. In ascending, the fingers can readily test the security of any ledge or crack, and estimate quickly its value as a hold. In descending, the feet go first and have, as it were, less power of discrimination. As far as possible, they ought to be guided by sight. In descending, as a foothold is of less importance than a handhold, the utmost care should be taken that the latter is really good, and the fingers should always grasp it in such a way that the value of the hold is not lost as the climber lowers himself down. For instance, a climber is descending a moderately steep slope with his back to the rock, and extends his arm to grasp a hold below. If his palm is directed downwards, and the hand bent back nearly at a right angle to the forearm, he is perfectly secure and comfortable until he begins to descend. Then begins the trouble. The moment the body is lowered, the elbow begins to bend and the hand to twist round. The grip is lost at once. The right position for the hand would have been with the palm directed forwards, while the fingers bent back grasp the hold. Then, as the body is lowered, the anchorage becomes firmer and firmer, while the arm may be kept rigid nearly the whole time. Such points of detail may seem too trivial to mention, but the writer has frequently noticed that beginners who fail to grasp the principles fail also to grasp their rocks properly, and it is after all on the careful observance of detail that the security of the party depends.

It must not be supposed that the two methods of descending rock which have been principally dwelt upon are each complete in themselves. A combination of the two is almost always necessary. On steep rocks the climber has incessantly to vary his method, now descending with his face to the slope, now seating himself on it, and thrusting the body forward while lowering the feet, now reaching out with the axe point and leaning right forward, as he may do in a narrow gully, until he is able almost to fall forwards against the opposite wall and alight on his selected handhold. Often a compromise between the two plans is the best. When the line of descent lies obliquely across the rock face, the mountaineer will find that by keeping his side turned to the wall he can progress best. The infinite variety of movement constitutes a great charm in rock climbing, but renders the task of describing it in words almost hopeless. The pencil may succeed where the writer is only too conscious that the pen fails.

Before passing on to consider the kind of rocks in detail, the writer desires to impress one principle on the beginner with the utmost emphasis, more especially if he be young, keen, and active. The man who desires to excel as a rock climber should before all things avoid the temptation to acquire a reputation for what is often spoken of as brilliancy. This quality is only to be acquired by much polish, and the polish should be the result of constant work. Too often the feats of those who hanker after a reputation as rock climbers are mere gymnastic efforts in the performance of which self-satisfaction is more considered than the safety of companions. Let a man endeavour to be considered thoroughly safe on rock rather than brilliant, for the latter term is often synonymous with showy. If he wishes to make himself a good rock climber let him not aim at becoming a mountain acrobat. So will he be more esteemed of those who best understand the business. It is better to be spoken of behind your back by the guides as a man who is always trustworthy on rocks than applauded to your face as a 'Herr' who goes like a chamois. The compliment is a doubtful one. Recollect that chamois are beasts that follow their leader rather tamely : they are chiefly concerned with their individual comfort and security, and lose their heads readily when in positions of unexpected peril.

Do not climb rocks either with a spring or with a jerk, whether going up or down. The slow, dragging action is the true and the right method. There may be moments when a jerk or a spring is required, but they seldom occur. The most difficult rock mountains are not necessarily those with passages of great intrinsic difficulty or danger, but those which it is hardest to climb throughout in a right and safe manner.

A really good rock climber progresses almost silently, whether his hold be good or bad ; he so applies and distributes his weight that he neither slips with a jerk nor dislodges a single stone. He is always firm at more than one point of attachment. He starts always with the proper foot. Directly he starts he progresses continuously, however slowly. He makes no springs. All his movements are deliberate. He seems to writhe up the rocks rather than to climb them. He does not move at all till hand- and foothold are placed as well as the conditions allow. While always ready for an emergency, for a slip on his own part or on that of others, he puts forth no more exertion than the particular passage demands. He depends on the concerted action of many muscles rather than on the violent contraction of a few. As you pay out the rope to him it passes through your hands as steadily as if it were attached to a weight sinking into the sea. He leaves the rock over which he has climbed in better order for those who follow than he found it. He clears away all loose stones. He has no occasion to try to improve his foot- or handhold, for he selected the best available at the outset, and at once placed his foot or grasped with his hand to the best possible advantage. He is not afraid to describe the position in which he finds himself truthfully. If he says he is firm, he is certainly so ; if he says he is insecure, look out ! And finally he is as good, as careful and thorough after twelve hours' climbing as when in the middle of his expedition ; as good, as careful and thorough when his peak is done and he is climbing down uninteresting rocks as when he is storming the last obstacle before the summit. Such are the characteristics, briefly and baldly put, of great rock climbers, of Jean-Antoine Carrel, of Michel Croz, or of Ferdinand Imseng, in the days that are gone ; of Melchior Anderegg, of Emil Rey, of Johann Von Bergen, or of Alexander Burgener in more recent times, to select a few names only at haphazard out of many that pass through the mind.

A will and a way

WHENEVER the route allows of it, a slightly zigzag course is the best to take in steep places. On rocks the nature of the ground may allow no choice, and the consequence is that the party are often perforce moving in a vertical line, one below the other. Now falling stones descend in a straight line, and these objects will acquire a very considerable momentum in a height of sixty feet or so, which may be the distance between the first and the last members of the party. Careful previous trial of hand- and footholds will prevent the dislodgment of stones : to stop them or to get out of their way when they are once started is quite another matter.

R

Many climbers of reputation, in describing their expeditions, will avow in the most ingenuous manner that they started any number of loose stones, and, indeed, with some mountaineers, as those who come late on the rope can testify, these misfortunes occur more in battalions than as single spies. It is largely a matter of attention and care. However active a man may be on rocks, and whatever the gymnastic feats he may be capable of performing, and, for the matter of that, whatever difficult rock peaks he has got to the top of, he is a bad rock climber if he is at all in the habit of dislodging stones. The difference between the professionals and the amateurs in this respect is immense, speaking generally. At the same time there are many amateurs able to climb on good, bad, or indifferent rocks for twelve hours or more at a stretch, without once dislodging a stone in such a way as to be of any danger to their companions. To clear away loose stones from the track, throwing them down out of harm's way, on the contrary, is good practice. Dislodgment of a stone generally means a slip. Directly a climber becomes tired and his attention to detail flags, the missiles begin to rattle about.

Stones are much more often dislodged with the foot than with the hands, and more often in quitting a step than in planting the foot on it. The climber may have misjudged the security of the foothold, and find it loose where it appeared to be sound. The greatest mistake he can then make is to attempt too quickly to change the foothold that has proved unsound. He will probably do this with some degree of spring or jerk. Neither of these actions is admissible for the most part in rock climbing. It is more essential in rock climbing than in any other department of mountaineering to bear the principle in mind that you must accept a slip made by yourself, for this gives the best possible chance of recovery. A man may have his foot lodged in a hold on which he depends for four-fifths of his security. To his distress, he finds his best step giving way. If he endeavours to shift his position suddenly he is in the worst imaginable plight. Unnerved at finding his selected foothold insecure, he is apt to leap from Scylla into Charybdis. Unable at the moment to gauge the strain that will be thrown on him by giving way, he plunges and sends a volley of loose stones flying down on his companions, rendering them more desirous of avoiding danger to themselves than of giving him assistance. A second or two is often quite sufficient to enable the whole party to be ready to meet the emergency.

The first duty of the man in difficulties is to call attention to the fact, a duty far too often neglected from a sort of vanity which is as foolish as it is out of place. Simultaneously he must endeavour to increase his security by means of his other fixed points. A forcible effort to strike off, as it were, from a badly placed foot, merely results in throwing additional weight on that which has already proved insecure. In climbing up, unless a mountaineer has made the mistake of placing his foot on a hold which gives way the instant it is touched, he can commonly maintain himself where he is, although his support is yielding, for a considerable length of time. Meanwhile the rope can be tightened above, or he can look about for a better position, or the man below can come to his assistance. If a piece of rock gives way the moment it is touched, little harm will result except perhaps from the falling mass. The climber will not slip badly if he has not placed much of his weight on the treacherous hold. It is the rock that gives way after a moment or two that is most to be dreaded, and even then it is scarcely of any significance if the climber will but bear in mind that it is of more importance to be deliberate in leaving the bad foothold or handhold than in leaving a good one.

To announce to the rest of the party that you are in a perfectly safe position—'ganz fest' in guides' phrase—when you can only just maintain yourself by an effort in a position of doubtful security, is an unpardonable error. The moment the rope is put on, the party becomes a chain, the strength of which is to be measured by that of its weakest link, and the man whose word cannot be depended on to describe his own position is a more dangerous companion than a man whose legs and arms cannot be trusted. Yet this selfishness, for it is none other, and this form of vanity in mountaineering, which deter a man from accepting assistance from others, are only too common. The rock climber has as much a duty to discharge towards the other members of his party as if he were crossing a snowfield fissured with hidden crevasses.

In very difficult places a far greater length of rope is required on rocks than on snow. A party of three or four on a presumably difficult rock peak unknown to them require at least 150 feet. They may not use it all; but they should be provided with this length. Frequently it becomes imperative when descending a long, difficult chimney, traversing a smooth passage, working up a narrow crack, or descending rotten rocks, that only one should move at a time whether guide or traveller. It follows that a large party on a difficult rock-mountain is undesirable. Five may be held a maximum. If the passage is say 80 or 100 feet in length, the party should unrope, and each member should be attached to the whole length and allowed to descend separately to a place of security. The process consumes much time, and on a cold day is very trying to the patience, but it is none the less imperative. At no time when the conditions of the rocks seem to require it, should this method be neglected from any ridiculous idea of beating records, as they are called, in time. Rock climbing, and for that matter, mountaineering generally, is not like quarter-mile racing, or bicycle riding. The whole conditions may vary from day to day, and almost from hour to hour. There is nothing more dangerous and more fatuous than the fashion which seems to be growing for quick climbing. A running-path or a road may vary so little in different places or under different conditions of weather, that the time offers a good criterion of the merit of a given performance. But on the mountain-side this can never hold good. The best climber is he who is secure and trustworthy from beginning to end, and the time occupied should not be considered an element in the case under any conditions whatever.

However skilled and experienced the various members of the party may be, it is the duty of each man, so long as he is roped, to behave as if any member of the party above or below him might make a mistake at any moment. It is common enough to read in popular accounts that so and so made an unexpected slip. Such a thing ought never to happen. Any man may make a false step ; and it is the duty of every member of the party throughout to expect that any of his companions may do so. The man who says that he never makes a slip with his feet, probably does so with his tongue. A false step is not an irremediable error, nor does it imply that the legs fly up and the body flies down. The beginner actually slips often, the young climber occasionally, the practised mountaineer very rarely.

The beginner will lose his balance more or less completely every time a rock gives way, while an expert may chance on an insecure ledge or a yielding handhold a hundred times in a day and never lose his balance once; and therein lies the cardinal difference between the two. With practice, the adaptation of the muscles to meet the particular emergency becomes merely an automatic action ; without practice an effort has to be made. The climber has but two things to bear in mind : one, that if he imagines he can climb rocks without ever making a mistake, he is alike destitute of wisdom and experience :

> Nay, deem not thus—no earthborn will
> Could ever trace a faultless line ;
> Our truest steps are human still—
> To walk unswerving were divine !

and secondly, that when he does make a slip he must accept it. If he jerks back he will but make matters worse than they need be. He should endeavour to get himself as much as possible into the attitude of glissading. The threatened fall is to be averted or checked not by the erring limb, but by the disengaged leg or the stick. No written words, however, can teach a man really how to correct a slip. It is one of the essential difficulties of mountaineering. Like the proper stroke at tennis or the cut at cricket, the faculty can be acquired by imitation, but then only by such as have real aptitude for the sport. Unless acquired, a man can never be a high-class mountaineer. When the climber is confident that he possesses the knack, he may go safely wherever a man may reasonably go at all on the mountains. If he is subject to the dangerous delusion that he is so skilful a performer as never to slip at all, he cannot safely go anywhere.

Grass-covered rocks, such as are constantly found in the neighbourhood of the high hotels, are much less easy to climb than people appear commonly to imagine. During the season it is of almost daily occurrence in the Alps for people who go out for solitary rambles on the mountain-side to find themselves involved in tolerably awkward predicaments. For some inscrutable reason people, professedly of sound mind, are in the habit of starting forth from their hotel for a Sunday ramble, shod in boots destitute of nails, and equipped with an umbrella or a walking-stick. Often when intent on botanical or sketching excursions they are no better provided. If they wish to cross a snow pass, such as the Theodul or Petersgrat, they engage a guide and carry an axe instead of an umbrella, and a rope over their shoulders in place of a yellow-backed novel under their arm. Yet the umbrella would be more valuable on the Theodul than on many of these mountain wanderings which involve scrambling over grass-covered rocks, and the ice-axe would be of more use on a grass mountain where there are no paths than on plain snow slopes. On a simple snow-pass there will be scarcely any deviation from the track which is recognised as the best, but on these grass mountains the cut direct is constantly being practised ; yet there are not so many easy variations on a fairly steep grass mountain of say six or seven thousand feet which a judicious person may take, and there is a right and a wrong way even up a grass mountain.

ROCK CLIMBING

At a height of six or seven thousand feet the soil clings but loosely to the rocks, for these are constantly wetted by springs. Tufts of grass when long offer apparently good grasp for the hand, but where they are longest there is most likely to be

Unstable equilibrium

water underneath, and consequently the roots have but a weak attachment. Again, the rocks at these lower levels have frequently been worn smooth in old times by glacier action. The nature of the rock surface is concealed by the grass, and the rambling scrambler constantly depends for his hand- and foothold merely on the soil loosely adherent to the rock beneath. Tufts of grass are at the best very insecure handhold; when wet they tear away readily, when dry they are extremely slippery, and even nailed boots give but poor security.

The inexperienced person who goes out for a ramble by himself will often attempt rock passages actually more difficult than any he would have to overcome if he ascended half a dozen of the reputedly most formidable rock peaks in the Alps. Few will deny that it is unwise to go alone on the high mountains; equally few seem disposed to admit that it is often just as foolish to scramble about alone and imperfectly equipped on the precipitous side of a grass mountain. On dry and slippery grass the axe is often wanted to make a step. In rare places it may even be admissible to strike the pick into the soil, and thus anchoring it, to draw yourself up; but if the axe constitutes the sole support, this practice is to be deprecated here as on rock. The pick or spike may be used to test the shallowness or degree of attachment of the turf. The rope is constantly required, and it is well to take a length always on any botanising or geologising ramble where no definite route is to be followed. The streams form a useful guide to the line of descent. The sound of the water enables the right course to be kept in mist; but if the rambler endeavours to make his way down the actual channel of the torrent, he is sure to be brought up from time to time by little cut-away cliffs, and the water-worn faces will be smooth and slippery. The handhold and the foothold should be tested even more carefully on grass-covered rocks than any other, for much is here concealed from view. The traveller in difficulty on the side of a grass mountain (and it is in descending that the difficulties are generally met with) will do best to make for any trees that he can see, even at the risk of a long détour. On tree-clad slopes the ground is firmer, and the roots thrust up above the shallow soil give good hand- and foothold.

ROCK CLIMBING

Grass, after dry and hot weather, is extremely slippery. This condition is most marked shortly after the snow patches have melted from the surface, leaving the dead grass of the previous summer exposed and flattened down. The utmost care is needed on such slopes when dry and hot, if the place is at all steep. As soon as the new blades begin to show through the going improves.

Glazed rocks.—However easy rocks may be intrinsically, they are all liable to one condition which may make them very formidable, and which is technically described by the term 'glazed.' Where these are likely to be met with has already been pointed out, but under certain conditions of weather rocks coated with a thin layer of *verglas* may be met with anywhere. In stormy weather at great heights, rocks previously in good order become glazed with extreme rapidity and form a serious obstacle to descending. The trouble does not merely lie in the fact that the rocks are slippery to the feet, and that the thin layer of ice cannot be dealt with by the axe : their surface is so cold that it becomes a matter of impossibility to cling to them with the hand. The mountaineer may sometimes find that a rock passage which he was able easily to ascend when he could use all four limbs, offers neither hand- nor foothold when he tries to go down it again. The best rule of guidance for dealing with glazed rocks is summed up in the advice:—Avoid them. Terraced or glacier-worn rocks when glazed will of course be much more formidable than those which are more splintered and broken up. It may be necessary to vary a route accordingly. In very cold weather the neighbourhood of any falling water or of any runnel trickling down the rock surfaces should be shunned. If the mountaineer is really caught in a trap, as at times he may be, owing to circumstances which no experience could have enabled him to foresee, and is forced to climb over glazed rocks, he is before all things bound to resist to the utmost the temptation to hurry. It cannot be expected that men, however good on rocks, however indifferent to low temperatures or regardless of frostbite, can use handholds efficiently when the cold is too great to even permit of their feeling what they grasp. Here again, as in any place that is at all difficult or dangerous, it may be absolutely imperative for the members of the party to proceed singly over the worst places, each one attached to the full length of the rope.

There is no condition that tests the quality of a climber more severely than extreme cold, and the rock climber, however slow or however ungraceful he may be, who is as safe on rocks that are glazed as on those that are warm, dry and rough, is a real expert, whatever his list of expeditions and whatever his 'times' may have been in achieving them. If the slope is of moderate steepness, say inclined at an angle of 25° or 30°, the climber who can really stand upright on rock (and such climbers are scarce) had best adopt that as the safest position. Ledges and cracks which are little more than finger-holds are often worse than useless, for the grasp unconsciously relaxes as sensation is lost from the numbing cold. Everyone has to be on the alert. A chance slip will be more readily checked if a man is careful to avoid throughout any positions in which, if he loses his footing, he is liable to turn on his back. In bad weather time is always a matter of great importance and speed essential, but this is gained not by hurry but by constantly moving. Notwithstanding the necessity for rapidity, it is wise for all the party to stop if anyone finds that he has lost sensation in his fingers and desires to regain it by clapping the hands together or thrusting them into his pockets; the last man, or the one who directs the expedition, should watch closely lest anyone should do this inadvertently. The rope is undoubtedly under such evil conditions of little use, and sometimes an actual source of danger to the party as a whole; still, in the writer's opinion the rule is absolute that it must be used. A place that is too bad to be traversed by a roped party, lest the slip of one should drag down all, is a place that should not be traversed at all.

Rotten rocks.—On some mountains the difficulties consist almost entirely in the disintegrated character of the rocks, which are then spoken of as 'rotten.' A climber with some practical geological knowledge will be able to tell beforehand where he is likely to meet with such difficulties. Thus crystalline rocks will be firm, but dolomite rocks or schistose mountains, such as the Matterhorn, the Ecrins and the Bietschhorn, furnish notable examples of rotten rocks. The stones strewn over and resting on the slopes and in the gullies do not constitute the chief trouble. Semi-detached fragments, loosely wedged in so as to give way readily, are far more serious. Mountains whose shape is pyramidal, owing to disintegration, will of course be most likely to prove rotten to the climber. Such peaks are often in the best order for climbing when there is a moderate layer of sound snow on their surface, provided that the stones below are not glazed. The irregularities of the face prevent the snow covering from sliding off, as it would if lying on smooth slabs. Snow-covered rocks give bad handhold, but on a rotten mountain foothold is distinctly the most important. Attention must be paid to directing the pressure, so as not to strain the hand- or foothold in the direction of its least resistance. The shorter the steps the better. The most tempting projections are commonly the most untrustworthy. Rocks will often be found so split up as to resemble rows of books standing on a shelf; the appearance is that of a vertical slabby or slaty cleavage, and is probably to be explained by the rocks having been subjected to extreme lateral pressure. Avoid such places : not only are the edges sharp, but the thin plates readily tear out or break off.

Under no conditions is a slow drag in moving upwards or downwards more important than on rotten rocks; often the action should be almost like that advised in crossing a doubtful snow-bridge. If the weight is shifted very gradually as one limb after another is moved, a loose fragment may indeed give way, but the climber has time to check himself, and if his steps are short and he is not spread-eagled, there will be no sudden jerk from the slip. The perfection of good rock climbing is for a party to go up and down a peak characterised by the rottenness of its rocks without ever dislodging a stone, save on purpose, or ever making a slip which the individual climber is not able immediately, and with safety to himself and to others, to neutralise.

Rapid progress is out of the question on bad and rotten rocks. The pace of a party has to be reduced to that of its weakest or its slowest member, and he must on no account whatever be hurried. Unquestionably slow progress is often very exasperating. To climb with extreme slowness in a cold wind down a long rock face, in a failing light and with stones rattling about, may beget impatience, even in the climber whose temperament is happily philosophical. But impatience on the mountain-side is very catching ; it begets hurry, and this means danger. People can find excuses for a man who goes slowly in deep snow, for that is a condition of things which appeals to the sympathy of all, but mountaineers are too often intolerant of a companion who wishes, not merely for his own sake, but out of regard for the party as a whole, to go very slowly on insecure though easy rocks. The state of mind may be human, but it is none the less foolish.

Scree.—The masses of small loose stones often found lying about the bases of steep gullies and ravines, and beneath the sharp cliffs of rocks, are spoken of in the north of England as scree, and the term is as applicable in any other mountain district. These stones are the result of the weathering and breaking up of the higher rocks. Where the disintegration is comparatively recent or is still going on, they will be destitute of any trace of vegetation, and are merely piled loosely together. When formed of very small stones, these shoots offer a very convenient way of descending, but if they are loose the ascent is a matter of extreme labour. Frequently it becomes necessary to traverse screes. The sensation of standing on the loose mass while each foothold gives, and the entire mountain-side appears to be shifting and sliding down, produces a curious giddiness in many people. Some persons experience great discomfort in crossing loose screes, although they are, in reality, perfectly safe. There is no need for the eye to choose the footholds, for these are all alike and all bad. The axe should be held above, and horizontally pressed into the loose mass as recommended when crossing fairly hard snow, and if the climber keeps his eye directed away from the immediate vicinity of his steps, the feeling of giddiness will usually pass off. While traversing scree, the more upright the climber stands the better ; indeed, he can hardly lean too far away from the slope. If he is unlucky enough to have to ascend he must observe the same rule. Sometimes it is almost possible in descending to glissade down. The great point is to keep the toes well turned up and the heels well down, as in going down slopes of deep snow. In descending it is best to lean rather towards the slope, keeping the axe about horizontal. Meanwhile a sharp look-out must be kept for any big rock that may chance to be in the mass or that may jut out, for such are very likely to trip up the unwary.

Terminal moraines, and sometimes lateral and medial moraines, are made up of huge boulders piled together in the most careless imaginable manner, thrown into confusion like so many spillikins in the child's game, so that it seems impossible to touch one without moving the whole set. The beginner finds the greatest difficulty in walking over this kind of ground, and watches his guides with envy and growing bitterness of heart as they skim over the obstacles, with their hands in their pockets.

Very little practice is really needed to enable a man to go at a fair pace, and with perfect safety to his limbs, over rough moraines. The foothold is entirely directed by the eye, and the result is that the trick is soon learnt. Big rocks that are flat at the top may seem nice to step on, but the base is more important to watch than the surface, and if the base is rounded or pointed the flat top is a delusion and a snare. Avoid stepping down into the deep holes. A slight movement of a big boulder of rock might pin the climber and severely bruise him. Keep on the surface as much as possible, and stick to the big rocks rather than to the little ones. It is better to step on to the sharper edges than to make for flat footholds, and to place the waist of the foot on the projections rather than the heel or fore part of the sole. There is much less liability to sprains and twists of the ankle or knee than appears at first sight, and it is often from fear of such injuries that an unpractised walker makes most of his difficulty. People do not as a matter of fact sprain their ankles when they have any fear of doing so. The great point is to keep on moving, and in this way long stretches of loose rocks can be passed over easily enough when once confidence is gained. There is a fine field in the Caucasus for practising this kind of walking, which is sometimes necessary for hours at a stretch. Those who visit Skye may find a certain corrie descending from between Sgurr Nan Eag and Sgurr Dubh down to Loch Coruisk, which, considering its size, is as exasperating as any to be found in any part of the world.

Experience does it

Dolomite rocks offer a fine field for those who take pleasure in difficult climbing. There is no need, as a rule, to make any deviation from the proper and accepted route in order to enjoy a tough scramble. These peaks and walls have their own characteristics. The rocks are extremely disintegrated, and hand- and footholds treacherous. Limestone worn smooth by the action of water is at no time very easy, and the climber will often be confronted with long smooth passages of rock, which will tax his powers to the utmost. The stones are peculiarly sharp, and progress up or down the intensely hot screes is by no means pleasant. Familiarity with dolomite climbing appears to breed enthusiasm, but, as in eating olives, the taste has to be acquired. The beauty of the scenery, the fantastic ruggedness of the rocks, and the comparatively low heights of the peaks may explain in some measure the attraction which this district has for many excellent rock climbers. Certain it is that a climber of some experience among the Swiss rock-mountains, or the crystalline 'aiguilles' of the Mont Blanc district, will probably find himself utterly at sea at first on these jagged, steep, little dolomite peaks. There are many tales of places quite inaccessible unless the climber takes off his boots. Undoubtedly by this device a much firmer hold can be obtained on smooth and slippery rock, provided the mountaineer is tolerably hard-footed and is not afraid of some discomfort. The ordinary mountain boot, studded with nails, is not the most convenient footgear for some kinds of rock. Still, in the Alps the variety of the ground traversed in a single expedition renders it, on the whole, the best covering. Too thick and stiff a sole increases the difficulty on smooth rocks : nails set closely together give less firm hold, but separated nails are more easily knocked out.

Fixed ropes.—Many places occur in descending where a spare rope may be fixed with the greatest advantage. Much time can be saved. If on difficult rocks the party propose to descend by the route taken in the ascent, the rope had better be attached on the way up. It will then be more certainly fixed at the right spot and will serve also as a guide to the proper line. The late Dr. Emil Zsigmondy, who was a most daring rock climber—too venturesome indeed to be imitated —brought this practice of fastening and utilising fixed ropes almost to an art. Most of his climbs were made without guides.

In fixing a rope to assist a descent it is often desirable to have the power to release it from its attachment when the last man is down. Economy is not the only consideration, for the spare rope may be required for use again further on. A loop can sometimes be jerked off a projecting rock to which it is fixed, but in very steep chimneys this is not always an easy thing to do. Mr. Whymper devised a ring (figured in his book) to overcome this difficulty : by means of a string pulled from below the noose could be slackened and the loop then jerked off the attachment, or else the whole length of rope could be pulled round the projection. Occasionally, on first ascents of difficult peaks, it has been found necessary to drive wooden pegs or iron stanchions into cracks and fasten ropes to them. These have usually been left in position, and some mountains are almost festooned in this manner. In the Alps now these devices are not often necessary. Permanent chains, iron rods, and even ladders, are only too abundantly supplied. It is well, however, before making use of any of these appliances to test them as carefully as can be managed, and especially to look to the secure fastening of the fixed points. To trust to a badly made rope that has been fixed on a mountain side exposed to the weather perhaps for years, merely on the strength of many others having made use of it, is to repose a trifle too much faith in those who have been before. Consider always that others who follow may trust to a rope they find fixed, and that the nature of the fastening may be invisible from below. A rope fastened round a sharp projection or secured to a stanchion should be encased in a leathern tube to prevent wear from friction. A gaiter will answer for the purpose, at a pinch. The climber who fixes carelessly a rope that he intends to leave, not only runs an immediate risk but incurs a heavy responsibility.

Special shoes are used in some countries for rock-climbing, the soles being made of plaited grass, like the Chinese slippers, or of twisted Manilla hemp or rope, and these are excellent for the purpose. The natives in the Pyrenees wear 'Aspargatas,' thin shoes made of jute with leather soles, or 'spadrilles' which have hempen soles. Mocassins would doubtless prove serviceable on smooth rocks, and those who climb for sporting purposes would find the silence of the footfall an advantage. The Caucasian mountaineers use a loose slipper made of one thickness of leather. Some of the native shoes are made in two pieces joined along the sole by leather thongs plaited together. These people generally stuff their shoes with dried grass, and thus shod will walk with wonderful ease and rapidity over smooth rock, but they are perfectly helpless on the smallest snow patch. All these shoes are, in fact, worse than useless on snow or ice, and stand but little wear on sharp loose rocks. If taken at all, they have to be carried and put on when the occasion demands it. Some wear climbing-irons on smooth rocks, and speak highly of their efficiency ; more practice is required to use these aids on rock than on snow.

Falling stones.—Apart from the small risks arising from the falling of stones dislodged by the climbers themselves, there exists, to a greater or less extent on all rock mountains, a distinct risk from falling stones. These unpleasant missiles must be reckoned with. Very little can be done to guard against them, and every possible precaution should therefore be taken to avoid their track altogether. There are two chief classes ; some are prone to fall when the sun is on the mountain, and others when the slopes and faces are in shadow and the temperature is below freezing point. The latter are far less frequent. Both may be met with on the same mountain.

Geological considerations will furnish the most valuable guidance.

Obviously, stones are most liable to fall on mountains whose formation renders them prone to rapid disintegration. On rotten schistose peaks they must be common. On compact crystalline rocks they must be rare. On peaked mountains, therefore, there will always be a source of risk that must be constantly borne in mind. Careful reconnoissance of a peak before attempting to climb it will clearly prove a valuable safeguard.

On stratified rocks consideration of the direction of the dip may enable the mountaineer to find a perfectly safe route up the most disintegrating mountain. Weathering may be as pronounced on one side as on the other, but the loose stones will be at rest on one aspect, while in a position of unstable equilibrium on the other. The illustration (fig. 7) will make

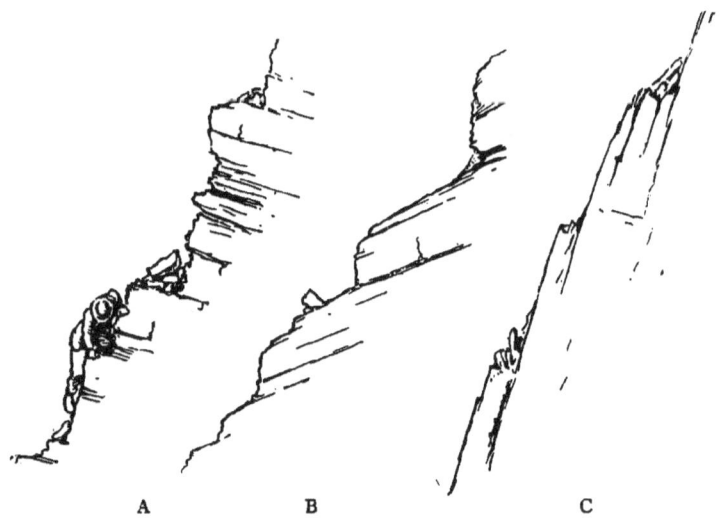

FIG. 7.—Stratification and surface

this clear at once. Moreover, the side free from falling stones will be the best to climb, as can be seen by comparing A and B. When rocks are glazed the difference of climbing with or against the dip becomes much more pronounced.

The stones that fall during the hot part of the day, when the sun is on the rocks, are principally such as have been broken away from the main mass and lie on snow, or are loosely adherent to ice. The stones absorb the heat, the attachment is dissolved, and the fragment started. At first it slides if the surface allows, and in all probability sets in motion other stones that lie in its course. The momentum increases rapidly, and the falling blocks, hurled down with terrific force, dash against the buttresses and projections. At times they fly to pieces as they bruise the unyielding boulders, and the fragments leap out into the air as when a shell bursts. Or they may knock out loosely-connected rocks, and in a few seconds the stone first started becomes one of a volley. On steeply cut-away cliffs they seldom touch the surface, and are quickly brought to rest at the bottom of the slope or plunge into the snow. But on more gently shelving slopes they strike frequently, and are therefore more dangerous. The rapidity with which falling stones start, when the sun's rays fall on the rock faces, is sometimes marvellous. On days when the clouds drift about, and sunshine and shadow succeed each other rapidly, the mountain walls seem alive in a moment as the mists part. Then, as the shifting curtain of vapour is drawn again and shadow follows, all is almost instantly quieted. The significance of this is not to be overlooked. On disintegrating peaks and slopes where stones are likely to fall, it becomes all-important to climb over the dangerous places while they are in shadow.

Stones may be dislodged also by the water that trickles over the surfaces as the upper snow is melted during the day. Slopes that become stone-swept when the sun is on them are never safe till again in shadow. It may often be prudent in descending to wait until the risk is reduced to a minimum. To sit cooling the heels in some sheltered spot on a rock ledge waiting till the sun goes down is rather trying to the patience, but the leader must not hesitate to insist on delay, if he thinks it wise. Smooth slabs and sloping terraces of rock covered with snow or ice, and situated beneath weathered and jagged ridges, are likely to give temporary lodgment to loose stones. The gullies which seam the mountain walls form constantly the natural trough down which stones fall. Indeed, by avoiding the gullies altogether and their neighbourhood perfectly safe routes can be devised up peaks and passes on which these unpleasant missiles are frequent. At times, however, the gullies give the only line of access. Careful consideration beforehand of the time of day when the gully will be in shadow enables the mountaineer to attack it with the least risk. Warm mists and hot winds, it must be remembered, will act as efficiently as sunshine in setting loose stones free, and indeed, on misty days, the uncertainty of the exact position and of the nature of the rocks above renders the risk greater than on a clear hot day.

The marks of falling stones are not very easy to distinguish, but a practised eye can often make them out from a distance. Keep a sharp look-out for marks on the rocks, when on faces where danger is suspected. The bruises when once pointed out will be recognised easily enough. Stones that fall when the temperature is low are principally due to the prising off of masses as the water that has filled the cracks expands in freezing. Huge blocks may be detached in this way. There is, therefore, an additional reason for avoiding any route that leads under faces and slopes over which water trickles.

It is not only in the higher regions that falling stones are a source of risk. Chamois when startled on loose ground send down showers of stones. On the lower slopes and grass-covered hills cattle, and particularly goats, kick loose fragments down with great unconcern, and it is never wise to walk up steep grass slopes in a line directly under a herd.

Large boulders poised on the ice near the snouts of glaciers are apt to slide off in the heat of the day. The enterprise of the Swiss leads them frequently to construct ice-grottoes under the terminations of glaciers, and these places are often visited also in order to see the source of the river as it emerges from under the glacier. Look carefully to all such places.

Notwithstanding every possible precaution, the mountaineer may find himself at times exposed to the risk of falling stones. They do not occur, as a rule, without some warning. The experienced ear recognises instantaneously the sharp crack of a falling stone as it strikes the surface. When the sound is heard each man should get as close to the face as possible.

The head is, of course, the part to be chiefly protected. When the rope is in use, and the party are on a dangerous place, every man should throughout be prepared for the emergency and note mentally at every fresh step the best shelter to make for. No jump is allowable, lest others be jerked out of their steps or dragged into an exposed position. Traversing a stone-swept ice gully is the most risky proceeding of all. Here there is no protection, and the climber has to trust to the chapter of accidents, and to the hope that he may not be included in that chapter. The leader should first cut the steps across, the others unroping if necessary at the side of the gully, so that the length required may be paid out. Then all must rope and cross with the utmost despatch. If stones do fall in such places, face them. Do not be in a hurry to get out of the way, for often the step in which you are standing may be the safest place. If a stone is falling with only moderate force, it can generally be dodged if watched. The advice has been given to parry falling stones with the axe, using it as a cricket bat or a quarterstaff. The counsel appears rather Quixotic, and when the stones have come from a great height the plan would be about as useful as putting up an umbrella while crossing an artillery range when practice was going on.

Some mountaineers boldly assert that the real risk from falling stones is infinitesimal. They point to the fact that very few of the many disasters that have unhappily sullied the records of mountaineering have been due to this cause. It may be so : we do not know the full explanation of all the calamities. No doubt the more sensational of the many Alpine accidents have not been due to falling stones. But surely precautions are not unreasonable on that account. The rules of mountaineering should not be framed merely to prevent the recurrence of disaster, but rather to obviate all risk as far as possible. No mountaineer who has once heard a falling stone whistle through the air close to him at a pace that renders it invisible till it dashes against the face below, will deny that he would come off second best if he got in its way.

Falling stones constitute a real risk in mountaineering, but one that good judgment renders largely avoidable. And no trouble should be spared to avoid the risk.

What has hitherto been said about rock climbing relates almost entirely to the Alps, employing that designation in a wide sense as extending beyond the mere confines of Switzerland. It is, however, in the Alps that difficult rock climbing will chiefly be met with and chiefly practised. Excellent scrambles may be obtained in districts like the Cuchullins in Skye, and there are many little rocky teeth in the Lake districts of our own country where the mountaineer may find abundant opportunities of testing for himself the soundness of the views expressed in this chapter. These small scrambles are but delightful episodes in a day's walk ; they have to be sought for, and just as much time as is thought desirable can be devoted to them. In regions like the Caucasus, on the other hand, where the expeditions are much longer than in the Alps, any serious length of difficult rock would prevent success on an expedition altogether.

'They're goin' very slowly!'

CHAPTER VIII

MAPS AND GUIDE-BOOKS

BY W. M. CONWAY

Topographers

OUNTAINEERING is a sport which can be enjoyed with equal keenness by men of widely different physical and intellectual capacities. It can be approached and pursued in many ways. Mountains have their characters like men, and each kind of mountain makes its own human friends. Some men climb for excitement, some for exercise, some for scenery, some for scientific enlightenment. A mountaineer's equipment must be determined by the nature of the climbing he proposes to undertake and the kind of enjoyment he desires to experience. This is true alike of the apparatus and other accessories that his pack should contain, and of the intellectual equipment that he may carry in his mind or with him in the form of maps or books.

It may, however, be broadly asserted that every intelligent traveller in the mountains needs to take with him the best existing map of the district he intends to visit. If he is a geologist he will, of course, not fail to provide himself with the best geological map that can be had. The mapping of the Alps falls to be done by four Governments, those of Switzerland, Italy, Austria, and France. The latest edition of the Swiss map (the Siegfried Atlas) is the model of what a map of a mountain district should be. The survey maps of other countries are less excellent in different degrees. For a general representation of a large part of the Alpine region, the Alpine Club map of Switzerland is much to be commended. Unfortunately more than half the Alps is not included in it. Small districts have now and again formed the subject of special survey. Mr. Adams Reilly's maps of the Mont Blanc range and the southern side of the Central group of the Pennine Alps are examples of what can be accomplished in a relatively short time by individual effort, properly directed and equipped. The latter of these two maps represents several glaciers and mountains with greater approximation to accuracy than is attained by the corresponding sheets of the Italian military survey. Other districts in the Italian and Eastern Alps have likewise attracted the labours, more or less detailed and successful, of individual students ; they do not, however, call for mention in a chapter which does not aim at containing either a history or a bibliography of Alpine surveying.

A very few years ago, the only map of the Caucasus was of the most inferior quality. A more or less accurate survey was carried to the limits of the cultivated area, whilst the upper regions were depicted in a fanciful style. The result of the repeated Caucasian journeys of mountaineers, chiefly English, has been to create, both in and out of Russia, an interest in the physical features of the range and a demand for their more truthful representation. To this demand the Russian surveyors have been quick to respond. Their new maps are of remarkable excellence, and, even in the glacier region, mountaineers will find them for the most part correct, except in some of the minor details of the upper snow-fields.

The main uses of maps to a mountaineer are four. They are useful for planning expeditions, for finding the way, for

MAPS AND GUIDE-BOOKS 265

identifying features in a view, and for recording routes that have been followed. There are few things in which the difference of intelligence between two climbers is more plainly manifested than in the way they use a map. This is a point which a beginner should bear in mind. The man who depends entirely upon local guides will of course only need a map for identifying his route. Nor will a map be of much service to the merely gymnastic climber—the man whose only pleasure in climbing consists in getting up difficult places—because each of the regular set of gymnastic climbs has to be accomplished by following some exact line of route in almost all its details, and such details cannot be identified or serviceably represented on any map. Most men, however, climb for the sake of seeing, at least as much as of doing, something. A map enables you to discover beforehand which is the best point of view to select for seeing the thing or the kind of thing you desire to see. It is customary to proclaim that all routes lead through fine scenery, and that all summits command incomparable views. The connoisseur knows better. There are degrees in these matters as in all others. Nature has built some mountains to be gazed from, and others to be gazed at. The youthful enthusiast generally confounds the one with the other. Nine times out of ten he will want to climb a mountain, the beautiful aspect of which has at some time charmed him. The glamour of what afar off looks beautiful, overcomes him. He imagines that *Dort wo Du nicht bist, dort ist das Glück.*

The man who studies his map carefully will avoid many a disappointment. He will also receive many an added pleasure. He will start on every expedition with a number of points to observe and of topographical problems to solve. He will thus come to attain a knowledge of range-structure on a large scale. Every climber must learn something of the architecture of individual peaks, but range-structure is a larger matter and is not so easily nor so necessarily learnt. The geographical importance of many a minor peak is only to be perceived by a man who has studied his map with care, or made himself thoroughly acquainted by years of mountaineering with all the ins and outs of a particular district.

Anyone desirous of discovering or inventing new expeditions, must, of course, avail himself continually of the services of a map; he can infer from it much about the structure of a particular side of a peak; he can also discover in what direction a favourable reconnoitring expedition may be made. But it is the Alpine wanderer, above all others, to whom a good map is the most valuable of friends. A stranger in any district may easily find out by enquiry which peaks around afford the most fashionable and delectable rock-scrambling; but it is quite another matter to select out of some possible threescore peaks and passes, mostly invisible from any hotel, the two or three that will best reveal the beauty and display the resources of a mountain group. The Alpine wanderer, who knows his map, and has some practice in the craft of wandering, can choose his expeditions with almost unerring certainty. He will regard the Alpine chain as a whole; as he moves on from district to district, year by year, or in the succeeding weeks of a single season, each expedition that he makes will link on to its predecessor, and lead up to the one that is to follow; so that ultimately he will often come to know more about a glacial basin that he has only traversed, than will some other man who has climbed many of the peaks around it.

It is not worth while to write of the use of a map for actually finding the way upon an expedition. Much depends upon the locality, the weather, the time of day, and other circumstances; more depends upon the climber himself, his experience, his memory, his power of inference, and what not. One man will continually lead wrong when the right way stares him in the face, whilst another will seem to have eyes that can see round corners. Path-finding is more an instinct than a science, and the man who possesses the instinct will discover methods of his own for developing it.

A map is supremely valuable for identifying features in a view, and nothing can take its place, except, of course, a companion with local knowledge. Yet it is remarkable how many climbers, not otherwise unintelligent, blunder in their application of a map to this purpose. Some are inclined to place too much confidence in the accuracy of the sheet they hold in their hands, and some too little. A few are hopelessly unable to recognise the horizontal projection of a peak whose elevation they have before them; and if this is true of mountains near at hand, it is still more true of distant mountain-groups. Though every near peak ought to be unfailingly recognisable to a man with a decent map, it is certainly by no means an easy matter to recognise a far-off mountain, with whose appearance one is not previously familiar. A prismatic compass is a great help towards such identification, and has other uses in connection with a good map, to which it may be worth while to make brief reference in this place.

In theory every mountaineer carries a good compass. Between a good compass and a prismatic compass there is not much difference, either in size or weight; but the latter is infinitely more useful. The equipment for its use consists of a piece of tracing-paper and a protractor, both of which can be carried in the pocket-book. You are on the top of a mountain (or at some point the exact position of which on the map you know) and there is some peak in view, near or remote, the identity of which you wish to discover. You proceed as follows. Take with the prismatic compass the bearing of some known point visible from your position, and then take the bearing of the point to be identified. Subtract the less from the greater, and that gives you the measure of the angle between them. With the protractor set out that angle on the tracing-paper; lay the tracing-paper on the map with the point of the angle exactly over your position, and make the proper line pass over the known point on the map; the point to be identified will, of course, lie on the other line, and will be assuredly recognisable if it falls within the map. If the map is cut up and mounted for folding, allowance must be made for the dislocation caused by the joints. Should you not know your position on the map, the prismatic compass will accurately fix it for you, if there are three points visible whose positions you do know. It is best to choose these points so that the middle one is nearer than the other two. Take the bearings of all three, and set out the corresponding angles about any convenient point on the tracing-paper. Lay the tracing-paper on the map, and move it about till the three lines you have drawn pass simultaneously over the three known points. The point from which the lines are drawn will then be immediately over your position on the map. There are many other uses to which you can put a map in conjunction with a prismatic compass, but a description of them belongs rather to an account of instruments than of maps.

Nor need we in this place discuss the question of mapmaking. A mountain map is made in the same way and by the same processes as any other. The positions of the various points have to be determined by the theodolite or the planetable, and the details must be either sketched in on the spot or taken from photographs. It may be assumed that, so far as the frequented mountain ranges are concerned, the theodolite has done its work. It is only in the filling in of detail that the amateur can now render any useful assistance. Mountaineers retiring from the more active forms of the sport might well turn their attention to this matter. Half-a-dozen lessons from a surveyor will enable them to take all the observations they require. They will find that the working out of their survey of, say, a single glacier basin after they come home will prolong into the winter months the keenest memory of the pleasures of a brief Alpine holiday. A week devoted to such a detailed survey of the Lys glacier, or of a portion of the Valtournanche range, would be well repaid, and the resulting map would form a valuable contribution to any Alpine Club's publications.

A climber gives others no ground for reproach if he neglects to use a map for any of the three main purposes thus far referred to; but there is one occasion on which such neglect shall not be forgiven him, neither now nor hereafter as long as Alpine literature is read. This occasion is when he sets himself down to record in print some route that he has followed, whether that route be new or old. Let him bear in mind that his description will probably be read, not on the mountain-side, but in a lowland or city chamber. His readers will have, in the first instance at any rate, to follow him not on the peak itself, but on its miniature semblance projected flat on the map. His written account must be comprehensible by aid of the map and that only, and should be written with the fact clearly in view. If the best existing map is inaccurate, the writer must point out and, if possible, correct the inaccuracy ; and as he should certainly climb (in the case of all new expeditions) with a map at hand, he cannot fail to discover on the spot any important error in the survey. When his account is written, let him imagine himself to be one of his future readers, and let him con his description with the semblance of ignorance in his heart and the map before his eyes. He will discover at once whether his account is clear, and, if not, in what points it needs amplification or correction. At least half of the published accounts of mountain ascents are topographically obscure.

And here, as a much-tried editor and reader of other men's accounts of Alpine expeditions, let me for once disburden my soul. Beware how you use the seductive words 'right' and 'left.' Do not speak of 'bearing to the right,' when you have not made plain the direction of previous motion. In describing a route up a glacier distinguish carefully between your right and the glacier's proper right ; [1] and in general speak rather of the east or west side of a glacier than of its right or left. The danger of confusion is still greater in the case of a gully ; the phrase 'rocks to the right of a gully' is necessarily ambiguous. Never content yourself with saying merely 'we followed a well-marked track' ; a track may be perfectly well-marked when found ; but to find it one wants clear directions about its

[1] Cf. p. 142, footnote.

position; however well-marked in nature, it may be totally omitted on the map. In case of a track leading up a branch of a glacier it is easy to fail in describing the particular branch, especially if you speak of it as the second or third 'branch to the left,' for one does not know exactly how to count. It is best to describe it as the 'tributary glacier descending from such a pass' or 'from between such and such peaks.' Do not say we climbed 'to the arête,' and leave it doubtful which ridge; all mountains have several. Be profuse and accurate in your references to the points of the compass; and take care you do not hold the map wrong way up, as has been done with remarkable effect by the writers of published descriptions that might be specified. If a peak has a south-east and a south-west ridge and no proper south ridge, do not say that you mounted by the south ridge. Do not describe a narrow glacier or snow-slope as a 'couloir,' nor a narrow valley (whether filled with snow or not) as a 'corridor'; no one knows what a mountain corridor is like. The hardest thing to describe clearly is a route up a great rock face. Such faces are almost invariably broken up by a few conspicuous ribs and gullies, and these can be counted on the spot from one point of the compass to another. Make it plain at what point you commenced the ascent of the face (between what two ribs, for instance), how the route was thenceforward related to ribs or gullies, and at what point any final ridge was struck; and be careful to distinguish between the true ridges of the mountain and the mere crests of its ribs. It is often convenient to describe your position at different times in relation to measured points (other than summits) or measured contour-lines marked on the map. Above all things make your notes on the spot. A piece of paper and a pencil are hardly less necessary as part of a proper equipment than a rope and an axe.

The numerous published accounts of early ascents of Mont Blanc may be quoted as instances of the vagueness that may remain in apparently elaborate descriptions of an ascent. Hardly one of them can be interpreted except on the spot or by a person who has actually made the ascent. Examples of perfectly lucid descriptions are Mr. Malkin's Diaries, recently published in the 'Alpine Journal.' Equally admirable are all Mr. A. W. Moore's accounts of mountain expeditions. The fortunate possessors of 'The Alps in 1864' own the classic model of whatever is clear, terse, and sufficient in Alpine topographical description.

Guide-books for mountaineers are built up out of abstracts of published accounts of ascents, corrected and, if need be, amplified by an editor from his own personal experience, from enquiries made on the spot, and from other information collected in various ways. Broadly speaking, the published accounts are the important factor; when these are bad, no good guide-book can be constructed out of them. It is, therefore, all-important for writers of such accounts to make them at once accurate and complete. The only times worth recording are those actually occupied in passing from point to point. No one wants to know how long a party dawdled in putting on the rope or eating their various meals. The times needed are those from the hotel to the sleeping-place, or foot of the mountain, or glacier, or edge of the snow, thence (generally speaking) to the beginning of the ascent of the actual peak, and thence to the top. The corresponding intervals in the descent should also be recorded. It is well to take the time at a bergschrund and when reaching a ridge, as well as at other fixed landmarks easily identified. As a rule, it is better in any published description to record the hour at which you started, the intervals of time occupied in the chief stages of ascent and descent, and finally the hour of arrival at your destination. Any pedant, if he needs must, can then discover the length of time spent in halts. And here it may not be out of place to observe that mountaineers in repeating ascents, and especially well-known ascents, would do wisely to note down and report to the editor of some Alpine periodical the more prominent changes which from time to time take place in the condition of mountains. Such changes are often greater than an inexperienced person would be prepared to expect. In the season of 1891, for instance, the Lyskamm was almost without cornices ; and at the same time a huge wave of snow, which a few years before had barred access to an obscure col, was replaced by a gently rounded saddle. It is well to know where such alterations may be looked for. Substitutions of ice for snow, over large areas, should also be recorded, and similar conspicuous changes. If mountaineers will pay a reasonable amount of attention to these matters, they will swiftly reap the reward of their pains in the increased number and efficiency of guide-books prepared specially with a view to their needs and convenience.

It may not be amiss in this place to introduce an account of an imaginary ascent of the Beispielspitz (14,000 ft.) as an example of the kind of description of a first ascent which is likely to be useful to after-comers. The description applies only to the portion of the mountain visible in the illustration p. 135, above.

. . . They mounted along the crest of the E. lateral moraine of the —— glacier till the lowest ice-fall was passed, and then they crossed the glacier horizontally above the ice-fall, and made their camp on the W. bank at a height of 7,500 ft.

Starting next morning at 2 A.M. by full moonlight, they mounted along old avalanche snow-beds, keeping the W. lateral moraine of the glacier close on their left hand. Thus in $\frac{3}{4}$ h. they emerged on the great snowfield above the second ice-fall. They crossed the snowfield in an E.S.E. direction, towards the foot of the great rocky buttress which descends southwards from a point high up on the E. ridge of the mountain, and divides the eastern from the central basin of the snowfield. Instead of taking to the rocks at the lowest point, they mounted over the snow of the central basin for a short distance, with the foot of the rocks close on their left hand. In 1 h. from the top of the second ice-fall they turned l. and took to the rocks, ascending them in $\frac{1}{4}$ h., by a convenient gully in their W. face, to the crest of the great rock rib. They followed this crest as closely as possible for 1 h., and then turned l., and for $\frac{1}{4}$ h. traversed the rocks into the ice gully E. of the rib. They crossed the gully and ascended the rocks on its E. bank as far as possible. They were then obliged to recross the gully, and take to the rocks on the other side. A short distance higher up the gully divided, one branch leading back towards the crest of the rib, the other ascending to the main E. ridge of the mountain. They, therefore, again crossed the gully below the point of division, and took to the rocks forming the E. bank of the E. branch. The crest of the great E. ridge was thus gained in 3 h. from the time when the gully was first entered. Two towers of rock in the ridge were turned in 20 m. by way of the rocks of the N. face; the snow-ridge was then followed for 50 m. to the summit, which was attained at 9 A.M. They spent 1 h. on the top.

In the descent the same route was followed. The snow-ridge was quitted in 25 m., and the top of the gully was reached in 20 m. more. The descent to the point where the gully was finally quitted took 20 m. They traversed in $\frac{1}{4}$ h. to the crest of the great rock rib, and, descending it and the gully below, reached the snow-field in 1$\frac{1}{2}$ hr. In $\frac{3}{4}$ h. they arrived at the top of the second ice-fall, and reached camp, 20 m. later, at 4.20 P.M.

Total times from camp and back: Ascent, 7$\frac{1}{2}$ h.; descent, 3 h. 55 m.

Alpine guide-books have developed *pari passu* with the development of Alpine travel. The history of both developments has been written by the Rev. W. A. B. Coolidge in his 'Swiss Travel and Swiss Guide-books,' to which the reader is referred for all information on the matter. We are here only concerned with the present condition and supply of guide-books for mountaineers, and the promise of improvement in them. There is no kind of book in the construction of which an author counts for so little and his readers for so much as a guide-book; and this is particularly true of a Climbers' Guide. A work of literature is in general the expression of some literary conception, and when the expression is complete the work is done. It cannot be tinkered at and added to from year to year. It does not grow. But a guide-book is a growing and changing thing. It is never complete. An ideal mountaineers' guide-book should be revised and re-issued every

T

year. There can be no end to the writing of it as long as Alpine exploration continues. The various Alpine Clubs publish their annual volumes and their monthly reviews, and all of them contain new matter that should be immediately digested and incorporated in a perfect Alpine Guide. Some fifty to a hundred new routes (many of them important) are invented or claimed every year. New huts are constantly being built; old ones keep falling down. Paths are made or destroyed. Even railways now have to enter into the calculations of the writer of a Climbers' Guide; the effect of the new Zermatt Railway, for instance, has been to make the Bietschhorn a Zermatt mountain, which before it assuredly was not.

Almost anybody, therefore, can write a guide-book for mountaineers provided that he is properly supported by the men who use his book. It is only the first edition that implies wide and patient research. If the foundations are well laid—a mere matter of patience for a man able to read the easy Alpine Club dialects of a few languages—the superstructure will be quickly and well, or slowly and ill, raised, according to the intelligence of climbers and their willingness to help. It is the men who abuse an editor for the blunders in his book (usually, but not always, other men's blunders) who are his best friends. An editor cannot, and doubtless does not want to, make personal acquaintance with every route included in his book; many routes must anyhow be recorded rather as warnings than for any other purpose. The editor must, therefore, largely depend for corrections and additions of detail upon information spontaneously contributed by those who have used his book and noted errors or omissions in it. For the most part, only accounts of first ascents find their way into Alpine periodicals. But first ascents are seldom normal. The times are generally slow (if that mattered); and, what is worse, the proportions of time spent over different parts of the route are seldom those afterwards maintained. Many moments are lost in discussing details of the way, or wasted in following a false scent. Again, the best route throughout is seldom taken in detail in a first ascent, and considerable improvements are commonly made by later parties. Alpine periodicals do not bring these facts under an editor's cognisance ; he is dependent for knowledge of them upon information directly supplied to him. For these reasons the development of guide-books for mountaineers is in the hands of climbers rather than in those of such persons as may be willing to write in their service. If the number of guide-book writers were much larger than it is ever likely to be, their work could not become of first-rate quality unless they were to receive adequate and continuous support from mountaineers.

The best system of structure and arrangement for a Climbers' Guide is not yet determined. It is easier to say what should be excluded from such a book than what it should contain. Clearly an author cannot expect climbers to carry in their packs, over lofty passes and difficult peaks, an account of his individual impressions and prejudices. A guide-book should comprise nothing but facts, or the nearest possible approximation to facts. As long as climbers congregate in swarms at a few great centres, they only require information about the ascents to be made from those places. But the Alps are wide, and consist of many groups. To pass from one group to another, the way lies often enough through valleys notable for beauty and sometimes memorable for historic events. The intelligent traveller needs information about many matters besides the mere route to be followed to the top of peak or pass, and it is the duty of a guide-book writer to supply such needs. Twenty or thirty years ago this was not difficult of accomplishment, and in Mr. John Ball's remarkable 'Alpine Guide' it was accomplished with great ability. Mr. Ball was essentially an Alpine traveller. He crossed the main chain of the Alps in all directions, and aimed rather at making himself acquainted with the mountain region as a whole than at accomplishing the ascent of a number of lofty or previously unclimbed peaks. His book, therefore, which must always mark an epoch in Alpine literature and exploration, was constructed by a man who had before his mind's eye the whole area of the Alps, and did not concentrate his attention upon small groups and the details of expeditions. The breadth of the writer's view is perceptible in every line of his work, and the structure of the book was determined by it. It describes the country as displayed to an active pedestrian traveller, moving on continually from place to place, and capable of overcoming any mountain obstacles that might arise in his route. It nowhere contains the kind of information required by a man settling down for a month's mountaineering about a single centre.

On the other hand, such guide-books as have in recent years been published, professedly for the use of climbers and climbers only, describe nothing but actual ascents to the summits of lofty peaks or passes, from some hotel or other habitable starting-point. The assumption, therefore, upon which they are constructed is that mountaineers are no longer travellers, but almost exclusively 'centrists.' The type of person for whom Ball's Guide was written still exists, and probably exists in greater number than will be readily believed by climbers who are only familiar with the great centres ; but such persons are no longer representative of the most active and advanced group of those who make mountaineering their serious sport.

The ideal guide-book must of course take chief account of the tendencies of the most active and forward of the younger men ; for the future is with them, and the sport of climbing will be developed in whatever direction they please to develop it. It may therefore for the present suffice, from the climber's point of view, to catalogue all the expeditions that have been accomplished, and to point out those that remain to be effected. But when that has been done for all the Alps, and done with the greatest attainable accuracy and completeness, the result will not amount to an ideal guide-book. That should certainly also contain the kind of information about valley-routes and grass-passes, about guides and hotels, and off-day walks around centres, about local points of interest, scientific and archæological, of which an intelligent person, even though he be a mountaineer, would desire to be informed. The kind of 'Alpine Guide' which should ultimately be produced is an epitome of the results of the exploration of the Alps in its widest sense. It is hardly necessary to observe that a long time must elapse, and many men must co-operate together, before the production of such a book can come within the range of possibility.

It may perhaps be worth while to take a concrete instance, and to consider how an ideal Guide to a single district should be constructed. Let us select the area of the Pennine Alps, contained between the Rhone and Aosta valleys and the routes over the Simplon and the Great St. Bernard passes. The work would fall into the following large divisions :—

a. An account of all the main valleys and the valley-routes, with any matter of historical interest concerning them.

b. An account of the villages and centres, of their accommodation for travellers, and of those shorter excursions to be made from them which are likely to be interesting to active walkers. Brief historical notes about the chief centres should likewise be inserted in this part, with a sufficiency of bibliographical references. Geological, botanical, and other scientific notes would likewise here find their place.

c. A list of all the club-huts and other places suitable for sleeping-out when mountain-climbing, and a minute account of the ways to them from the various hotels and railway-stations.

d. An account of all known routes to the peaks and passes of the range, with full references to the literature concerning them.

It will be seen that the existing volumes of the 'Climbers' Guide to the Pennine Alps' are only commensurate with the last, and, in part, of the last but one, of these four sections. The first and second will be in the main covered by the revised edition of Ball's 'Alpine Guide.' The new edition of Murray's Guide traverses much the same area.

There still remains one important question to be considered in connection with the fourth of the above indicated divisions.

In what, for the purposes of a Climbers' Guide, does a difference of route consist? The question is one that has only arisen in recent years. As long as there were plenty of unclimbed peaks the attention of mountaineers was directed mainly to the attainment of all summits and the traversing of all obvious passes. It was not till the more important peaks had been climbed and passes crossed that the detailed exploration of the higher regions of the Alps can be said to have commenced. A mountain that had been climbed from the east was then attacked from the west. Ultimately numerous different ways were discovered by which the summit of almost every mountain can be attained. The pleasure of novelty and the flatteries of fame betrayed some of those whose misfortune it was to come late into the field, to attempt an undue multiplication of routes. Ascents were published as novel which in reality were only the most insignificant modifications of routes previously accomplished; or secondary points, humps on the ridges of well-known mountains, were made the goal of special expeditions, and baptized with all the circumstance of independent peaks. A limit must, of course, be set to this kind of overproduction, and the good sense of the large body of climbers may be trusted to enforce it.

It may be broadly stated that, for guide-book purposes, there can seldom be more different routes to a mountain than the total number of main ridges and faces the mountain possesses. A face that can be climbed at all can generally be climbed by many lines of ascent. These are not different routes, but mere varieties of a single route. A ridge may often be reached from both sides, but if so, in all probability the two methods of approach together form a pass rather than, in the true sense, different routes to the peak. Of course no law can be made that shall apply universally, but lucidity in these matters is best attained by adhering as closely as possible to some simple convention. It is easier for the writer of a guide to do this than for the editor of an Alpine periodical, mindful of his sensitive *clientèle*.

MAPS AND GUIDE-BOOKS

In discussing the question of guide-books for mountain climbers we have confined our attention to the Alps, because thus far, if we except perhaps the Norwegian mountains, about which the present writer is wholly ignorant, there exist practically no guide-books to any other ranges of snow mountains in the world. The time, however, has surely come when some attempt should be made to write a guide-book for Caucasian travellers. The lack of such a work is the greatest existing hindrance to the progress of travel in that splendid region. Caucasian literature is already so bulky that a novice may well hesitate to attack it. A journey that requires some months of preliminary study will not be lightly undertaken as a holiday amusement by busy men. An intelligently constructed guide-book would do much to smooth away the difficulties, resulting largely from mere ignorance, that prevent many an active mountaineer from visiting a range, of whose beauties and interest he has in these latter years been willingly compelled to hear so much.

Ein Rückeimer.

CHAPTER IX

MOUNTAINEERING BEYOND THE ALPS

BY DOUGLAS W. FRESHFIELD,

Hon. Sec. Royal Geographical Society.

Carduus Rhodanogletscherensis

HÉOPHILE GAU-TIER, in a letter written from Zermatt, moralises in the following romantic manner on the return of a young Englishman from an ascent of the Matterhorn :—

Quoi que la raison y puisse objecter, cette lutte de l'homme avec la montagne est poétique et noble. La foule qui a l'instinct des grandes choses environne ces audacieux de respect et à la descente toujours leur fait une ovation. Ils sont la volonté protestant contre l'obstacle aveugle, et ils plantent sur l'inaccessible le drapeau de l'intelligence humaine.

In this passage we have the poetical and—since poetry is the highest truth—the true view of 'Mountaineering.' At least it expresses much that was understood in that word by the men who thirty years ago added it to the English language. The Alpine explorer of that date found his reward in something beyond gymnastics; his pursuit called forth moral as well as physical energies; he broke the charm of the unknown, he subdued the hitherto unconquerable. He used the title 'Mountaineer' not in the common sense, but in that in which it was employed by the Westmoreland peasants who refused it to Wordsworth because the poet, as a rule, kept to roads or beaten paths.

Every mountain, it may be argued, is *new* to the man who has not climbed it before. One might go further, and urge that no mountain is the same on two different years, or even days. Dull indeed must be the traveller who does not feel a thrill, as peak or crest sinks under his eyes, and he looks out beyond on some tract seen for the first time, some region of 'strange forests and new snows.' Few sensations can be more stimulating or delightful than the first glimpse—over the hard line of the white ridge, that has for hours confined the view—of the saffron-tinted snows that line a horizon a hundred miles off, or of some vast vaporous space, whether it be the Plain of Lombardy

Islanded with cities fair,

or the Steppes of Scythia, broken only by the flashing lines of the rivers that flow towards the Caspian. Yet the full flavour of mountaineering in its highest perfection is undoubtedly reserved for the first-comer of all, the adventurer to whom success is uncertain up to the moment when the circle of the horizon is complete about him. And this zest—the zest of the early attempts on the Matterhorn, so eloquently described by Dr. Tyndall and Mr. E. Whymper, is no longer to be enjoyed in the Alps. A few persevering wanderers may still come home annually from Tyrol or Dauphiné, murmuring 'a poor thing but my own' of the secondary, or more probably sixth-rate, summit they have discovered and scaled. But these sorry remainders are nearly all gone. Recent climbers have had to fall back on the desperate resort of going the wrong way everywhere; and—since the wrong ways lead after all to the old goals—the device begins to be admitted even by those who practise it, except perhaps by a few of the more simply muscular, to be but a humorous pretence, a trap to catch an afterglow from the sunny days of fresh adventure.

In another way the Alps have in the last quarter of a century lost not a little of their primitive charm.. Thirty years ago Alpine mountaineering was still rough work; even in the Bernese Oberland it involved lying in hay-barns and caves, or under the stars, eating 'strange flesh' and munching last winter's bread; it took men away from all recollections and echoes of city life, from their clergyman and their tradesmen, from newspapers, telegrams, and the daily post. In the Central Alps we have changed all this; or rather the Swiss have changed it for us. Travellers have become industrial objects; the statistics of the 'Fremden-Industrie' are scanned annually by a nation that lives on it. Even the artistic Parisians see nothing vulgar in the 'vulgarisation of summits.' French savants are engaged in struggling against nature and common sense in the endeavour to erect a Climbers' Shelter on the actual summit of Mont Blanc. The old mountain resort, 'The Crown' or 'The Eagle,' has been converted by the fairy wand of the Messrs. Cook into a barrack blazing with electric light, ringing with telephones, and furnished with a cosmopolitan crowd, a British chaplain, London newspapers, and local Society Journals. Crowning injury of all: a railway up the Vispthal enables the Hairdressers' Society of Vevey to spend six hours at Zermatt, and Transatlantic tourists to see 'the glacier region in sample'—I quote textually—in a single afternoon!

Even when we fly further into the mountains civilisation nowadays remains with us. The cave, sacred to many memories of years—and perhaps friends—departed, has been superseded by a Club-hut, with berths like those of a sleeping-car and as many rules and regulations, where we lie packed between

Germans who limit the ventilation and fellow-countrymen who in their holiday humour are likely enough to put forward the boorish side of our national middle-class character.

I should be the last to dispute that even in the Alps it is still possible, for those who know a little topography, study 'The Alpine Guide,' and are not slaves of elevation above sea-level, to escape into more primitive regions. If it yearly grows more difficult for travellers who are forced to take their holidays in August and September to avoid what Shelley called 'the polluting multitude'—this difficulty, like others, is to be overcome. I trust nothing I may have written here will be taken by any would-be mountaineer as an excuse for not knowing the Alps; or as anything more than an incitement to know something beyond them. If I have spent three summers in the Caucasus, I have spent some thirty in the Alps, and I am convinced that a fairly comprehensive knowledge of that noble chain is one of the first requisites in an explorer of more remote ranges.

There are not a few who have this knowledge, and it can be no wonder that among such climbers and in the last few years there has been a growing tendency to seek new haunts and fresh experiences, to open the map of the world in place of the map of Central Europe. Let us do the same here, and reckon over what mountain regions may best tempt the mountain lover who has duly served his apprenticeship in the Alps.

In a work of this kind there must, perhaps, be some limit fixed. And I understand that, with an exception in favour of the British Isles, our editor for his present purpose confines 'Mountaineering' to glacier regions. Fifty years ago, or less, this limit might have been an indistinct one. In an article on 'Glaciers,' published in the year 1848, I find this surprising statement : 'Although the towering summits of the Andes and the Himalaya mountains penetrate far above the snow-line, yet no true glaciers are to be found.' Humboldt was mainly responsible for this popular delusion. Thanks to Hooker, Richard

Strachey, and Godwin Austen, in the Himalaya, to Whymper and others in the Andes, we know better now; at least the world does, for there are still some Rip Van Winkles left. An eminent 'scientist' in the employment of the Indian Government assured me only the other day, as the result of his personal observation, though there might be snow and ice in the Himalaya, it was all stationary, and there were no glaciers.

Bound by this restriction to snowy ranges, I must pass by Corsica and the Carpathians, the Apennines and the Atlas, Syrian, Greek and Bosnian heights, though elsewhere I might have much to say on the pleasure to be found among them by the true mountain lover.

The Sierra Nevada of Spain cherishes one small glacier, but its level summit-ridges, though they reach 11,280 feet, and out-top the highest of the Pyrenees, are singularly wanting in the variety most climbers desire.

Despite Mr. Packe's and Count Henry Russell's eloquence, the Pyrenees have hitherto singularly failed, among Englishmen, to attain the vogue to which, as a paradise for lovers of fine scenery and flowers, and a gymnasium for rock-climbers, they are entitled. Their glaciers are comparatively small, their peaks, with some exceptions, such as the Pic du Midi and the Canigou, are blunted. In their upper region they resemble rather the western end of the Oberland—the Diablerets district, than the Pennine or Dauphiné Alps. Their valleys are deep, and there is a lack of high halting places and mountain centres, such as attract the gregarious tourist. But the very depth of their valleys tells in favour of the grandeur of the scenery; the beechwoods of Luchon have a grace hardly known in the Alps, the range is studded with picturesque tarns, the valleys on the Spanish side of Mont Perdu are extraordinarily fine in form and colouring. From the climber's point of view, if few of the peaks are difficult, there are difficulties enough to be met with among the cirques and limestone cliffs that form their buttresses. The peasants of the Pyrenees are habitually bolder rock climbers than those of the Alps, and they have invented special foot-gear for traversing the limestone ledges.

A young climber might do worse than spend a summer—or several summers—among these mountains. He would not, it is true, learn icecraft, but he might gain experience that would be of use to him farther afield. Each great range has its distinct character, and he would learn to know something beyond the Alps.

Next to the Alps and the Pyrenees, the nearest snowfields to our shores are those of Norway. Many rock climbers have lately betaken themselves to its heights, and have been well rewarded. Norwegian mountains differ in character from the Alps. The familiar comparison to a table represents the typical Norwegian group. The snow and ice take the place of the cloth, the peaks stand up round the edge, lofty relatively to the valleys, low relatively to the extensive *névés*. The glacier passes are long and tedious, the climbs short and often arduous. Less mountain craft, on the whole, is needed than in the Alps, and Norway can hardly take the place of Switzerland as a training-ground for ranges loftier and more arduous than those of either. It will be visited for its own sake, and will doubtless always have, as it deserves, its own devotees. But when the last word has been said there seems—at least this is the impression of a reader—to remain a sense, that from the mountaineer's point of view the peaks want individuality, that the summits of the Jökulls are, after all, Lesser Alps, while what we most of us look tor are Greater Alps.

Iceland has its volcanoes—but volcanoes are an inferior type of mountain, and very much alike all the world over. And the long distances, monotonous landscapes and execrable climate of the northern island are serious deterrents to those who have revelled in the infinite variety of Alpine beauty. It is a country for sturdy and dogged explorers, to whom hardship is its own reward, and beauty a secondary consideration.

When we turn to Asia, or rather to the confines of Europe

and Asia, our eyes necessarily rest first on the link between the mountains of two continents, the Caucasus. There, in the latitude of the Pyrenees, towers a chain 700 miles long, loftier than the Alps and equally iceclad. Up to 1868, it is true, British geographers of repute had made bold to state that 'its peaks are mostly flat (*sic*) or cup-shaped, and the existence of glaciers is uncertain !' In fact, the peaks bear the same relation to Alpine summits that those do to the blunter ridges of the Pyrenees; while—if the exceptional Aletsch glacier is put aside—the Caucasian glaciers surpass those with which we are all familiar, both in the extent of their snowfields and the magnificence of their ice-falls. The first exploration in 1868 of the chain by mountaineers was followed only at considerable intervals by single journeys. In the last ten years, however, it has become a recognised field for climbers, and in the last five summers there have been some fifteen different exploring parties at work climbing and photographing above the snow-level. To the well-known attractions of the Alps the Caucasus adds those of strange people, noble forests and a rich and varied vegetation. But it has also, it must be allowed, drawbacks beyond its eight days' distance from London. There is risk of fever on the Black Sea coast and in the lower parts of the Mingrelian valleys, and some climbers and many guides have found the mountain diet produce serious indisposition. Again, there are other difficulties to be encountered besides those peculiarly attractive to the mountaineer; and the climber pure and simple, who holds every hour wasted that is not spent above the snow-level, is apt to grow impatient of delays caused by impracticable horsemen, broken bridges, and pathless forests. It is, I must add, an entire delusion to suppose that the Caucasus is not full of rock climbing, and of problems for rock climbers. Skill and steadiness and judgment in ice-falls and on snow and ice slopes—a branch of mountain craft which has fallen somewhat into the background among Alpine climbers and guides of late years—have been peculiarly called for in this new country. But when the great peaks have all fallen, rock climbers will find plenty of rock pinnacles, of Drus and Charmozes, to exercise their nerve and muscle upon.

The vast ranges of Central Asia, the Altai, Transaltai, Karakorum, Hindu Kush, the unmapped and hardly named chains of Tibet and eastern China, promise an inexhaustible field for mountain exploration which should last until the human race is extinct, or has grown too effete to enjoy an out-door exercise. The Himalaya and the great chains that lie immediately between it and the Altai have attracted as yet but one party of *mountaineers*, in the European sense of the word. They have been of course visited, explored, and partially mapped by a multitude of Anglo-Indian travellers, sportsmen and surveyors, some of whom may be disposed to claim the title. I do not forget the travels of the Schlagintweits, of Hooker, Godwin Austen, Strachey, Tanner, Younghusband and others. But in a work on Mountaineering we can hardly class as mountaineers mountain travellers who were not skilled in the use of rope or ice-axe, and who, as far as I know, did not scale a single snow-peak a second-rate Swiss guide would consider difficult. The Anglo-Asiatic traveller or surveyor is apt to think a rope is only useful on a flat glacier, and seldom ventures beyond the point to which his native companions will lead, or at least follow him. Even within these limitations, however, very much valuable information has been collected, and it has been sufficiently proved by Mr. Johnson in Kashmir (see 'Alpine Journal,' vol. xii. p. 58), and more recently by other officers of the Indian Survey Department, that men can not only breathe, but work both with their limbs and brains, at heights varying from 18,000 to 22,000 feet.

Russian and Chinese ranges may be left to the coming century and future editions. For the moment let us turn to Africa. There by a curious paradox permanent snows are found only on the Equator. The great volcano of Kilimanjaro is now easy, if costly, of access. The glaciers of Kibo, its chief summit, have not yet been fully described. And on its secondary peak, Kimawenzi, there seems to be plenty of rock-scrambling. The Masai tribes who live round the more northern volcano, Kenia, which, to judge from sketches, is a fine rock-pyramid, a volcano in the last stage of decay, have, I fear, not yet been brought to a proper knowledge of mountain worship ; while the yet more remote range discovered by Mr. Stanley in his last journey, Ruwenzori, seems still beyond the beat of those who are not personally conducted by the interesting Austrian naturalist, for whom a trip to the coast was so philanthropically arranged by the directors of the British East Africa Company. I may add that whoever goes out to these mountains should be a naturalist as well as a climber, unless he is content to create widespread disappointment on his return.

Let us now leap the Atlantic and turn our gaze on North America. Here on British territory, in the Selkirks and Cascade Range of the Pacific coast, a very noble field has been opened up for explorers in the last few years by the completion of the Canadian Pacific Railway. It is true that the peaks, so far as they have yet been measured, appear to be only of moderate height (12,000 to 13,000 feet) and of no immoderate difficulty.[1] But the extent of ground covered by glaciers is immense. There is wild life enough in the forests at their base to satisfy the sturdiest youth, with a happy absence of those terrors to the average Briton, foreign officials and an unknown tongue.

Farther north on the frontier of Alaska, Mount St. Elias, towering in an unbroken sheet of ice, the loftiest on our planet, 18,000 feet above the stormy waters of the Northern Pacific, challenges assault, and offers the exceptional attraction of a fair chance of drowning before the assault can be delivered. Farther north still, into what Charles Lamb called the 'icy privacies of the Pole,' we need not extend our survey. Dr. Nansen's feat in traversing the southern corner of Greenland is hardly likely to excite much emulation. A 'shining tableland' hardly represents heaven to the mountaineer. And

[1] The heights assigned to Mount Brown and Mount Hooker (15,000 to 16,000 feet), peaks standing north of the railway, are probably exaggerated.

surely the prime attraction of mountain snows lies in their incongruity. The richer the vegetation at their feet, the greater the fascination of the glaciers. There can be few greater delights in travel than the swift change from the snows of Monte Rosa to the vineyards of Val Anzasca, from the glaciers of Ushba to the azalea thickets of Suanetia. In a country that is all snow and ice, mountaineering must cease to be a recreation, and become a monotonous task. Not but what Arctic explorers might in several instances have done better than they did had they had a little Alpine experience in the art of travel and transport over rough ice.

Turning south, beyond Mount Shasta and its neighbours, there is little from our point of view to be noticed in the mountains of the United States. Its citizens, with praiseworthy patriotism, make the most of the few infant glaciers that may be found on the Rocky Mountains. They have even discovered and photographed an embryo bergschrund. But they cannot pretend to offer a field for rope and ice-axes.

Mexico has several snowy volcanoes which serve to amuse climbers who are in their neighbourhood for other reasons.

In South America great things have been done, and much more remains to do. Mr. E. Whymper has fought with all the difficulties of travel and mountaineering among the Great Andes of Equador and prevailed. The descriptions of the country, the climate, and the mountains he gives, will, however, hardly prove tempting to the general climber. A bleak uplifted tableland dotted with broad-based volcanoes, on whose glaciers the sun seldom shines, peaks which are rendered dangerous rather than difficult by the atrocious conditions of weather, ranges without spurs or valleys, forests, lakes, or rivers : these may offer a field for the resolute explorer or the scientific worker, but they can present few inducements to the holiday-maker and the pleasure-seeker.

Farther south, among the loftier, and probably not less accessible peaks of Peru and Chili, Dr. Güssfeldt has broken ground. This region deserves a fuller exploration. But the

German traveller was hampered by the failure of his climbing companion, and no one has yet followed him with better success on the slopes of Illimani.

The explorer who is sufficiently enthusiastic for glaciers to encounter ceaseless bad weather, may anchor a yacht at the Pacific end of the Straits of Magellan and climb the peaks, shrouded in perpetual mists, that overhang them.

The Antipodes—to say nothing of the South Pole—are still left. Of what there may be in the heart of New Guinea we are still ignorant. That magnificent myth, Mount Hercules, had unfortunately nothing better to rest on than the wit of a speculator and the credulous haste of publishers and critics. But in both islands of New Zealand there are snowy peaks. The summits of the North Island are volcanic : the highest, Ruapehu, reaches 9,700 feet. To the climber the ranges of the South Island, the so-called Southern Alps, are likely to be the more attractive. There, indeed, a glorious field is open for the mountaineer; a range that, with a snow-level of 5,000 to 6,000 feet, reaches 12,000 feet and has the glaciers of the Alps, the forests of the Caucasus, and the fiords and waterfalls of Norway, all brought into the closest juxtaposition. The Southern Alps may be too remote for the vacation tourist ; but they are at the doors of a young and rising nation of Englishmen who, as their cities grow and prosper, and town-work increases, will turn —are, indeed, already turning—to mountain exploration as the noblest form of sport.

So far I have written mainly from the point of view of the Long Vacation mountaineer, who makes travel subsidiary to climbing. But I desire also to take this occasion to emphasise the advantage that an acquaintance with mountain craft gives the traveller who looks on an ascent as only a secondary object. If our African travellers had known how to cut a step, or to judge the approaches to a great peak, it seems improbable that Kilimanjaro would have been left to a German, or Ruwenzori be still unascended. If those who provide funds for our Indian Survey Department had taken the advice offered them some years ago by practical mountaineers, it would be by this time able to produce mountain maps that could be compared favourably with those made by the Russian Staff in the Caucasus; and it would have learnt better than to neglect the advantages of an intelligent use of photography in survey work above the snow-level. Captain Younghusband's recent exploring expedition in the Karakorum and the Pamirs was undertaken without a camera in its equipment, and the photographs with which he illustrated his narrative were lent him by his Russian rival Captain Grombchevsky ! There are, indeed, not a few regions where the traveller will be better equipped for his work if he possess some small skill in picking his way among snow mountains, some power of describing accurately their features. The number of otherwise educated men incapable of recognising in nature or a photograph the distinction between glaciers and snow ravines, or describing a mountain landscape with any approach to distinctness, would be astonishing did we not remember that the faculty is of very modern growth, and that only two generations ago such men as Humboldt were without it.

Again, as Sir Joseph Hooker has pointed out in his 'Himalayan Journal,' the word 'perpendicular' is one of the most abused in the language. The descriptions in a very fascinating book of Himalayan travel, the late Andrew Wilson's 'Abode of Snow,' were generally given quite a wrong significance. It was a convenient commonplace with the critics to write that the author had 'conquered difficulties that would have dismayed the Alpine Club.' In fact, he was an invalid, and had only gone where he could be carried ! His energy was, perhaps, the more splendid; but his travels were not 'mountaineering.' A climber with equal powers of description —no doubt a rarity—might have had a very different story to tell.

Let us assume the field of exploration chosen. The next question to arise will be the season to select. What is true of the Alps and Pyrenees is true also of Norway and the

Caucasus. July and August are the months for the great peaks, June and September are in many respects preferable for travel and ordinary snow-passes. The seasons in the Caucasus vary extraordinarily, particularly in the snow-falls; all is snow one year, where the next bare rocks will be found at the same date; one summer the *névés* have a firm surface all day, the next the traveller flounders to the knees on every snow-pass. Let no lover of nature forget the two glories that await the early and late comers in this country, the blaze of the rhododendrons and azaleas in June, the jewel-like glow of the woods in the last days of September.

For the Central Asian ranges, speaking generally, midsummer is the best season. This rule includes all north of Kashmir. On the Himalaya within the range of the monsoon the traveller can use September, but what is the climber to do? In October, when the skies clear, the cold is already settling on the heights, and frost has driven the herds from the upper pastures. How far do the constant summer rains of the outer spurs of the hill stations reach the snowy crest? The question is one I hesitate to answer. Probably a general answer would be misleading. Sir Joseph Hooker certainly gives good reason to think that in Sikkim the sky clears with every mile of advance northwards behind the first snows. Different districts have undoubtedly very dissimilar snowfalls. A Himalayan group bearing to the plains of India the relation the Oberland peaks bear to the plains of Lombardy, that is partially cut off by intervening heights, enjoys probably broken, but not hopeless, weather during the summer.

For Central Africa Dr. Meyer and Mr. Johnston found July the best season. In the Equatorial Andes Mr. Whymper spent the first six months of the year. But the weather seems to be uniformly villainous, and he never had a clear view from the top of any of the great Andes. South of the Equator, in the South American continent, as well as in New Zealand, the mountaineering season is the reverse of the European, that is, February and March.

In North America the Alpine season should be selected—July being preferred, as the days soon shorten. For Iceland, too, the summer closes early, if it ever begins.

What are the chief requisites for the composition of the exploring party? First, as to its head. He should know the Alps well, and in some detail. He should have a clear conception of the objects of the journey, and the best means to attain them. The more autocratic he is, and is allowed to be by his comrades, the better for their chance of success in exploration. On this account, at least one successful mountain explorer has preferred to travel with only such companions as he had a right to command, and if results rather than pleasure are the first object of the journey, there is much to be urged in favour of this plan. The leader should have wandered up and down among European mountains without *local* guides, perhaps with one or two friends and a guide from another district, so that he may be accustomed to study the ground, to cultivate a sense of locality, to give a prompt decision when called on in critical moments.

Next to the qualities of the leader the most important consideration is the constitution of the party; for, whether the party be made up of guides or fellow-climbers, mountaineering is of necessity a social form of enjoyment. Happy are those who can set out with old and tried friends! Wherever it is possible, before undertaking any long expedition, climbers should test in a preliminary trip their powers, tempers and sympathy. Identity of pursuits or hobbies is by no means to be desired. Diversity, given reasonable mutual forbearance, will produce far better results. One man is handy in the multifarious work of camp-life, another at the cooking-pot; one is master of the camera and the plane-table or theodolite; another has a microscopic eye for reading the landscape, remembering the details of a glacier, or identifying a summit. One has a taste for rocks, another for plants or insects. Some men are incapable of a topographical observation, others of doing up a parcel. Neither defect is necessarily fatal to a traveller as an ingredient in a successful combination. A strong party is a unit with the strength of the combined qualities of its members, but without their individual shortcomings.

As to number, three to five including the guides (if guides are taken) is best. Six may be held the maximum. Four is a number that has the advantage of lending itself to temporary severance. But it is one too many for comfort in a seven-foot Whymper tent on a long journey. For a night now and then four may do well enough.

There are now a certain number of Englishmen trained in the Alps capable of dispensing, so far as climbing goes, with the professional aid of guides. By guides I mean here not men of special local knowledge, but Alpine peasants with a general experience of mountains and hereditary powers of climbing, developed by constant use. There is, however, the very serious consideration to be taken into account that in many countries the inhabitants, whether on account of superstitious scruples, or from fear of the unknown, or from idleness, cannot be induced to act as porters in snow and ice expeditions on ground unknown by them. They are apt to frustrate an ascent by their obstinacy in sticking to familiar tracks. Thus in the first ascent of the Leila peaks in Suanetia we had the pleasure of seeing our provisions going off up the wrong glacier, because there was a pass known to our native companions at its head!

The traveller who makes up his mind to 'mountaineer without guides' beyond the Alps must recollect therefore that it will often mean to mountaineer without porters; and this, to men not weight-carriers from their youth up, must be a very serious consideration. Mr. Mannering's recent book on his explorations in the New Zealand Alps shows how, even in the case of hardy young colonists, climb after climb was rendered ineffectual by the handicapping of the climbers. Every man takes into the mountains a certain store of physical force; the more he exhausts on the lower glacier, the less he has left for the struggle on the final slopes, the oftener he will fail to reach the peak, and the less power of observation he will have when he does succeed in reaching it.

One of the surest marks of a young and inexperienced traveller is an appetite for needless hardships. Alpine climbers are peculiarly liable to this distemper, since in their brief *exeats* from civilisation they have naturally and harmlessly sought to make the contrast as complete as possible by shirking no hardship or exposure. What may be permissible or even meritorious for a few days becomes pernicious when the journey is one of weeks or months. I could point a moral from my own Caucasian experiences, but I prefer to quote an Asiatic explorer, Captain Younghusband, on this point :—

Let me give my opinion, that more roughing than is absolutely necessary should always be avoided on principle, for it not only makes one less physically fit when the time for real action arrives, but also, if continued for month after month on long explorations, degrades the mind, gradually obscures the brightness of the intellect, and makes one forget that one belongs to a civilised portion of the human race.

Should the traveller desire to follow up the question of 'the rarity of the air,' to provide fresh material for the discussion of the greatest height attainable on mountains by human beings, it will obviously be of peculiar importance to him to preserve his organs from previous overstrain. I omit purposely in this chapter any remarks on this very important and interesting topic because, depending as it does so much on physiological and medical considerations, it seems to belong rather to the editor. One point, I may suggest in passing, must not be overlooked. Mountain air does not affect the same human frame at the same height in any uniform way in different places and on different days, and this quite irrespective of training.

An exploring party without guides must consist not only of step-cutters and weight-carriers : its members must be ready to work hard in camp after the proper day's task is over. But if it is laborious to have no guides, it is generally irksome to travel with them. Picture a party of English domestics torn suddenly from the comforts of the housekeeper's room and the convenience of the tradesmen's calls, and thrown into the backwoods of Canada or the Caucasus. It may give some idea of the deplorable gloom and helplessness of the average Bernese peasant taken from the guides' room and the well-provisioned club-hut, and set down among strange people in an unfamiliar land. He has to suffer many real hardships; he makes the most of them by his shiftlessness, and he spends his spare time in adding to them imaginary terrors. Anything novel serves his turn as a bogey; natives who wear knives and pistols must needs be murderers; the fresh mutton is reckoned unwholesome to a digestion accustomed to salt meat and rancid cheese; the wine, slightly flavoured with goatskin, is poisoned ! Home-sickness soon attacks him severely. Often he makes himself really ill by sleeping in wet clothes and eating unripe fruit. Sometimes, as Louis Carrel on Chimborazo or one of my guides on Tetnuld, he gets badly frost-bitten by perverse want of attention to boots and gaiters. In the end his unfortunate employer, after dragging him for weeks through the mountains in a semi-crippled condition, is likely—just as he is beginning to feel some real anxiety for his companion—to have the great, but hardly unmixed satisfaction of seeing him not only perfectly recovered, but very superfluously elated, at the first town he comes down to.

Having drawn this picture, let me make haste to say that it is far from universally true of Alpine guides. I have had the services of a François Dévouassoud. The two Maurers, the Carrels, and other Alpine peasants have known how to accommodate themselves to strange circumstances, and done excellent service to their companions.

My general advice to explorers in want of guides would be :— Do not apply to a great centre such as Grindelwald or Zermatt ; select an unspoilt man, one who likes going, who loves adventure more than pay, one who has been taken up and down the Alps a good deal, frequently therefore called upon to lead over ground new to him. Individuality may overcome race and locality, but, unless you know your man, look for him rather in the Italian than the Northern Alps, and if you take a Teuton, take a Tyrolese rather than a Swiss. Guides of Latin race, whose fellow-villagers are accustomed to wander far as workmen or emigrants, do better, as a rule, than the more stay-at-home and stolid Teutons. There is a peculiar narrowness in the Swiss character, due probably to the absence of the civilising and mentally enlarging influence exercised in larger realms by the conscription with its consequent transference of young men over wide distances. This defect is brought into strong relief in travel. I dwell on it with more confidence since I have the support of a distinguished German traveller, Dr. Güssfeldt, in my view.

In many mountain regions an interpreter is a necessity to Englishmen. Few of us are fluent in Russian, or Spanish, or Hindustani; and in the Caucasus Russian is not enough. There the traveller frequently loses opportunities, or makes mistakes, by not being able to communicate with Tartars or Georgians, except in a language they dislike and imperfectly understand. If you can find a man who speaks fairly the needful tongues, the humbler he is as a rule the better. He will be more entirely at your orders, and more assiduous in his special duties. If possible, the offices of interpreter and cook should be united in one person. Be patient; conversations are not to be carried on, or orders given, in places where time is no object, with the brevity that prevails within the sphere of Bradshaw. It is human to lose one's temper from time to time, it may often be judicious to appear to do so; but it is expedient to prove that as a rule one is not unreasonable. Do not show a perpetual suspicion in small matters of expense, and always be ready and prompt to give praise as well as blame, wherever it is due.

Much of the success of exploration depends on a power of dealing with the people of the country, of enlisting their goodwill and services as hosts, purveyors, horsemen, or porters.

Now the people to be dealt with are often extremely difficult to manage. They have nothing, perhaps, to fear, and but little they care to gain from the traveller. Money is not a necessity to them, nor is time a commodity. They have no clubs or newspapers, and are as fond of 'sitting on the wall' as Eton boys. Talk is their delight, and it is as a subject of endless conversation that they most value their visitors. The experienced traveller will endeavour to discover the most influential spokesman, and through him he will negotiate. He must never, with Orientals, make the blunder of assuming that the price first asked is the value they really set on their goods or services, and break off in a huff at its unreasonableness. On the contrary, he must smile blandly and play the game according to the rules, by himself at first offering far less than he means to give. When hiring porters he should have, as far as possible, his luggage ready divided into packages of a reasonable weight. Whenever it is convenient he should arrange to pay by the journey, not by the day. On the march he will find frequent opportunities of securing a certain respect and liking, now by himself taking the lead in soft snow, now by lending in bad weather a waterproof covering, best of all by presenting the morey to buy a sheep, or some suchlike luxury, when the men have had a long day's tramp. Presents, such as knives, razors, whistles, opera-glasses, snow-spectacles, &c., will always be appreciated. They should be given as rewards, not as bribes. Barbarians or semi-civilised people are children—without children's courage in face of the unknown—and may be best treated on the same principles. But, as with children, their respect is easily won, and once won may be long kept.

It is no easy thing to create glacier guides, or glacier porters, even out of hardy mountaineers. A peasant may be a terribly fast rock-climber and think nothing of snow-passes, yet he is brought to a standstill where danger or even difficulty hardly suggests itself to the Alpine climber. A crevassed *névé*, any slope which requires big nails or step-cutting, are too much for him—on hard snow his hay-stuffed sandals handicap him. Time alone will bring about a change in his point of view; time and proper footgear. A snow-wall of the character of the Strahleck will stop men who have no nailed boots. I once brought some Caucasians down such a wall, but they arrived at its base head foremost, and were not at all encouraged to try a second. The most practical step towards manufacturing native guides out of the raw material to be found in a new field of mountaineering would be to take out socks, boots, and gaiters, such as they could wear. There would be prejudice at first—just as there is prejudice among English mountaineers against crampons—but it would be overcome by experience.

The party constituted, its leader should set himself to prepare in detail his plan of campaign. For this purpose he must study the best maps and books procurable. This can most easily and thoroughly be done in the library and map-room of the Royal Geographical Society, at 1 Savile Row, W. The traveller bent on exploration must remember that he will have for the first time to do without a guide-book. And since he cannot expect to carry many volumes with him, he will do well to make up his own skeleton guide by a number of extracts and memoranda drawn from all the best material available.

The journey should, as a rule, be planned so as to be adaptable to circumstances, or even reversible. No man should make himself a slave to a programme, which better knowledge may advantageously modify. But, on the other hand, the explorer who is without a definite scheme is liable to be carried away by every change of weather or mood, by every blast of local, and often misleading, advice. One of the first difficulties in travel is to sift and discriminate the hearsay information pressed on you on the spot. It is essential to discover your adviser's standpoint before you can decide how much he can help you, or how far he is likely to mislead you. The first party who went to the Caucasus were assured that their plans were impracticable by counsellors whose ideas of travel were limited by post-roads, and of mountaineering by the powers of native hunters. The best adviser of all is a map—where one exists ; and even a very defective map may in judicious hands be most valuable in settling a general plan. Its very defects and omissions are often suggestive.

In using a large-scale map the intelligent traveller soon learns to make a certain allowance for the idiosyncrasy of the office, or even the individual responsible for it. One survey—that of the British Isles, for instance—will make property its primary object, and be less careful on heights not appropriated to human uses. You cannot tell at a glance on the Scotch or Welsh maps, as you can on the Swiss, where precipices are likely to occur. Another will be very accurate as to forests and bridges, but careless as to details not likely to affect military movements; such was the now superseded 5-verst map of the Caucasus, where the snow region was represented by smears, and ranges misplaced several miles. In some surveys, as in a few of the Dauphiné sheets, a surveyor will become recognisable by work below the level of his comrades. The notorious misplacement of the sources of the Exe and the streams of the Doone Valley in our own Ordnance Survey is credibly reported to have arisen from their distance from a public-house, and the personal habits of one of the surveyors who, at the beginning of the century, had charge of the district.

The journey that is most full of interest, and fruitful to persons of ordinary intelligence who love mountains not only as a gymnasium, is one arranged so as to allow of excursions from a successive series of centres, which for the sake of scenic completeness should lie, whenever practicable, on *both* sides of the chain. No one, for example, can form any idea of the Central Caucasus who has not visited the southern slopes as well as the northern. While the heavy camp moves from point to point, the high expeditions can be made in light marching order. There are some climbers who prefer to go to one centre only, since thus they get more scrambling in a given time. These are scramblers, not mountaineers, and when, on their return, they talk with confidence of the general characteristics of the chain they have visited, when they dogmatise on their narrow experience, they are apt to mislead others as well as themselves. One of the keenest of the first generation of mountaineers, my friend the late A. W. Moore, used to urge that no one knew a mountain group properly until he had not only crossed the pass at the head of each valley, but also followed each valley down to the plain.

Arrived at some central point suitable for a camp, the first business of the prudent explorer is to insist on a general reconnoissance from any convenient minor summit. I think it needful to insist particularly on this point, although it has been referred to elsewhere in the present volume. For more than half the failures of pioneers have arisen from a neglect of this preliminary precaution. The worst place to judge of a peak or pass is on its own slope; the parts necessarily obscure the whole. Climbers are tempted to take what is the wrong line, because it is the easiest for the next hour or so. It is only positive knowledge of what lies beyond that will counteract this temptation, to which Alpine guides are quite as prone as amateurs. If you cannot reconnoitre, preliminary study of a predecessor's photographs may in some cases supply the needful information. Take all the photographs you can get with you, and have them always handy for reference. Your map may be correct as to the position of outlying peaks, and if you have photographs of the snows from two of such ascertained positions you can work out the topography of the district. Even distant photographs are useful. For instance, a photograph of the Kinchinjanga group from Darjiling will in competent hands serve to locate and piece together photographs taken at the base of the mountains.

Do not assume that the conditions of snow and ice are the same all the world over as in the Alps. In moister and warmer climates (e.g. the Caucasus and New Zealand) the snow has a trick of peeling off underlying ice that is terrible. I have seen whole slopes cracking and hissing as the avalanches raced down, like drops on a window-pane, to the great trench at their base. I have found a strong torrent flowing down an ice-gully at the head of a *névé* 3,000 feet above the summer snow-level.

Look to the particular character of the rock in the portion of the range you attack. One side of a valley may be solid granite, the other rotten schist. Otherwise—*experto crede*—while crawling for hours under volleys, any single missile from which might be deadly, you may have occasion to reflect on the vital importance of geological studies to mountaineers.

Above all things, in dealing with peaks and passes greater than those of the Alps, start early ; I would almost say—if there is a moon—start late, before midnight. For you can hardly be too far advanced at dawn. Everything will be in better condition ; your chances of a view greatly increased, of a night on a ledge lessened. It is disagreeable, no doubt, to walk in the dark ; but it is far more disagreeable, remember, at the end than at the beginning of a long day's work. This point needs to be emphasised, since the erection of huts has done away with the need of very early starts in the Alps, and guides have in consequence learnt to wait for daylight to rouse themselves. It was not so twenty years ago. It follows, naturally, that the meal before starting should involve as little cooking as possible, that the packs for the day should be got ready over-night. A self-cooking soup-tin between every two travellers meets every need, and causes no delay. On an ascent it may often be prudent to leave bits of stick as landmarks ; if they are cleft at the top and a piece of cardboard pointing the direction fixed in the cleft, their utility will be increased. Some plan of signalling both by day and night should be agreed on between the pioneers and the party left below in camp.

I am relieved by the preceding chapter on Equipment from dealing with that matter in any detail. Study it, together with Mr. Whymper's lists, in the 'Hints to Travellers' of the Royal Geographical Society, and the list of implements and instruments taken by Mr. Conway in 1892 to the Karakorum, which may be seen at 1 Savile Row. According to my experience nothing protects from rain so well as a mackintosh, no wraps are so serviceable as a light Scotch plaid, a comforter, and a knitted helmet. Self-cooking soup-tins, kola chocolate, a good store of tea, are all of great advantage.

The note-book should be always at hand and frequently used. Notes on the spot, even if lost afterwards, are of use. For the making of them forces the eyes to observe and the mind to record. The sketch-book will now generally be replaced by the pocket-camera. Yet those who can sketch or make even rough notes of local colour should not neglect to do so. But, first of all, let the mountaineer beyond the Alps keep his eyes open and his intelligence on the alert to notice all the novelties, natural and human, that surround him in the valleys as well as on the mountains. Do not let him—as some do—turn himself into a one-eyed Dervish, and because many generations have passed by and not seen the great mountains, make it his peculiar pride to see and care for nothing else !

The following extracts from hints I have printed elsewhere may suggest some headings for the note-book and indicate the classes of observations, one or more of which may be serviceably undertaken without destroying the pleasure of a holiday journey or making a toil of a pleasure, even by those who do not pretend to a 'scientific aim' and would perhaps regard such an aim as a tyranny.

Nomenclature.—The nomenclature of peaks is always a difficulty. The same difficulties arise in all countries. In the first place, the mountain people often speak a different dialect or a different language from the surveyors or travellers. Again, unless in the case of very prominent or isolated peaks, they frequently do not give any distinctive name to the individual summits. Or if a peak is conspicuous from two valleys it has two names, one of which has to be preferred. Sometimes a whole chain is named from the pasturage at its base, or from the valley that lies beneath it. In such cases, when the hunters of the locality, generally the only authorities, can give or agree on no name, a peak may fairly be called after the glen or pasture nearest it, and from which it is best seen. Again, when a peak has two names, the one given to it on the side from which it is most conspicuous should prevail. A name given to a block or *massif* may always fairly be applied to its loftiest point, as was done in the Alps in the cases of Monte Rosa and Piz Bernina. Where no local name is forthcoming, a name may suitably be found in some characteristic of the peak. Personal names should only be introduced on rare occasions, and then after careful consideration. One main point to keep in view is that the Government survey is, or ought to be, the final authority, and that, when its decisions are altered, much literary and practical inconvenience is caused. The orthography of place-names, particularly in the Caucasus and Asia, is a frequent source of difficulty. If every valley has a different name for the peaks, every hamlet has a different pronunciation. When a place-name has found its way into literature it is vexatious and frivolous to vary it because your particular informant's pronunciation is not represented. The system instituted by the Geographical Society, and now generally adopted, should be followed, and any new names brought into agreement with its rules.

Geology.—Apart from their primary objects, mountaineers may find opportunities far exceeding those of the ordinary tourist for observing and collecting facts and specimens illustrating the nature and products of their field of work. E.g. they may collect (carefully labelling in each case the height and position from which they are taken) specimens of rocks, particularly of the highest rocks among the snowfields. Bits of the prevailing rock (not crystals and spar, which look pretty) are most useful. They should, as a rule, be broken off *in situ*, and not taken from moraines. But a collection of stones from the terminal moraine of a glacier, specified as such, will be valuable as showing roughly the geological structure of the ranges forming the glacier's basin.

Mineral springs, hot springs, and their temperature, where

they issue from the ground, may be profitably noted, together with any evidences of volcanic action, recent or ancient.

The snow-line and timber-line (that is, the upper limit of forests) should be noted, with their variations according to rainfall, exposure, prevailing winds, &c. The heights of the lower ends of glaciers should be observed, and marks (with a date) placed on the rocks opposite, showing the exact position of the end of the ice, so that its future movements of advance or retreat can be ascertained. Erratic boulders should be reported on, and unmistakeable ancient moraines; rock surfaces scraped and polished by ice (*roches moutonnées*) are also worth notice, though less positively recognisable.

Botany and Natural History.—Rare specimens of the flora, particularly from the *highest* rocks, should be taken. Bulbs or seeds of any unusual species should be obtained whenever possible. Any uncommon objects in natural history (beasts, birds, or insects) may, of course, profitably be collected.

Photography.—Photography is invaluable not only for topographical delineation, but also for obtaining types of the population, their dwellings, domestic furniture, antiquities, monuments, &c.

Anthropology.—Measurements of natives, made according to the principles set out in the Royal Geographical Society's 'Hints to Travellers,' p. 222, would in many cases be valuable.

Local Traditions.—An interpreter of intelligence may aid in collecting evidence of old customs, religious or legal; forms of worship, and modes of tenure of property. Also by copying down local ballads and traditions.

In the Caucasus at this moment there is an opportunity in this respect, which will be lost as civilisation and schools spread. Old burial-grounds, especially those where ancient objects have been found, should be noted.

Instruments.—Information as to instruments used for

survey purposes and instruction in their use may be obtained from the Map Curator of the Royal Geographical Society at 1 Savile Row, on any day (except Saturdays), between 10.30 A.M. and 5 P.M. The Society also provides instruction to intending travellers in botany, geology, zoology, and photography.

'Sic itur ad astra'

CHAPTER X

CLIMBING WITHOUT GUIDES

BY CHARLES PILKINGTON

A Rotifer

BEFORE dealing with a subject which, in its day, has been a burning one, it is well to consider what those best qualified to give an opinion have said on the general question. It is not unfair to sum up the decision that would probably be endorsed by the majority, as follows :—
'Guideless climbing is to be strongly deprecated as a general practice. Under certain conditions it is reasonable enough.'

What those conditions are, and what reservations must be made, it is the purpose of this chapter to point out.

That no definite laws have been laid down is not surprising, when the great variety of expeditions, and the varying circumstances under which they are made, are taken into consideration ; for instance, the risks to be avoided, but prepared for, by men who contemplate going over the Petersgrat or Adler Joch, or climbing the Morteratsch and Breithorn are very different to those which may be encountered by the men who would attack the Dru or the Weisshorn and cross the Jungfraujoch or Moming Pass; again, those who stop long at one place, and make few expeditions, study their mountain well, learn all about it, and wait for thoroughly good weather, set themselves a much easier task than those who, in a more ambitious manner, go about from place to place, ascending peaks and traversing passes about which they know little, as often as their legs, or the weather, permit.

Also, what a contrast the same mountain may present on different days, nay, even on the same day! That steep and narrow ridge of rocks, its handholds and ledges small, but firm and not too far apart, so interesting and pleasant to climb when the rocks are dry and warm, can be very different when the wind is howling over it, seeking to tear you from its glazed and slippery crest; your half-frozen fingers scratching the falling snow out of the hidden cracks and crevices; your feet numb and cold, scraping and kicking to find the narrow steps; your eyes half-blinded by the biting sleet, and your body covered with frozen clothing. Step-cutting up the ice slope above is simply a little healthy exercise on a bright warm day, when halts can be made at any time to 'admire the view,' or change the leader if the work is prolonged. But it is quite another matter when the clouds are between you and the sun, and a biting wind is drifting the snow dust and ice chippings around. The leader may keep himself from freezing, but the glazed and slippery weapon in his hand is very different to the well-balanced ice-axe he so gaily wielded on the Gorner icefall a few days before. Meanwhile, the others, if wise, will also cut and improve the steps, lessening the leader's work and increasing his speed; but they must reserve their force, for their turn in front will come, and they have to see that the heavy, ice-covered rope is kept neither too tight nor too loose, and all have to work together at the full stretch of their mental and bodily activities in spite of the arctic confusion raging round

them. And when at last the top is reached and you are all cowering under a rock for a few moments' peace from the strife, someone must untie the string of the rücksack to find and unpack the frozen provisions that you must eat to live.

Even then it is not over. There is the weary trudge over the snow basin, with the lengthened rope and tired limbs; there are hidden crevasses to be detected, the best route through the ice-fall to be hit off, the place to leave the glacier, and the faint goatherd's track, if there is one, to be found, that will take you down to the alp below. Having traversed the peak, your energies and all your mountaineering skill are still wanted to bring you out of the lower difficulties before the night closes in.

It may be urged that the advent of bad weather might have been foretold, and a retreat made earlier in the day; but the best weather prophets are often wrong, and the most prudent party may be caught by a storm in places where it is more easy to advance than retreat. So remember, that all these things may come to pass even on mountains of ordinary difficulty, and before arranging to make a series of ascents without guides, or undertaking one of the more difficult expeditions, where the party may be exposed to all the dangers of mountaineering, think everything well over and see how far you are justified in undertaking the risk. And remember too, it is not only your own risk, but that of the three or four men composing the party, who are naturally dependent on one another.

It is well that each member of your proposed party should be a good and pleasant travelling companion : it is absolutely necessary that he should be a steady and safe climber, have a thorough knowledge of all the details of mountaineering, and enough practical experience to enable him to make use of his knowledge. For each man must be able to take his own share of the work, whether it be carrying the provisions or making the road. The success of some particularly difficult climb may depend on that member of your party who is specially experienced or proficient in some special branch of mountaineering ; a particular peak may be won simply because one man has an unusually long reach, for instance; but the success of a long, varied, and arduous climb, or of a series of expeditions, depends on the strength and skill of the entire party ; each man coming to the fore on different occasions, and all working harmoniously together like a well-trained football team.

Do not undertake any serious climb till you know each other's capabilities well : the mere knowledge of them is not enough : you ought to have had practical experience of them also. It is most annoying to be misunderstood, and find that the man above gives a pull when he is only asked to keep the rope tight, makes a spring when you are expecting him to steady you with the rope, or lets it coil about your feet in a difficult place when he ought to have gathered up the slack as you moved forward. These small inconveniences, caused by want of consideration for the others, the result of ignorance or inexperience, if multiplied indefinitely throughout the day, mean much loss of time and energy. On difficult places they mean not only a much greater loss of time, but possibly danger.

Watch a party of experienced men descending amongst broken rocks. They let themselves down the easy parts as slowly as over the more difficult places, and never quicken their pace until all are on easy ground ; each man climbs for the whole party, not for himself alone.

If the members of your party have travelled much together, the value of each man's opinion is known, and discussions as to route are quickly brought to a close. They ought very rarely to occur in actual work, when it is better to keep your breath for the work before you than to waste it in talking, which leads to thirst rather than to a definite conclusion.

Some of the party ought to have a good share of that topographical insight which gives a broad rough knowledge of the lie of the land, the directions of the ridges and valleys, and the general contour of mountain forms; should understand steering by map and compass, and have practised it often on less dangerous hills both amongst clouds and in fine weather, and ought to possess to some extent that sort of instinct, call it what you will, which enables you to retrace your steps, on a difficult and confused mountain-side, even if the clouds come down; and to find the knapsack or *cache* of provisions which may have been left behind during the ascent. The best guides have this instinct highly developed, but amateurs can cultivate it also, for it is greatly the result of observation. There can be little doubt that it can be readily acquired amongst our own hills, where the climber is forced to study his surroundings for himself and with reference to his route. It will soon become a habit rather than an effort. The study can be advantageously continued when climbing with guides, but if we never climb without them we may not think of cultivating it at all.

Very careful mental notes should be taken even on the easy lower ground, and any prominent object should be connected in the mind with the route. In taking these notes, look back as well as forward, for you thus fix the view on your brain as you will see it on your descent. A note-book and pencil may be of assistance, but it is often better to rely on your memory, which is a book that can be kept open in the worst weather, leaving both hands free for climbing. Constant and diligent practice of this mental photography will make a marvellous difference in your power of observation, and afterwards, if you go with guides, you will cease to wonder at their skill, though you may still admire it. The occasion will doubtless arise when your judgment will be better than theirs, for even they are not infallible.

To cultivate this faculty of observation and the power to apply the knowledge acquired, it is absolutely necessary that every member of a guideless party should have passed a thorough preliminary training on the hills at home, and be accustomed to act on his own responsibility. It is one thing to know how a difficult cliff or ice-fall should be attacked; it may be quite another to lead the attack under those adverse circumstances which are so often met with in mountaineering. It is better for the same reason that each man should take a share in leading, so that he may be able to do so when the necessity arises.

There ought to be a good iceman in the party, not merely a step-cutter, of whom more than one will be required, but a man who knows where to expect hidden crevasses, who has some idea from above, as well as from below, which is the best route through a complicated ice-fall; who knows when a snow slope is safe, and when ice is to be expected beneath the surface of the snow; who can tell when the steps should be sunk, at any cost, through the loose covering into the ice beneath; and who will also have had sufficient experience of his own knowledge to prevent unnecessary labour in this direction.

One qualification more, and the most important of all. Every man of your party must have had a long experience under good guides. They are the best teachers, and walking behind one of them for a month in the Alps is a better lesson than anything that even they could impart by speech or writing. The teacher, however, must be a man of wide experience, and good alike on ice, rocks, and easy ground. Especially should he be a good iceman, for it is in this department that amateurs require most training. The best models are of the old Oberland and Chamonix schools.

Whatever may be the subject, it is well to be taught early by competent masters. A good style is easily acquired if there is nothing to unlearn. Therefore, early in your mountaineering career place yourself under a good guide, watch him carefully and practise what he teaches, both at home and abroad.

It is no argument to point to some particularly good amateur and say that he has not had this training. Examine his case carefully, and it will be found that, either he is brilliant on rocks, but a dangerous man to follow on snow; or, if he is good on both—a very rare occurrence—he will tell you that it is the result of many years' hard work, and that, had he had the advantage of lessons under good guides in his early years, he would have much sooner attained his present skilfulness.

Travelling constantly with good guides induces many to rely solely on the skill of the professionals, taking no notice of the surroundings except from an artistic, botanical, geological, or geographical point of view; some indeed only notice the summit or other distinctive feature, with regard to the hour and minute at which it is reached. This bad habit should be especially avoided if you contemplate going without guides, or exploring unknown mountain regions even with guides.

Note how careful your tutor is not to bustle himself or you in any way, how slowly ponderous his movements are in the early morning, how solemnly he places his feet in the flattest parts of the rough Alpine path. He seems to take no thought about anything, except to avoid knocking the nails out of his boots. But if you watch carefully you will see that he is

'Will it hold?'

observing everything that is taking place around, and that if the country is new to him, he carefully scans the backward path, so that he may recognise its distinctive features on his return. Do not get impatient if he is a little slow in kicking unnecessarily large steps up the moraine, or on the few yards of ice leading up to rocks above the Bergschrund; it is a most excellent fault, rarely found in bad icemen. Imitate the practice, and it may save you from a nasty fall some day, for too often vigilance is only thought necessary on the more dangerous places. An inexperienced amateur is very likely to fall into this error, and he must remember that when without guides he must allow a greater margin for safety than when with them. You will find that your guide can go fast enough should the necessity arise, and that in crossing a dangerous couloir his steps will be cut several sizes too small for the comfort of the party.

It is not a guide's business to teach you how to climb on your own account, though the men are generally ready enough to tell all they know; he may also be clumsy in imparting information. The student must therefore strive to get all he possibly can out of his teacher. Mountaineering is a difficult subject, with many branches always producing fresh questions to be solved and new difficulties to be overcome.

There is one other point to be considered in making up your party. Is it better to take a fourth man or not? Four is usually a better number than three for companionship, is easier over deep snow, safer over large snowfields, and would be useful in case of a sprain or other accident to one of the party. On the other hand, three move much more quickly over difficult places, over all rocks, and are safer also from the dislodgment of stones by the leader; in fact, are more easy to handle; and it must be remembered that on some occasions, speed means safety, as, for instance, the crossing of a doubtful gully, racing a storm, or getting out on to easy ground before evening sets in. On no account should two men undertake a long glacier expedition, and even on a rock peak the rule that no party should consist of less than three members should not be broken.

It is better, at any rate at first, to avoid the great centres, such as Zermatt, Grindelwald, or Chamonix. It is true that around them there are better paths through the lower alps, and that on the big peaks the huts are high up and comfortable, that it is easy to get provisions and porters up to your bivouac, and, in short, the conditions are more luxurious in every way. But, on the other hand, there is then a greater disinclination to turn back from an expedition when the weather becomes bad, or when one of your companions becomes temporarily indisposed, or, what is still more annoying, when you have made a mistake in the route, over the lower and easier part of the mountain, and have not time to retrace your steps and begin afresh. You know that most of the guides and some of the visitors will put down your return to incompetence, though nothing would have been said on the subject had you been professionally led. The moral courage to turn back in case of doubt is one of the most important qualities in guideless climbing, and it is rarer than it ought to be. The stronger your party is the easier it becomes, for the knowledge of your own powers naturally makes you think less of the admiration or sneers of incompetent critics.

A little episode, within the writer's knowledge, is to the point. A guideless party had returned early to their hotel, at one of the great centres. The weather had been doubtful, and they had given up their expedition. The guides and loafers sitting about the door turned up their noses and scoffed at their want of pluck or knowledge. One of them, a first-class man, came up afterwards and said, 'It is bad for us that you go without guides. I know that you can go on the mountains as we go, and that you are like chamois, but still most gentlemen would be safer with guides. But it is not so with you; for I see you know when to turn back. I think you were quite right: with many guides you would have gone on, and it would not have been wise to have persevered!'

On a snow mountain there may be many steps cut ready for you, and the way up may be marked, as in a map, by a dotted line. Worse still, there may be others on the mountain, and unless you start early, not only will you have to bear the remark that, 'It is easy to climb without guides if you follow in other people's steps,' but your expedition will lose its main charm, the pleasure of finding and making your own way. If another party is sleeping out for the same expedition, you will probably all start together; then, if the other is a fast one, follow quietly and enjoy the joke as well as they, for you are not entering into a practical competitive examination with the guides. If the others are not fast, take example from what the best guides do in such cases; let those who know the ground best lead up the easier slopes of the mountain, and when the peak is well in view, take the lead if you care to do so and are being delayed by the others. It is best, however, to be modest. Racing is an extremely dangerous amusement, which destroys nearly all the pleasure of an expedition, and leaves you hot, tired, and perhaps discontented with yourself or your companions.

The weather in the rather lower districts is generally better than at the great centres; a bad storm also does not usually render the peaks dangerous for a week afterwards, as in the higher regions, and expeditions can consequently be more often made. There are no guides to tell you whether a mountain is in a fit state to be climbed or not, and few climbers will be met whose movements might give you some idea of the state of things above. You must judge of the snow entirely for yourself, and you will have the courage to put your opinions more boldly to the test, because the risks are not so great, and there are fewer critics in case of failure.

Out-of-the-way districts are distinctly the best for beginners; there they must find the way through the meadows and forests, and on the upper land below the snow, and learn by practical experience that part of mountaineering which is not climbing. It must be remembered, however, that amongst these lower groups of mountains there are many peaks to be found more difficult than the Matterhorn, and as dangerous as the Schreckhorn or Dent Blanche.

Although all the different branches of mountaineering have been dealt with in other chapters, it may be advisable, at the risk of repetition, to draw the attention of those for whom this one is written to some of those precautions which they especially should observe, or might possibly overlook.

If you are in a district new to you, it will be wise to reach the hotel, chalet, or bivouac, from which you are to start for your ascent, fairly early the previous afternoon. Even if you know all the details of a previous ascent, one of the party at least should explore the way through the lower ground, either by walking some distance up it, or by examining it from an opposite slope, even if you are not able to reconnoitre the mountain properly. In deciding on the lower part of the route, be sure to keep in mind whether you propose to start before it is light or not, as in the former case simple ground is to be preferred, even if a trifle longer. If there is a stream over your path, find a good place for crossing: avoid bushy ground and long grass, as it is better to keep dry as long as possible, and if it is likely to be damp put on your gaiters before starting. Further up it is safer to keep near the stream or glacier which drains the snowfields, as you are less likely to be cut off; if, however, you think you will be forced away on to a buttress or face, try and mark a point whence you can traverse back again, without much loss of distance or height; but remember, that from below a point that appears separated by a deep dip is often connected by a level ridge with the mountain mass. It is well to have your programme decided on before you enter the district, and to take every opportunity of studying your peak carefully, with reference to its ascent. A route can often be traced from a summit five or ten miles distant.

On approaching a glacier avoid its snout if there are many stones upon it, especially giving it a wide berth if there are smooth and ice-worn rocks below, for the stones may fall at any moment, and will travel a long way; as for instance below the Rosenlaui glacier. There seems to be no infallible rule for threading the labyrinth of a broken ice-stream, but often the more broken parts are the easiest, for the smooth ice or snow cut with long lines of wide crevasses may be impossible. The rocks on either side may give you a passage when all else fails, and a rocky mound dividing a glacier is often an easy way; only remember that if the glacier is stony and the ice above steep, it may occasionally be raked.

On the upper snowfields keep a watch on either side, for crevasses concealed in the centre are often exposed there, showing enough of their direction to give you warning, and in the afternoon especially it is well to prod deeply in doubtful places.

In rock climbing avoid places where ice and stones may fall from above, keeping to the ridges rather than the hollows, as being both safer and easier.

It is sometimes well to leave small cairns as guides for your return. The last man should build them. One or two stones are enough, and they should be erected in safe places where they will not be knocked over by anything you may send down, and where they will be visible from above as you descend upon them; fix their positions as carefully as possible in your mind, by often looking back at them during the ascent.

A face of rocks when seen in front is very deceptive, and an apparent precipice can sometimes be scaled with ease.

A clean white line through the rocks is a delight to the climber, but see that it is not likely to be swept by falling ice or stones before venturing into it; dirty gullies are generally icy, and therefore laborious in step-cutting, as well as dangerous, and the rocks will be the quicker and safer way. If, however, a gully is safe, take it by all means; it is a straight road, but look well to the condition of the snow, and its steepness compared to its width, for a narrow couloir is safe at an angle that would be dangerous on a broad one, or on an open snow slope. If there is any doubt avoid it, or keep to the side with one hand on the rocks, or at any rate let one of the party be within reach of them, and prod well where ice is suspected. A steep open snow slope ought always to be entered on with extreme caution, as there is less to keep the snow in position than in a gully; do not be content to test the upper layers only; note carefully the condition of the older snow below, which forms the connecting link between the surface and the solid ice beneath. In ascending a steep slope of soft snow, drive the axe handle deeply in as a support, and do this before the step is taken, and not while you are stepping.

If you are forced to ascend an ice slope, think chiefly of giving your blows with good aim and swing, rather than with a short strong stroke; but see that your balance is good and your feet well home in the steps, or else have one foot in the step and one knee on the ice, and be sure that the floor of the step slants slightly in towards the slope. You must learn to cut with either hand, and, like a woodman, will soon find that if right-handed it is easier to cut from off the left shoulder, for the hand nearest the axehead must constantly slip backwards and forwards along the handle, directing rather than forcing the blow. Watch a blacksmith after his blow is delivered; his right hand, firmly grasping the tail of his hammer, is dragged backwards till the head is almost in his left hand; he then easily throws it over his left shoulder and brings it down again, slipping his left hand on to his right as the blow descends, thus delivering it with the whole swing of the handle.

On most steep snow ridges there is an overhanging cornice of snow on one side or the other. Every climber ought to know that cornices exist, and that they may cease to exist at any moment if they are used as a path; yet many lives have been lost by walking on them, and very many men have been placed in great and unnecessary danger without having the slightest suspicion of it. The upper surface of the cornice, being usually flatter than the slope of the mountain, offers a tempting path, and men are lured unconsciously towards it, especially if the slope below is hard, requiring the use of the axe. People are too apt to underrate the overhang, which is not always apparent when close upon it. Never steer so near the debatable ground of the breaking line of the cornice as to seek for it by sounding with the ice-axe. The writer once saw and felt an immense mass break away beneath his feet, the leading guide being actually standing on the part that fell, although he had driven his axe-handle deep into the snow at each step. It is probable that the break lay more or less along the line of the holes made in sounding. Some warning is given by the loud crack when the break occurs.

Care should not be relaxed when the summit is reached. A long halt is usually made there. See that all the rocks used as seats are firm, and that the snow stepped on is safe. The axes should be driven well into the snow or lodged in a secure corner of the rocks, and the sacks and provisions must be placed in safe positions, and unless the top is broad and safe, the rope should not be taken off.

In descending below the snow-line proper, large patches of snow in excellent condition for glissading are frequently met with; they save much time, and are a great help. If, as often happens, they fill the bottom of a small steep valley or gully, be careful where rocks show through, and especially near the end, where the stream may have made a large tunnel; an awkward fall, if nothing worse, may be the result of gaily sliding over it. But it is not only in depressions and valleys that this danger may occur; some of the lower slopes of snow, especially those which are rapidly wasting away in early summer, close above the highest pastures, may cover some slight rocky pitch, and although there may be little evidence on the surface, a large chasm may have been formed by the melting of the snow from the face of the rocks. It is all the more dangerous from the fact that the travellers, elated with victory, may be hurrying downwards glissading and running, thinking only of the green valley below. A watch should be kept on either side, for the break is usually more or less continuous, especially in limestone districts, and should such indications be observed, the pace should be moderated till the danger is past. If, however, the hole is seen too late, throw yourself down on the snow and shoot over it sitting, but never lose command of yourself, or you may rush into other dangers below.

Have everything ready for a start the night before ; it is so easy to forget things in the cold wretchedness of a dark morning. Make your knapsack as light as possible, but carry plenty of provisions to last you, should you be later than you expect ; a famished party is a weak one, and most accidents have come at the end of a hard day when men are tired, hungry, and careless. Never throw away a chance. Think of every danger that can possibly arise, and be prepared for or avoid it. Keep the rope on too long, rather than too short a time. As an instance of the great care taken by a good guide, the writer was once breakfasting under the shelter of a large stone with Melchior Anderegg, when the latter gave a loud and apparently unnecessary 'jodel' : asked to explain, he said he wished to frighten away the chamois which were likely to be on the face of the mountain, for they would see us when we left the rock, and we might be hit by the stones dislodged by them, as we skirted under the face of the peak.

Much has been said about the comparative skill of amateurs and guides, but a few words here may not be out of place. Everyone, I suppose, admits the natural superiority of the first-class guides even to the most experienced amateurs, and in the first class the writer includes a very large number. But what of the residue, some of whom are employed to go up the most dangerous mountains, and to traverse the most difficult passes ; who are reputed to be first-rate climbers, capital carriers, steady and safe? The traveller is told that they have been up the Weisshorn, Matterhorn, and Gabelhorn numbers of times, and judges them accordingly. The amateur attacks these very peaks, having never climbed them before, and he is also judged by results ; but the amateur is handicapped to the

Y

amount of the other's local knowledge. The second or third-rate guide must be taken to a new district and a peak he has never previously seen, and asked to lead his party to the summit. Some of them will do it, but many will fail entirely. The writer has seen them perfectly useless, frightened at the responsibility, and occasionally apt to take one into dangerous places. Physically they are well able to stand the work, and they are very useful men as second guides, and on arduous rather than difficult climbs better than the best amateur would be; but often they are not such good mountaineers. It is not their fault; they seldom have the intelligence and power of observation, nor have they had the wide and varied training under experienced teachers that some amateurs have had. As to third-rate guides, they can only be recommended as porters on such places as the Tschingel or Theodul Pass.

But, because third-rate professionals undertake difficult expeditions, it must not be thought that anyone may attack a dangerous snow mountain. The amateurs who have been compared with the guides are only those who have had a long training, several seasons at least, under the best of guides. A man may be a brilliant rock climber, strong and courageous, but unless his energies are tempered with knowledge, his brilliancy and strength will only lead him to danger or disaster. It will be a sad day if at any time in the future the warnings of those who can speak with authority are disregarded, and men who have had little experience on snow begin to climb without guides, or who, having climbed much, even with the best guides, have thought more of 'times' than of the way their leader surmounted the difficulties.

To sum up the question in a few words: every guideless party must consist of three or more members who know each other's climbing well; each must have had a long experience of lower hills in all kinds of weather; must have climbed in the high Alps for at least four seasons with good guides; must have the moral courage to turn back if advisable; must be able to take his share of the work, and lead the party in case of need.

CLIMBING WITHOUT GUIDES 323

The pleasures of an expedition carried out by your own unaided efforts are far greater than if you had been profession-

Letting him down gently

ally led, in spite of the anxieties and labour you incur. It is the act of overcoming the difficulties, the mental climbing as well as the physical, the charm of leading, or following

Y 2

when you share the responsibility of the leader, that give such a zest to the work; and when all is over how much more clearly and freshly it all stands out in your memory than if you had followed guides. The hopes and fears, the anxieties and troubles, are all forgotten, or if remembered are only thought of as a fitting frame for the rocky peaks and snowy plains of your holiday.

Caller Herren

CHAPTER XI

HILL CLIMBING IN THE BRITISH ISLES

BY CHARLES PILKINGTON

πᾶ βῶ; πᾶ στῶ;

MOST of us have probably felt our first mountaineering aspirations awaken when in sight of British hills, and though no great effort, and no climbing, as the word is understood in this chapter, is really needed to reach their summits, they afford endless opportunities for the gratification of our climbing instincts and for the cultivation of all those faculties, habits, and virtues, which collectively form the art of mountaineering.
In addition to a climbing qualification a man must have a general knowledge of mountain forms, and be fairly able to choose a route and find his way through unknown ground, if he would be called a mountaineer. To attain this distinction it is not necessary for him to leave his native land, though if he would graduate in the more advanced standards he must study the higher subjects and become acquainted with the glacier world.

In speaking of the hills of Great Britain the writer wishes to show how mountaineering can best be learnt by those who cannot travel far afield, and how those who are able to do so may improve the many opportunities that are thrown in their way at home.

The aspirant who has not the advantage of living near mountains, and whose visits to the hills are rare and of short duration, may follow the brilliant example of the cricketer who bowls endless balls at a stick on the lawn, or of the salmon fisherman who diligently throws his fly over the garden pond, tenanted by the garden toad alone. The cricketer gains little if he hits the stick, and the fisherman would be annoyed if he caught the toad ; but they have their reward, and so will the climber if he as diligently makes use of any material that lies ready to his hand. A quarry or a chalk-pit may supply his need, an ordinary cross-country walk will train his legs, strengthen his ankles, and accustom his eyes to localise objects. A ruggedly built railway bridge has before now supplied a difficult rock chimney ; a well-known climber has been seen through a telescope solemnly kicking steps up a steep clay-bank under an overhanging grassy cornice ; while a little practice with a woodman's axe or even a blacksmith's hammer may lead the way to the use of the ice-axe.

Then come the days that can be spent amongst the hills, perhaps not more than three or four at a time, but they are as enjoyable as many spent in the Alps, probably because the time is so short and the contrast to the immediate surroundings of our ordinary life so great. It is true the weather may be bad, but we are not so dependent on it here as in Switzerland, and there is an astonishing amount of fair weather in most wet days to those who are out all the time. These are the opportunities for the beginner to acquire knowledge of his craft, of which the learning is almost as pleasant as anything that may come afterwards.

At first it is better that he should have someone more experienced than himself in the party. He will more quickly learn to judge pace and distance, will be shown what can be climbed and how to climb it, the practical use of the map, compass and rope, what he ought to take notice of with reference to his route, and will pick up as many little devices and methods in a few days as he would have taken months to learn by himself. But mountaineering does not consist in knowing how all these things should be done; it is in being able to do them, and to do them safely also. It is a great mistake for a beginner to follow a good man up a place too difficult for his own unaided efforts, even if firmly held from above; it teaches him to scramble not to climb. If he really wishes to learn he must realise what his powers are, and keep within them, always remembering that there is danger in every cliff, and that the true art of climbing is to reduce that danger to its smallest possible limit. Besides, learning mountaineering is not only the training of the body, but the cultivation of the mind to take in the various shapes and directions of mountain masses, the details of moor, crag, and valley, to gain the knowledge required to overcome the difficulties of the way, and to cultivate those other qualities of perseverance, courage, determination and prudence, which are so necessary to the safe conduct of a difficult expedition.

The choice of routes even over easy ground requires much practice, and unless a man has often handled a map and compass and worked out the route for himself, he will find his theoretical knowledge of little use in case of need, when the wind and rain are driving over him and the clouds are thick around. Climbing in cloud and mist is a most difficult and interesting branch of mountaineering, and nothing but long practice will enable a man to lay out a route and keep to it when the range of his vision is limited to a few yards. Mist seems to have a confusing and terrifying effect on the mind to which it is unfamiliar. The feeling that you and your companions are on a wild and unknown mountain-side, blotted out from the rest of the world, alone, no one knowing where you are, and miles away from any human aid, has a deadening effect on the brain, most dangerous when all the faculties are required to be on the alert. A mountaineer must have studied the hills in all their varying moods, and be familiar with all their aspects. He must know them well enough not to be flurried or dismayed by any change or unforeseen danger, and be able to decide and act quickly on his own judgment. To cultivate his capabilities for responsibility in case of emergency, necessitates his climbing on easy hills ; for if a novice undertook the conduct of a difficult glacier expedition, a disaster would inevitably occur should adverse circumstances arise and quickly terminate his period of responsibility.

It is possible to get a long walk on a five-acre field, but it is much pleasanter to have it spread over twenty miles of beautiful country. In the same way, rock scrambling can, to a certain extent, be learnt on any crag, but it is much more profitable and enjoyable to practise it amongst many hills and varied scenery, when its study can be combined with that of many other branches of the mountaineering art. Our hills seem especially suited for the safe education of beginners. The expedition need never be too prolonged, no dangerous difficulty need be undertaken after the climber is tired, neither need his energies be taxed to the utmost for hours at a time, thus avoiding two of the greatest sources of danger in the Alps. If he is wise he will not at first undertake the more difficult expeditions, but should he have done so inadvertently and find the route beyond his powers, let him have the moral courage to turn back at once. Frequent practice over easy rocks will gradually fit him for more arduous climbs, and he can return again at a later day with increased experience and a stronger party.

Many an experienced climber, although he does not despise our British hills, ignores the climbing that they can give him, and looks upon them simply as places where good training walks can be had amidst beautiful scenery. But unless he has actually tried the more difficult rock work, he can have no idea of the great obstacles that may be encountered even in the summer. In the winter months, when the snow lies deep in the gullies and the rocks are icy and cold in the shade, these ascents may be dangerous and often impossible, but he will find plenty of others that will give him as much excitement

HILL CLIMBING IN THE BRITISH ISLES

as he may care to take. Although the alphabet of mountaineering can be safely learnt amongst our hills, it must be pointed out that if treated with undue levity they may become extremely dangerous. The writer would especially urge on those who have had some experience in Switzerland, that though short pieces of difficult rock work are far less dangerous than long climbs of extreme difficulty, such as are met with in the Alps, where the bodily and mental strain lasts much longer than the first period of interest and excitement, yet on the other hand, the muscles of most men, who snatch an odd day for climbing at home, are not so well trained, nor is the general condition of mind or body at all equal, from a climbing point of view, to what it is after a week's work. When out of practice, men have not got the power of stepping freely down rough rocks, cannot balance themselves on narrow ledges, and may find, in the middle of a protracted struggle up a smooth slab, or when hanging on to a small crack, that they have overrated the strength of their hands and arms. Some of the expeditions demand the utmost skill and experience, and there is no ever-watchful guide in front to say that this rock is not fast, and that ledge not to be trusted ; every doubtful hold must be carefully tested by yourself, and nothing taken for granted. Experience of Alpine mountains will carry you through, no doubt, but do not trust to that alone ; though most of the dangers met with in the Alps are wanting, there are others, not usually met with above the snow-line. For example, the snow in winter is often very loosely adherent to the rocks. Small patches of snow-covered grass give a very treacherous foothold, and even an ordinary grass slope becomes exceedingly slippery during a short snow-storm, though the snow may melt almost as quickly as it falls. Moss now frozen into a green ice often grows over the only convenient cracks, the soil and rocks, frozen only on the surface, may give way when apparently firm, and on one occasion the writer remembers retreating from a face of rocks, because they were completely covered to windward with beautiful feathers of hoarfrost, some nine inches long, such as are often seen on the Llanberis side of Snowdon. The gullies are often in a wet and greasy condition, rendering them exceedingly difficult, and any tufts of grass growing in such places are always to be distrusted. Tufts of heath and fern are very tempting, but must not be relied on in dangerous places; a man must surely be a fool who allows his safety to depend on the strength of a root of heather. It may be thought that these warnings are too simple and useless, but is there any climber who has not been tempted, nay more, is there one who has not to some extent yielded to temptation in this respect? He thinks, just one foot more and I am up. Then comes the struggle up the rock, with the finger nails convulsively digging into the grass above. Let us hope that it holds, and that his folly will be repented of at leisure.

A steep slope

One great advantage that our hills possess is, that they are always accessible and can be climbed at any period of the year. May and June are the best months for most purposes. The air is fresh, the days are long, and the weather generally fine.

This applies especially to the western coast of Scotland, where the months of July and August are often wet and misty. August and September may be better for some of the exceptionally difficult climbs, for the rocks are generally warmer then than earlier in the year, and there is less greasy moss or ooze adhering to damp places. The end of October or beginning of November shows the mountains and valleys in their richest and most varied colouring, with the exception perhaps of those districts, belonging more to the shooting man than the climber, whose chief beauty lies in the purple heather. In February and March the snow is usually in its best and most Alpine condition, but this depends much on the season. In the high central regions of Scotland, April is also often a favourable month. If snow climbing is the object, the rule seems to be, delay your visit as long as possible, but beware lest you wait till all the snow is gone.

It is not to be wondered at, nor is it any discredit to the guides of this country, that they are able to do little more than show the way along an ordinary mountain path, or take a party on a recognised expedition. This is in fact as it should be, for as our hills are not dangerous, it is better that amateurs should study mountaineering for themselves, taking examples only from the best masters of their craft. Still it is a pity that more of our native guides, many of whom are strong enough for anything, do not accustom themselves to carry packs. It would be a great convenience to be able to send a few knapsacks over an easy pass, while taking a longer and more difficult expedition on the hills ; but an English dalesman or Scotch gillie will look very seriously at a burden that a Swiss boy or girl would laughingly run away with ; also, now that so many men carry photographic apparatus into all sorts of places, a mountaineering porter would often meet with employment. Again, it is never well to ascend an out-of-the-way hill alone ; a knee can easily be twisted or an ankle sprained ; and what if this happened in the bogs of Rannoch or the solitudes of Carnedd Llewellyn? If accident left you companionless in a mountaineering centre, it would be very convenient to be able to engage a man who took an intelligent interest in the hills, and who was able to climb rocks fairly well. There are a few, quarrymen mostly, who are able to climb difficult rocks, and perhaps two or three who know something of the uses of a rope, but the beginner will be very foolish if he engages a local man to take him up any difficult place, for however good a climber he may be, he will not understand how to render much assistance to his employer. Most shepherds and many Scotch gillies are capital men on a mountainside, and can hold their own against the world over grass, heather, or broken ground : they can climb rocks too, but as their occupation does not necessitate it, they are not brilliant cragsmen. The only professional cragsmen in our islands are those who collect seagull eggs from cliffs above the sea. The writer has had no personal knowledge of their performances, but it can hardly be called climbing, as the youth or boy is generally lowered over the cliff by a rope held by his companions above.

As to the kit required, the less of it the better. A strong woollen shooting suit, well worn and properly nailed boots, thick stockings, a metal pocket flask with drinking-cup, a hat that will not blow away, woollen gloves and silk handkerchief, a map, a compass and plenty of provisions, are all that is necessary for the day. A strong stick, alpenstock, or ice-axe must be taken, as the occasion may require ; a rope also is necessary, always in winter and often in summer, if anything more than an ordinary ascent is contemplated. As many men object to flourishing a rope in the open daylight, a rücksack may be added to carry it and also the provisions, for even the hardened sandwiches of a British hotel are scarcely proof against a steep rock chimney. A complete change of clothes is also to be desired, but it is much pleasanter not to carry them, especially as your heavy luggage ought to include extra shirts and stockings, and all those little odds and ends that seem to become more necessary and numerous as years go by.

As the easiest way of showing how an ordinary excursion amongst the hills may be made a lesson in mountaineering, and to draw the beginner's attention to those things which he ought to notice, we will localise the advice and go to the hills themselves. The 'Lake District' is about half-way between London on the one side and Edinburgh on the other. Dungeon Ghyll is a good starting point, as it is easily reached and has the advantage of being close to the foot of the mountains. We will ascend Bowfell by the north-east face of rock, and then follow the ridges on to Scafell and down to Wastdale Head.

The October morning is wild and dull, with occasional clouds drifting over from the Scafell direction, and every now and then forming under the Bowfell cliffs. We start through the trees behind the house, and up the valley, a pleasant walk with the crags gradually emerging into full view. The top ridge can occasionally be seen through the clouds that keep forming and dissolving under its shelter, and after a short consultation we choose a line of attack that will land us on its crest a little to the north of the summit.[1]

The path now bends to the left, and mounts quickly. We have marked from below the spot where we intend to leave it, but find that, as usual, the inexperienced eye has chosen a place too low, and we must follow the track some distance further before quitting it for the lower slopes of our mountain. These slopes are fairly steep, so that it will be well for us to start very slowly, pressing steadily and silently upwards. It is not necessary to be dull or morose, but there is a time for everything, and the first steep slope of a mountain is not the place for a political discussion. In ascending a mountain it is usual to go uphill, but some men walk thoughtlessly about, often taking a few steps downwards, because they have been admiring the view, talking or thinking of other things, instead of keeping their attention to the work before them. They get tired and annoyed, and at the end of the day will think more of getting

[1] If the reader has a one-inch ordnance map of the district open before him it will enable him to follow the route described.

off 'those wretched hills' than of the effect of the evening lights upon them. By attending strictly to business, not only will you have more time to look about you when a halt occurs, but you will be able to take in quietly the beauties of the view as you walk along, for it is one of the essentials of mountaineering that you should notice and remember everything, and you must carefully practise this dual effort of your brain.

The ground before us is broken into little ridges and valleys. Generally the first give the better foothold, but occasionally the bed of a little streamlet provides a series of convenient steps. When the route lies up steep grass, it is well to zigzag much, so as to change the direction of the slant pretty often, as it is easier for the feet and better for the boots. The clouds lift a little and give us a view of our proposed route up the rocks, so we take the compass bearing of the summit ridge as nearly as we can, and note the distance of the top from the point where we hope to strike the ridge, as it will help us to find the cairn should the mists be thick above. A slope of scree is now entered upon; we must not rush at it, but put our feet down carefully and firmly, and a little practice enables us to choose the firmest footholds almost mechanically. When the slope becomes loose and shaly we must quietly submit and let our feet sink, for struggling and scrambling is useless as well as tiring. It will not do to scoff at these small details, for their sum will be considerable before Wastdale Head is reached. A steep gully filled with moss and grass has next to be ascended, and here it is better to kick an occasional step and see that each foot is securely planted, before we actually trust our weight to it, for it is annoying to keep slipping back and having to do the work a second time. If there is any doubt about the firmness of a step, whether it be on rocks or grass, in ice or snow, it is well not to use it too much as a fulcrum, but, with the help of our hands and stick, gradually transfer our weight with a quiet, upward swing. The rocks at the side of the gully now offer us an easier way, but care is still necessary, for though easy, they are seldom climbed, and are in their normal condition loose, and now wet and slippery with the rain and mist. The clouds sweep down upon us, but we keep steadily upward and reach the crest at last. As we turn to the left a dark mass looms through the clouds, but knowing that its apparent size is a misty delusion we pass it, still keeping within touch of the crest, for near it the top must be. It is found at last. Taking our bearings with the map and compass, and noting carefully the direction of the wind, which will be fairly steady over the lofty ridge along which we have now to return, we hurry down, as it is cold and wet. We keep the wind on our left cheek and feel the cliffs on our right hand, as it is easy to stray down to Eskdale on the south and west. We soon pass the spot where we first struck the crest on our ascent, and shortly afterwards find ourselves distinctly rising again and appear to have reached a summit. A look at the watch tells us that we are on the northern end of the Bowfell ridge, for we have not had time to reach Hanging Knots, even if we make due allowance for the fact that we have been going downhill most of the time ; and the map also shows us what we had not noticed before, that the ridge here takes a sudden western bend towards Hanging Knots. With a resolve to be more careful in future we turn slightly to our left, but we quickly lose ourselves again ; a timely break in the clouds reveals to us a beautiful glimpse of upper moorlands, a glorious mixture of greys and yellows, the colours intensified by the wet and their silver frame of fleecy cloud. The compass tells us that this is Upper Eskdale, and as we gaze at the wonderful picture the clouds break above us on the right, and show the pass of Esk Hause over which they are swiftly sailing. We have skirted fully 300 feet below the summit of Hanging Knots, instead of passing over it ; however, it does not matter now that we have found the pass.

The route from here to the top of Scafell Pikes lies over an upland stony wilderness, upon which we enter with some misgivings, for the clouds have again come down upon us, but we know that our direction should be west-south-west ; that if we at any time steer to the left hand we shall drop into Eskdale, and if to the right, we must strike the track leading to Styhead Pass, or, if the mistake is made later, the stream leading from there to Wastwater. We walk in single file, so that the last man may correct the leader's variations from the true direction; we look at the compass every few minutes, and at last, after many a twist and many a turn, reach the great cairn on the highest point of England.

Before starting downwards we again take a good look at the map and compass, for many mistakes are made by leading off in a hurry, and here the desert of loose stones is the same on every side. Making due west so as to hit the Wastdale side of Mickledore's broad ridge, we turn to our right under the Scafell cliffs, and follow the water to Wastdale.

Our walk may have been a good lesson in the art of mountain travel, though there was little rock climbing in it; but as this branch of mountaineering has been dealt with elsewhere in this volume, the writer would here only draw attention to a danger which exists more at home than abroad, viz. the danger of 'fancy climbs.' Near Snowdon, in the Lake District, and to some extent even in Skye, those who would do some fresh thing have been driven to climb places of extreme difficulty, which are now known and named, and their ascent has become the ambition of many a climber. The novice must on no account attempt them; he must content himself with easier expeditions, such as the Crib Goch ridge of Snowdon, the 'Broad Stand' of Mickledore, and, as he becomes more experienced, with one of the ordinary ascents of the Pillar Rock or the north ridge of Sgurr-nan Gillean. He may console himself with the reflection that most of these fancy bits of rock-work are not mountaineering proper, and by remembering that those who first explored these routes, or rather created them, were not only brilliant rock gymnasts, but experienced and capable cragsmen.

A word of advice to him even after he has qualified himself to attempt them. He should not take a novice with him;

BRITISH HILL WEATHER.

the main difficulty should be attacked in the morning, so that there may be no need to hurry or scamp his precautions, as the pace will of necessity be very slow on the difficult parts of the climb. The rope should be hitched over or round a rock as often as possible, and care taken that all slack rope is gathered up.

It is quite impossible within the limits of this chapter to give a list of all those places in Great Britain where climbing is to be found. All that can be done is to draw the beginner's attention to some districts where mountaineering can be profitably studied in most of its branches, and leaving him to find his own routes, both on the map and on the hill-side, lead him to learn gradually the difficult but pleasant lesson of mountain topography.

To commence with the English hills. Many think that nowhere else is such a beautiful variety of hilly scenery to be found in so small a space as in the Lake District, where the hills though grassy have many a precipice and rocky ridge, and whose pleasant valleys are conveniently dotted with little inns. From any place in it there are many capital walks, such as from Coniston to Wastdale Head, passing either to the south or north of Harter Fell, according to whether you cross the Scafell range by Burnmoor or Mickledore, and as the way has to be worked out almost entirely from the map, it is a valuable lesson in the art of path finding.

The finest walks however are along the watersheds. The man who has never tried a long tramp over the main ridges of a hilly country can have no idea of the charm there is in being for hours as it were on the summit of a peak, and having the pleasure of alternate ascents and descents thrown in at the same time. The views are lovely, the air exhilarating, boggy tracks of moorland are avoided, and many a little unknown piece of rock work can be found to enliven the way. For instance, start from Wastdale Head up the Pillar Mountain (ascend the rock or not), and then to Black Sail Pass, Kirkfell, Great Gable, Green Gable, Brandreth, Grey Knots, and down to Seatoller or Buttermere. If you have to return to Wastdale Head, strike down direct from Great Gable, or continue the walk to Scafell Pikes, returning by Mickledore and Scafell. Or take the ridges of Skiddaw and Blencathra, or of Helvellyn and Fairfield. One of the advantages of these walks is that they can be made as long or short as is desirable, but the enthusiast will gradually discover the best things for himself, and a walk invented by the climber is often more interesting to him than one of the more recognised expeditions.

The best centre for rock climbing is Wastdale Head, as the group of hills above it includes the Pillar Mountain, Great Gable, and the Scafell range. Of course the Pillar Rock is one of the first objects of a novice's ambition ; there are three recognised routes to its summit, one from the west and two from the stone shoot on the east ; find them out for yourselves, but remember to take a rope, and have someone in the party who knows how to use it.

The Scafell group has beautiful crags on every side, and opposite to it Great Gable sends down a splintered and almost invisible ridge called Great Napes, on which stands a sharp and extremely difficult needle. The Dow crags near Coniston, and the faces of the Langdale Pikes hanging over Dungeon Ghyll, can be climbed in many directions of varying degrees of interest and difficulty. In fact, it will be your own fault if you leave the district disappointed with the scrambling you have found in it.

The hills of North Yorkshire and Lancashire, of Derbyshire, Northumberland and Durham, have long stretches of little known moorlands, which are capital places for learning the use of map and compass ; hidden away in their recesses are certain crags, which will give many a practical example of the way in which large blocks of limestone will break away when apparently firm, a very necessary lesson to those who would explore the dolomitic mountains of Alpine Europe. There are many places on our long coast line that boast of fine rocks, and the white cliffs of Old England have proved themselves too much for many of her sons. Chalk is a very troublesome material to climb; it is loose, breaks away, and if wet, either with rain or the draining of the land above, forms a sticky paste, which, lodging between the boot nails, renders them of little use. Doubtless many of those who get into difficulties are only shod for a walk on a promenade or pier, but chalk cliffs must not be treated carelessly even by the well equipped.[1]

In Wales, although there are many beautiful mountains, the climbers' interests ever centre round Snowdon and the neighbouring peaks. If Ben Nevis is the king of our mountains, surely Snowdon is the queen; her graceful cone rising high above, but seemingly so dependent on her steep supporting ridges. It is true that there is an uninteresting western flank, but she has evidently, in her benevolence, allowed the glaciers which shaped her sides and arranged her ridges to leave it thus, so that even those who cannot climb may share the pleasures of her crown. There are several paths to the summit, but the most interesting climb is the ascent of the extreme end of the Crib Goch ridge, passing along its narrow crest and round the cirque to the top of Snowdon. It is a fine walk, and if snow and ice lie on the rocks, or a high wind be blowing, care must be taken in places to avoid a slip. A beautiful descent from Snowdon is to the east, over the summits of Lliwedd, leaving the ridge at the point most convenient for walking back to the hotel. The north face of Lliwedd can be climbed right up from the base of the great central gully, or by the buttress to the right, but the ascent ought only to be attempted by a very strong and fully equipped party. The precipice of Snowdon offers a dangerously steep and slippery ascent direct from Glas Llyn to the top. The cracks and ledges are small, far between, and filled with much earth and grass, which has often to be scraped away, to give even a moderate hold; if many tourists are on the summit, tokens of their presence are apt to fall, and no climbing skill or precautions will save you from a well-

[1] The coastguardsmen have often to rescue tourists by means of ropes.

aimed gingerbeer bottle. This climb is therefore not to be recommended. There are several difficult ways of reaching the Crib Goch ridge from the north, and the great buttress called the Parson's Nose is a particularly fine piece of Cyclopean anatomy.

At the head of the Ogwen valley, Tryfan sends down several fine clefts and buttresses into the wild hollow of Cwm Tryfan, up which many an interesting route may be chosen. The central gully is the most difficult; the climbing on either side of it is easier. From the top a most interesting ridge leads to Glyder Fawr; no better one for a beginner exists in the British Isles; it really gives rock work if you keep to its crest, is never too difficult, and always safe. The long range of Cader Idris is bold and striking on the northern sides, and Arran Mowddwy is well worthy of a climb from the east. It is useless, however, to enumerate other routes. If you want to climb rocks go to the Snowdon district; if you want to learn general mountaineering go anywhere, and take such walks as from Festiniog to Beddgelert over Moel Wyn and Cynicht, or from Bethesda to Capel Curig over Carnedd Davydd and Carnedd Llewellyn.

The McGillicuddy Reeks, near Killarney, look as if Erin had given some thought to the climbing education of her children. The Mourne mountains form a lovely little group, and if you happen to be fishing in County Galway, a little scrambling can be had on the picturesque Twelve Pins of Connemara, when the wide views over wastes of water and bog will compensate you for the easiness of the way. There are many detached masses in the south and west, and wandering amongst these out-of-the-way groups of hills is always interesting; it is unknown country, and the knowledge of mountain craft gained in finding your way amongst them, will be of great assistance on more serious climbs.

The greater part of the Highlands of Scotland is composed of sweeping moorlands and grassy or stony hills, beautiful in shape and colour, but not of those forms which especially attract the climber. There are, however, brilliant exceptions. Take a meteorological map of the country, mark the districts most deeply shaded and which are said to have a rainfall of 80 inches and upwards ; there you will find plenty of work to do. It is true that the lofty range of the Cairn Gorms, the wild hills of Torridon and Loch Maree, of Ullapool and Assynt lie beyond this magic circle ; yet within it there is as much good climbing to be done as in all the rest of Great Britain, and certainly sometimes more than one's share of the 80 inches rainfall. If you object to rain, why go there in the rainy season? The months of May and June are nearly always dry and fine. The inns being not so crowded give better accommodation, and there is less danger then of being stopped on the hillside by indignant gamekeepers. Do not blame them. The fellow-feeling that you ought to have for your brother sportsmen should make you careful not to disturb his deer forest in the autumn, nor to examine his grouse nests if you accidentally come across them in the spring ; he also in his turn should never stop you unless for some good reason. A mountaineer is often heard to complain that the mountains are overrun with his fellow-men, and that he finds it increasingly difficult as time goes on, to revel in 'the vast solitudes of nature.' Thoughts like these should prevent him from forcing his way through private grounds and desecrating another man's solitudes, when there is a lonely and more appropriate way near at hand, even if a trifle longer, and should cause him to carefully bury such parts of the débris of his luncheon as will not be eaten by the birds or foxes. Let him efface himself as much as possible, remembering his own professed feelings, that the grandeur of the hills is increased by solitude, and, as he must needs be conspicuous on a mountain ridge, let him be doubly anxious to be modestly unobtrusive when on lower ground. The Scottish Mountaineering Club have taken up this question, and their example should have a good influence on some of those tourists who march about like an invading army regardless of season and place.

The Scot believes himself to be of a favoured nation, and so he is in the matter of hills. The inhabitants of his largest city can in a few hours by steamer reach the head of Loch Long, or the more distant isle of Arran, whose sharply pointed hills are the characteristic feature of so many beautiful views on the Frith of the Clyde, and whose far-famed glen will rejoice the heart of any mountaineer.

Close to the great tourist route of the Caledonian Canal towers the great mass of Ben Nevis. Certainly he has an observatory on his summit, and nothing serious happens there without its being known at once in London and Greenwich. He has suffered many an indignity from science and fashion, but still remains, though not one of the most beautiful, yet solid and majestic, the king of all our mountains. His tremendous northern cliffs fall steep and rugged into the depths of the glen below, and few of us could boast of having gained the highest point of our native land, if the only way thereto lay up those beetling crags. Glen Coe is near at hand on the same route, with its wild corries and bold and seldom explored buttresses, fit sentinels of the rugged solitudes of Etive. The whole length of the coast from here to Loch Inver is one long and variedly coloured panorama, and the mountaineer who wishes to visit Skye or the hills of Western Ross-shire cannot do better than take the steamer from Oban. There can be few lovelier sails than this, or easier ones either, for the waters are nearly always land-locked, and smooth compared with the wretched tossings of the English Channel. Unfortunately the ordinary lines of steamers do not give one the chance of seeing all the beauties of Loch Hourn and Loch Duich, and the wild mountains at their heads which invest these Fiords with all the beauty and mystery of this western coast ; but enough remains, and what mountaineer would turn aside on his first visit, when the serrated ridge of the Coolin Hills beckons to him over the purple headlands of Sleat and Strathaird.

Sligachan, in Skye, is the rock-climbing centre *par excellence* of the British Isles. Every black Coolin is worth ascending for the climb alone, if you take it from the right (i.e. the wrong) direction, and some give a good scramble even if taken from the easiest side. The steepness of these volcanic rocks is most remarkable; the hard 'gabbro' of which they are composed has resisted the action of the weather better than the softer rocks which once surrounded them. Their weathered surfaces are rough and crystalline, giving capital hold for hands and boots. The mass is traversed by veins or dykes of all sizes; where these veins are of softer material than the 'gabbro,' their faster weathering has produced deep cracks and wall-sided gullies, filled with loose stones; but where the vein is harder than the containing rock, it stands up as a slab or pinnacle, often of the most fantastic shape. Both the ridges and gullies thus formed are often exceedingly rotten, and supply an endless field for the study of loose stones; so let the leader remember that his friend is below him, and let the last man remember that falling bodies gather momentum, and that if stones do fall, the nearer he is to the man who sends them the better.

The usual ascents are now comparatively free of loose stones, as for instance those of Sgurr nan Gillean, by either of its three ridges, climbs varying in difficulty from the easy eastern to the fascinating climb up the northern crest. But many of the other peaks present to the climber ridges, faces and gullies, in the glorious primeval rottenness characteristic of a virgin peak. The Alpine climber will find an additional interest in the district from the remarkable indications of former glaciers; in some places it would seem as if they had disappeared but a few years previously, and occasionally, whilst climbing some smoothly rounded buttress, one almost expects to see the ice itself creeping through the next depression.

Always take a rope; 40 to 50 feet is generally enough; it is a precaution that may save you the disappointment of having to give up some particularly interesting piece of rock work for an easier route. There are also some clefts which cannot easily be ascended if anything is carried on the back, but the sack and water-gourd can be sent up by the rope without any danger of destroying them, and a large water-bottle is a great comfort on these dry and barren rocks.

Perhaps the finest example of a Coolin Peak is Sgurr Alisdair,[1] with its shattered ridges, loose stone shoots, black precipices, and lovely views over sea and land. The 'Inaccessible Pinnacle' of Sgurr Dearg is the Matterhorn of the district, and the east ridge is now safe enough for a man with a steady head, but it is very narrow with a sheer drop on each side ; great care should be taken in the descent, one man only moving at a time, and let the last man be steady and keep the rope tight. The western ascent, not nearly so fine, is very short, and has, moreover, one difficult and dangerous step, where no help can be given to the leader, and where, if he fell, he could not be held by the rope ; once up, however, he can, if required, give very effective aid to those below. One other ascent may be mentioned as a typical one. Gain the top of Bidein Druim-nan-Ramh from Harta Corrie, either by the gully separating its peaks or by the rocky face where the eastern ridge loses itself in the peak ; the passage from one summit to the other is very fine ; then keep to the ridge over all the tops of Mhadaidh, and descend by its wonderful west ridge into Corrie nan Creiche. If this walk is not long enough, continue along the ridge from Mhadaidh over Greadaidh, and so on until you have had enough for pleasure and then descend towards the west. The compass is not always to be trusted amongst these hills, as the rocks are magnetic. The maps also are incorrect, and though the new one-inch ordnance is an improvement on the old one, the names of nearly all the mountains are illegible and one or two are misplaced.

The hills of Sutherland are more remarkable for the beauty of their sweeping moorlands than for the boldness of their forms, but the weird masses of sandstone that seem to be

---

[1] Name omitted in new one-inch ordnance map.

HILL CLIMBING IN THE BRITISH ISLES 345

stranded amongst the bogs of Assynt will raise alike the curiosity and ardour of the climber. The most wonderful of them is the long ridge of Suilven, which, seen from Loch Inver, appears as an inaccessible and almost perfect sugar-loaf. To the south, just beyond the county border, are the little known Coigach Hills, conspicuous amongst which, for its fan tastic shape, is Stack Polly with its ragged top. Further south, above the head of Little Loch Broom, rise the Teallach Hills. Seen from either side the range is very fine, and from a distance of forty miles to the north the writer has seen bands of snow from top to base, even towards the end of June, showing the steepness and depth of the northern gullies. The Black hills of Torridon have many splendid precipices of terraced sandstone, and many a sharp and shattered ridge, but, like all the other sandstone mountains of the western coast, they have no tops. They are, in fact, magnificent ridges rather than peaks, and it should also be mentioned, are all strictly preserved deer forests.

The central ranges of Scotland carry much more snow than the hills of England or Wales. Large quantities remain in some places far into the spring, and having time to settle, become somewhat like Alpine snow, requiring much kicking and step-cutting. It is quite impossible to give any list of the climbs that can be made in this district. Roughly speaking, the finest peaks lie on or near the watershed of the river Tay, from Ben Lui on the south to Glen Coe and Ben Nevis, then across the country to Ben Alder, and away past Dalwhinnie station, to the Cairn Gorms in the east. The bold outlines characteristic of the range are well seen from the railway. Much of the district consists of strictly preserved deer forests, but this need not trouble the climber. There is plenty of other land, and these hills must obviously be much more interesting to him at a time of the year when the snow, which lures him upwards, has driven the usual inhabitants into lower feeding grounds. The guardians of the 'Sanctuaries' will be found in a more amicable frame of mind, than at a time when a thoughtless visit might cause much loss to the sportsman who had paid beforehand for his stags, a loss which, if repeated constantly throughout the shooting season, would drive him away in turn, to the great loss of the inhabitants, who get far more money from the shooting tenants than they do from pedestrians. After the train to the north leaves the upland, it sweeps down with a rush into the valley of the Spey, and there, far away to the right, the giants of the east block in the moorland, and pressingly invite us to stop at Aviemore, to make a pass through their blue recesses. What a pity it is that these Scottish hills are so far away from some of us, and that we are induced to climb in the Snowdon and Scafell district again and again, instead of making raids into new country and fresh conquests across the border.

In conclusion, let us sum up the lessons that the mountains of the British Isles can teach us. That they can give healthy exercise, and cultivate in us the power of appreciating the beauties and grandeur of nature, has long been known to the many, but apparently only the few have hitherto recognised what it is the purpose of this chapter to point out to others—namely, that they form a good and safe training ground where men may learn and practise nearly all that is necessary in the art of mountaineering. Amongst them we may learn the proper use of our legs; the balance of our bodies, and so to regulate our movements that distances may be traversed and heights scaled with the least possible expenditure of force. We can learn to discriminate between the real and apparent angle and difficulty of a steep mountain face, how to judge of pace and distance, and to steer by map and compass even in the worst weather. We may learn to climb difficult rocks, to avoid dislodging loose stones, and to guard against those dangers that are peculiar to grassy mountains. We can practise carrying a pack, and to a great extent learn the use of the ice-axe and rope, and something also of the varying conditions and appearance of snow. We can cultivate perseverance, courage, the quiet uncomplaining endurance of hardships, and last, but not

HILL CLIMBING IN THE BRITISH ISLES 347

least important, those habits of constant care and prudence without which mountaineering ceases to be one of the finest sports in the world, and may degenerate into a gambling transaction with the forces of nature, with human life for the stake.

A general depression

CHAPTER XII

THE RECOLLECTIONS OF A MOUNTAINEER

BY C. E. MATHEWS

An I Glass

CANNOT think it unnatural for an old mountaineering hand to compare the Alps as they were forty years ago with the Alps of to-day. The means of access to the great playground have changed ; the people one meets with are of a different type ; everything, but the grand old mountains themselves, seems to wear another aspect ; how many of us ardently wish that the changes were in all cases for the better.

If we consult that interesting archæological curiosity, a Continental Bradshaw for 1852, it will furnish ample food for reflection. Forty years ago no railway even approached the Alps. The Great Eastern of France took the traveller as far as Strasburg, from whence he could proceed to the terminus at Basle, where a new line was in course of construction to Zürich, and another from Dijon to Lyons. Berne and Aosta could be reached only by diligence or carriage. Alpine travellers went from Dijon through Besançon to Neuchâtel ; and through Dôle, over the Jura Mountains, to the Lake of Geneva. There was no railway within a hundred or a hundred and fifty miles of the various centres of Alpine attraction. In these days a man may leave London at eleven in the forenoon, and breakfast the next morning at Geneva or Lausanne.

The establishment of a direct and rapid communication between Paris and Turin and the construction of the Cenis tunnel brought the first great line of railway within measurable distance of Geneva ; and a junction with Culoz was soon effected. The St. Gotthard line, a rare triumph of engineering skill, now brings Lucerne in direct contact with Milan, and happily does not deface the charms of the beautiful valley of the Reuss. The line from Geneva to Martigny, carried on in due course to Sion, and thence to Brieg, will some day be continued through a Simplon tunnel. To such lines, necessary for military and commercial purposes, although they pierce the heart of the Alps, no reasonable objection can be taken. Alpine scenery is on so large a scale, that when a railway is completed and its banks clothed, if the valley along which it runs is sufficiently wide, the eyesore is far less serious than might have been anticipated. Lovers of the picturesque may bear with equanimity the sight of the line which now runs from Geneva to Cluses, and which is shortly to be extended to Sallanches. The line from Ivrea up the Valley of Aosta may also be tolerated, if not approved. It is the construction of railways up the lateral valleys, and extending into the very fastnesses of the Alps, that fills us with sorrow and alarm. It is a scandal to the Republic that a line should have been permitted between Interlaken and Grindelwald, or between Lauterbrunnen and Mürren, but to mar the exquisite beauty of the Wengern Alp is a crime. Alas ! for those who hailed with delight the extension of the Rhone Valley Railway from Sion to Visp, on the ground that it saved the traveller many hours of dust and discomfort. How little they thought that Visp in its turn would be made the starting point of a line to Zermatt, and that the once unfamiliar shriek of the locomotive would startle the climber on the Weisshorn or the Dom.

So with the high roads. It is no longer necessary to leave

the diligence at St. Gervais, and rattle in the little char-à-banc from that village to Chamonix. The high road now runs through that great caravanserai, and is continued over the Tête Noire to the valley of the Rhone, and the once lovely mule tracks of the Brunig and the Furca, the Bernina and the Albula, are seen no more.

Well ! we have our 'improved facilities for communication,' and the rush of tourists has begun ; the Alpine regions, but not, I am glad to think, the Alps as we understand them, are visited by constantly increasing hordes. The small mountain inns are dying out. Who does not remember that typical little house, the Krone at Pontresina, some thirty years ago, where Herr Gredig welcomed many an early mountaineer? Enormous hotels now choke that once Arcadian village, where the London season is further protracted, Parisian cookery advertised, and electric lighting laid on.

The new régime has extended from the villages to the mountains. Except in the remoter districts men no longer camp out in the open, or in holes in the rocks, share a hayloft with a shepherd, or a chalet with a maker of cheese. Huts are constructed all over the Alps containing a stove and similar luxuries, and climbing is made easy to the modern mountaineer. The old chamois hunter, who thought himself well paid with five or six francs for some more or less difficult pass, is replaced by the local guide with a tariff of twenty or thirty.

The explorer, the geologist, the student, the lover of natural scenery, the mountaineer, now mix with those who seem only to think that they may as well be there as anywhere else. While good manners are not always cultivated as they once were, they are still sometimes taught, as two ill-bred English tourists once discovered to their cost, when they learnt from a justly angry German professor that England was most admired where Englishmen were the least known ! The changes which facilities of travel bring, have not apparently added greatly to the number of Englishmen who go to the Alps solely for the pleasure of climbing; and although there were giants in the old days, we are all proud to think that the calibre of the climber has not deteriorated. What manner of men were those who founded and developed our amusement, and who may rightly be described as the makers of mountaineering?

Among these ever pre-eminent must stand the name of JAMES DAVID FORBES. He was no mountaineer in the modern sense, and never accomplished any very difficult feats: he was certainly not an athlete, and would probably have broken down under the stress of what would now be considered a really arduous expedition. He was pale, thin, and had indifferent health, but his expression was singularly sweet and winning, and he had the beautiful and refined manners of the old school. He was a remarkable instance of the hold the mountains have over men of rare intellectual endowment. His early career was of unusual brilliancy, while his manhood was spent in scientific investigation, being a Professor of Natural Philosophy in the University of Cambridge. The first

J. D. Forbes

of the modern Alpine pioneers, he began to peer into the secrets of the mountains when the Alps were an unknown land. The perusal of his fascinating 'Travels in the Alps of Savoy' (first published in 1843) sent many a novice to the great playground. Before commencing his glacier investigations, he read every line of the celebrated work of De Saussure, and used to say that the endeavour to follow the great historian of the Alps in his own country, and to meet him on his own ground, was an attempt which he felt to border on presumption. He travelled in almost every district in the Alps; visited the still inhospitable regions of Dauphiné in 1839 and 1841, afterwards ascending the Jungfrau, and crossing the Col du Géant and the Col d'Erin.

Wherever he went, the secrets of Nature were probed 'with clear eyes and an open heart': nothing was taken for granted. His mathematical and scientific knowledge was brought to bear upon disputed facts, leading him to propound what is now generally known as the viscous theory of glacier motion. To him are owing the first accurate survey and map of the Mer de Glace. After visiting Norway in 1851, his health failed and rendered further Alpine investigations impossible. Only the early climbers know what happy results are attributable to his works, precepts, and example. He it was who suggested to Mr. Adams Reilly the survey of the whole chain of Mont Blanc, and Forbes often spoke to me with great interest and delight about the *chef d'œuvre* of that accomplished mountaineer. He was intrepid and active in the days when climbers were few : when old, no one who sought his assistance or advice ever went empty away. He used to say that more was to be learnt from the hills than mere mountaineering, and though he had a lurking fondness for a man who could do his twenty hours' hard walking without flinching, would add—and with infinite truth—'You will enjoy and learn more by doing half as much in the same time.' He was the father of scientific climbing, and his death in 1868 brought sorrow to the hearts of many. If ever there was a good and true man it was 'Icy Forbes.'

Next in order among the great Alpine explorers comes the name of one whose recent death has left a gap which can never adequately be filled. Probably few of the many men of intellectual eminence who have found their best form of recreation in the Alps have been as many-sided as the late JOHN BALL. He was a mountaineer and a statesman, a man of the world and a man of science, a traveller in many countries and appreciated in all. Under-Secretary for the Colonies in Lord Palmerston's first administration, he was one of the keenest observers of natural phenomena, had high literary

THE RECOLLECTIONS OF A MOUNTAINEER

attainments, and certainly no one ever possessed a more thorough knowledge of every part of the Alps. He was the first president of the Alpine Club, and I well remember not only the respect but the enthusiasm with which he was always greeted by the then younger generation of Alpine men. His early exploits date back to what seem pre-historic times, for it was in the year 1845 that he made the first passage of the Schwarz Thor from Zermatt to San Giacomo d'Ayas, accompanied, or rather impeded, by a Zermatt peasant, in days when regular guides in that region were unknown.

John Ball

He edited the first series of that remarkable work 'Peaks, Passes, and Glaciers,' first published in 1858, and between 1863 and 1868 brought out 'The Alpine Guide,' which is of course now somewhat out of date, but which has been rightly described as the 'Encyclopædia Alpina,' and 'the first of Alpine classics.' It not only chronicled the work of previous explorers, but contained countless suggestions of which younger men at once availed themselves. Probably no guide-book was ever more clear in grasp or comprehensive in detail, more stored with general and special knowledge ; the whole being lighted up on almost every page by a graceful and charming literary style. After visiting Morocco and the Great Atlas, he made a long tour in South America ; his botanical discoveries in the lower

Andes, Valparaiso, and Brazil being recorded in his 'Notes of a Naturalist.' His interest in scientific discovery was so intense, and his mind so stored with knowledge, that no one could be in his company without feeling its subtle influence. On one occasion a periodical was lent to him containing an article wherein the writer, with apparent gravity, discussed the question 'Whether life was worth living.' It caused him to place on record the 'contemptuous pity he felt for the man whose mind could be so profoundly diseased as even to ask such a question—as if a soldier with the trumpet call sounding in his ear, should stop to inquire whether the battle was worth fighting.'

The various theories of the motion of glaciers, the excavation of mountain lakes and Alpine valleys, and the distribution of plants were very dear to him. His charming speeches, which so many of us remember, combined a rare courtesy with true kindness of heart. A witty poet, whose works in these days are too little read, has told us that

> Some minds improve by travel, others rather
> Resemble copper wire or brass,
> Which gets the narrower by going farther.

Of Ball it may truly be said that he improved the minds of others by travelling, for wherever he went an interesting record was made. Professor Bonney paid him a just tribute when he wrote that 'He seemed never hurried, never ruffled, incapable of bitterness, and still more incapable of anything base. When men of his nature die not only friends but many others also are the losers, because an influence for good is taken away from the world.'

ADAMS REILLY was another of the heroes of old days. Cultured, unselfish, and sweet-hearted, he inspired the strongest and most lasting personal attachments. His Alpine ambition, like that of others, was fired by Forbes' celebrated work. He began climbing in 1860, and a few years later made the acquaintance of Forbes, which soon ripened, and they became fast friends. When the health of the great pioneer failed, he said to Reilly, 'You have the necessary capacity, youth and health—will you carry out the task that I have endeavoured to begin?' Reilly accepted the charge, and went to work at once with his theodolite, and after two years' continuous labour achieved the distinction of 'presenting for the first time to the eye of the tourist and the physical geographer a correct and skilful delineation of the most remarkable and most elevated mountain chain in Central Europe.' This was followed some five years later by a complete survey and map of the southern valleys of the chain of Monte Rosa, which were so accurate that they are still essential to modern travellers. After Forbes' death Reilly wrote the Alpine portion of his biography, a work of the deepest personal interest to him, for there was no man for whom he had a more chivalric devotion. No mountaineer was ever better known in the valley of Chamonix ; he had been seven times on the summit of Mont Blanc, and there were few of the guides and porters of the district who had not at some time or other been employed by him. Reilly was unselfishness personified. I remember an occasion when he met an absolute stranger at a friend's house high up amongst the mountains. The stranger being taken ill, was ordered home without a moment's delay. Reilly generously offered to accompany him, and took care of him as far as Geneva, but there being no improvement, accompanied him to Paris, then to Calais, and at last to Dover ; and when he had delivered the sick man to the care of his friends—and not till then—did he return to the Alps. Such an act of sterling kindness must be recorded, for how many mountaineers are there of whom such a tale could be told? There is more than one man now living, who, when suffering from accident in a foreign land, has been nursed by Reilly with a beautiful tenderness, and who knows what it is to pine

For the touch of a vanished hand,
And the sound of a voice that is still.

Few of his friends will ever see the runic cross, carved with

Swiss flowers, that marks his resting-place above the waters of Lough Derg ; but as long as men love the Alps, his work and his name will never be forgotten.

Another name must be placed in the front rank of the early explorers—that of THOMAS WOODBINE HINCHLIFF ; a man with an irresistible personality and a temper so genial and delightful that nothing seemed ever to disturb it. On one occasion, having suffered severely from a gun accident, which rendered necessary the amputation of some of his fingers, he declined all anæsthetics and smoked his pipe calmly during the operation. He was a mountaineer of the old school, and feats were not in his way. The woods, the meadows, and the flowers charmed him as much as the rocks and the snows. He enjoyed a fine climb with all his heart, but seemed equally happy on a quiet day. Success and failure seemed alike to him, possessing as he did an infinite capacity for enjoyment which any man might envy. On leaving the Bar and devoting himself to travel and literature, he said that by selling his wig, his gown and his books, he had realised more than in the seven previous years of vanity and vexation of spirit. He made one of the earliest ascents of Monte Rosa, and then climbed Mont Blanc. His 'Summer Months among the Alps,' published in 1857, became immensely popular. This work, and that of Mr. Justice Wills, ' Wanderings in the High Alps,' which had appeared the previous year, gave a great impetus to mountain exploration.

Hinchliff died at a comparatively early age : a loss felt alike by climbers and by guides. I well remember that on reaching Zermatt almost the first question one asked was ' Has Mr. Hinchliff arrived?' What pleasant hours we spent in his company ; with what unfailing modesty he gave his advice to those who were younger and less experienced than himself ; with what cheerful hilarity he joined in our festivities ; no man was ever more welcome at Zermatt, or the Riffel, and when he left, some of the sunshine seemed to go with him.

A. W. MOORE was another of the Alpine immortals. The

THE RECOLLECTIONS OF A MOUNTAINEER 357

son of a director of the East India Company, he entered into its service as a boy just after leaving Harrow, and for the rest of his life served with unswerving honour and fidelity either the Company or the Crown. In 1885, after many years of laborious duty, he resigned with a pension, but returned two years later, on his appointment as Political and Secret Secretary at the India Office. He left England for Monte Carlo on account of ill-health. But he was dead when kindly friends who went to nurse him arrived. His mountaineering achievements were as brilliant as his official career was distinguished. It was not till 1860 that Moore, then a youth in his twentieth year, first visited the Alps, but he soon made his mark, making the ascent of several great peaks for the first time, and was one of the memorable party who first crossed from Courmayeur to Chamonix over the summit of Mont Blanc. He spent two seasons in the Caucasus, and shared with Mr. Freshfield and Mr. Tucker the honour of being the conquerors of Kasbek and Elbruz. Moore was one of the early pioneers of winter exploration, and crossed the Finsteraarjoch and the Strahleck not only at that season, but by night, reversing with marked success the usual hours devoted to Alpine expeditions.

He had a rare genius for topography, and probably had a wider geographical knowledge of the Alps than that possessed by any of his contemporaries. His frankness and his good temper were proverbial, and his loss was widely and keenly felt. I had the good fortune to take him on his first winter visit to Wales about thirty years ago, where we played at Alps on Tryfan, Snowdon, and the Glyders, and had happy times together at the good old hostelry at Pen-y-gwryd when Harry Owen was still young. Moore was always brimming over with good-natured fun. Mr. Whymper, in his 'Scrambles amongst the Alps,' says of Moore that on his first visit to La Grave he reported upon the resources of the wretched little inn at which the party had to stay, informing his companions that there was nothing stable about it except the smell. Once, when we were in a remote part of Wales, he could not, or he would not, understand that people could exist within a day's journey of London in absolute ignorance of the English language. The constant answer of 'Dim Saesoneg' to questions put to peasants on the wayside puzzled him extremely. He believed that they were shamming, and on one occasion resolved to put the question to a severe test. Encountering an old woman he went up to her, and taking off his hat and making her a profound bow, he said, 'Fairest of creation, I love you to distraction.' Of course he received the invariable answer 'Dim Saesoneg,' and after that he gave it up. On another occasion we climbed Cader Idris in the depth of winter, arriving on the summit at five in the afternoon, just in time to see the sun fall into Cardigan Bay, and to find ourselves alone and in the dark. Moore was very short-sighted, and declined a descent to Dolgelly under these circumstances. For some hours we waited on the ridge, slowly freezing, when, by the aid of a reading-lamp which Moore suddenly recollected was in his knapsack, we crept slowly down and reached comfortable quarters exactly at midnight. These are occasions on which it is easy to test the various qualities of a comrade, and rarely have I found a man more honourable, more true-hearted, or more sincere.

He was a clever, even brilliant writer, and his journal of the 'Alps in 1864,' which many of his old friends will remember to have read in manuscript, and of which a limited number of copies were subsequently printed, remains one of the rarest and the most coveted of all the books of Alpine adventure. Old friend, we miss you sorely. The gods must have loved you well, or you would not have died so young.

Of eminent foreign mountaineers who have joined the majority, the name of Signor QUINTINO SELLA will ever hold a foremost place. He was born amongst the mountains, and in the intervals of his multifarious duties as Financial Minister, and at one time Premier of Italy, he found in the Alps almost his only relaxation. He was not only the President of the Italian Alpine Club, but its founder, and was held in esteem and even in reverence by its many members. While an ardent politician, he was also devoted to literature and physical science. His sons and his nephews enjoy a splendid mountaineering reputation. When I was at Courmayeur with Mr. Morshead in the year 1879, Sella was there, and we all meditated one of the most charming of all Alpine excursions, the passage from Courmayeur to Chamonix, over the summit of Mont Blanc. Sella started a day earlier than we did, for he was then advancing in years and had to use great caution. He spent the first night of the expedition in what was then the new hut on the Aiguille Grise, and occupied another day in constructing a rude shelter some two hours nearer the summit of the mountain. On our arrival at the new hut, we found that he had sent one of his porters to request the loan of an extra blanket, which fortunately we were able to supply. Starting very early the following morning, long before dawn we overtook Signor Sella on the open glacier, and were greeted by him—though we could not see each other's faces—with the greatest kindness and cordiality. We started together, but his pace was proportioned to his years, and he begged us to proceed. A few hours later, in descending from the summit, we again met the distinguished Italian just as he was clearing the rocks at the foot of the Calotte. He looked fatigued, and having a succulent West India pine-apple amongst my stores, I begged his acceptance of a thick slice of it. No refreshment could ever have been more intensely enjoyed, and he could not have been more profuse in his acknowledgments had this small gift paid off his National debt. He spent that night at the Grands Mulets, reaching Chamonix the next day very little the worse. On his return we spent a most delightful evening in his company, and when he left us we felt that we had parted with a gallant mountaineer, and a most able, courteous and kindly gentleman.

CHARLES HUDSON must not be forgotten. He was certainly the most accomplished and expert amateur of his time, and in mountaineering adventures he was a very Paladin of chivalry. He was the first who ascended Mont Blanc by the new route of the Aiguille du Goûter without guides in 1855; and a few years later, having climbed to the Dôme by the ordinary route, he completed the ascent by the Bosses du Dromadaire, a grand excursion in which more than one mountaineer now living sincerely wished to have been able to precede him. He was one of the party on the Col de Miage in 1863, when Mr. Birkbeck fell more than seventeen hundred feet, and assisted in nursing him during the period of grave anxiety which elapsed before the arrival of Dr. Metcalf from Geneva. Hudson volunteered to go from St. Gervais in search of the doctor, and he walked the whole distance, forty miles, and back again in less than twenty-four hours. His was a beautiful and unselfish nature. He was said to have been as surefooted as the born mountaineer, but the race is not always to the swift, and he perished in the sad misadventure which marred the first ascent of the Matterhorn in 1865.

WILLIAM LONGMAN also rendered admirable service. He used to say that he laid no claim to the character of a mountaineer; but there was no man who had a keener love of the mountains, and his position as an eminent publisher brought him into kindly contact with many a climber who desired to introduce his experiences to public notice. The publication of 'Peaks, Passes, and Glaciers,' and the great success which attended it, were due alike to his foresight and his enterprise, and the unfinished history of 'Modern Mountaineering,' on which he was actively engaged during his last illness, forms an abiding monument to his memory.

FRANK WALKER also deserves the most honourable mention. There was no man better known or more universally popular among the climbing fraternity. He visited the Oberaarjoch and crossed the Theodul in 1826, and in later years climbed all the great peaks in the Swiss Alps. He was one of the party in the first passage over the summit of Mont Blanc by the Brenva Glacier, and when sixty-five years old he made one of the early ascents of the Matterhorn. His mountaineering experience was unique, extending as it did over a period of fifty years, and it may truly be said of him that he 'bore without abuse the grand old name of gentleman.' Perhaps it may be permitted to one who was on the Jungfrau with him 'two-and-thirty years ago,' to record that his gifts and his powers have, together with his house, descended to other members of his family, and that there is no home in these Islands where succeeding generations of mountaineers have received a more genial welcome or enjoyed a more generous hospitality.

JOHN FREDERICK HARDY, too, was one of the makers of mountaineering. He was in the first English ascent of the Finsteraarhorn, had ascended Etna, travelled in Greece, and was the first on the summit of the Lyskamm. Though a University Don, he had a boy's power of enjoyment. Once, arriving on the summit of the Strahleck Pass, wholly dishevelled as to his attire but profusely decorated with Alpine flowers, he encountered the astonished gaze of a young Undergraduate whom in his capacity as proctor he had a short time previously severely admonished for certain irregularities in his University costume ! In later years Hardy lost all the activity of his youth, and, becoming subject to those physical disadvantages which are also recorded of Hamlet the Dane, took the rôle of a confirmed ' Centrist,' and was known for many years as ' The king of the Riffel.'

Such are some of the leading men who have created and developed the craft of mountaineering, and who, unhappily for us, are no longer able to answer to their names when the muster-roll is called. How is it that the mountains appear to have fascinated chiefly those persons whose general intellectual endowments have been so high ? It is certain that the English makers of mountaineering had already climbed high in the paths of literature and science before they prospected in the new field. How is it that men like Pratt, Bell Scholar and Senior Classic, should have gravitated to the mountains? or Francis Maitland Balfour, of whom Darwin said that 'As an investigator he was of like rank with Cuvier'? From the dead let us turn to the living. It is interesting to notice that such men as Tyndall and Tuckett, Wills and Stephen, Whymper and W. Mathews, Lightfoot and Ellicott, Montagu Butler, Taylor, Leaf, and a host of others, Bishops, Deans, Senior Wranglers, Senior Classics, men of letters and men of science, have sought and found amongst the glories and the beauties of the Alps the most true rest and the most perfect form of recreation.

Before passing to some of the great guides who were created, or at least developed, by the growing taste for mountaineering, we must not forget two who stood as it were midway between amateurs and guides, and who for many years rendered essential services to both.

The kindly figure of ALEXANDER SEILER, proprietor of the great hotels at Zermatt and the Riffel, will be seen no more. When Forbes visited Zermatt in 1841, he lodged with Herr Lauber, the village doctor, whose house had become a little hostelry. Mr. Ball stayed there four years later, and also A. T. Malkin, Ulrich Studer, Dr. John Forbes, Engelhardt, John Ruskin, and other early explorers. In 1852 the Mont Cervin Hotel was opened, Lauber subsequently selling his little inn to Seiler, and it gradually developed into the Monte Rosa Hotel, to which was added a little wooden 'Dependance' on the Riffelberg. It was in 1856 that I first made his acquaintance and that of his charming wife. The Monte Rosa was then not nearly half the size that it is now, but from the earliest days men realised the meaning of the words 'Shall I not take mine ease at mine inn?' when they received the ever-courteous and obliging attentions of the best host and hostess in the world. Other and more commodious hotels were in due course started at Zermatt, all of which Seiler ultimately absorbed; but the Monte Rosa was the mountaineers' true home, and none of them would willingly go elsewhere. The old guard will well recollect the special attentions which they invariably received, the bottle of rare wine which found its way into the bedroom but never into the bill, the warmth of the greeting on arrival, the little breakfast or dinner on departure when they sometimes induced Seiler to partake of the hospitality he had so considerately provided. Zermatt can never be the same place without the man who, though innkeeper, was a warm personal friend, whom worries disturbed not, neither did success spoil. Perhaps this worthy couple were seen at their best in time of trouble or disaster. How many there are who cannot recall without emotion their sympathetic kindness in times of gravity or sorrow, and who still think of them with gratitude and affection.

FRANÇOIS COUTTET, known as 'Baguette,' will be missed at Chamonix almost as much as Seiler at Zermatt. He was a guide before he became hotelkeeper, and was accomplished in both capacities; mountaineers always entertained for him a hearty regard; careful, shrewd, and persevering, he both deserved and achieved success; he too is gone, and another link broken in the chain which bound us to the times that are past.

Whatever success a mountaineer may have achieved, he owes entirely to some one or other of the Alpine guides. Men sometimes climb now without professional assistance, and rightly so if they are sufficiently qualified; but some competent guide must first have taught them more than the rudiments of the art. No amateur, however accomplished, has the quickness of perception, or the fertility of resource, that are to be found only in those who are 'native and to the manner born.' Guides, like the rest of mankind, are good, bad, and indifferent, but of the best of them it is impossible to speak too highly. They have been evolved by the necessities of the times. The germ of the guide was the chamois hunter. Half a century ago guides properly so-called did not exist, except, perhaps, in the valley of Chamonix, and even these, with rare exceptions, knew little or nothing of any district but their own.

In 1856, when exploring the Val de Bagnes with Mr. W. Mathews, we had the assistance of Benjamin Felley, a chasseur of Lourtier. The same year, in making the passage of the Col du Mont Rouge, we sent for Bernard Trolliet, 'le premier chasseur de Bagnes.' Few of the Swiss peasants except the chamois hunters knew anything of the higher ice-world. As mountaineering became popular, competent guides were always in request, and many of them soon became masters of their craft. They began to understand the perils of ice and snow, and to recognise when climbing was safe and when it was dangerous. They learnt the mysteries of séracs, they found out how to circumvent or evade crevasses, and how to give avalanches a wide berth. In short, they learnt the business of taking the traveller safely up his peak and bringing him safely down again. As the years rolled on the guides were organised and subjected to regulations, many of which, particularly at Chamonix, were of the most absurd description. Ultimately at all the great centres—Chamonix, Courmayeur, Grindelwald, and Pontresina—more or less competent guides could always be secured. Some of these men have achieved a splendid reputation. They have become not only the guides, but the friends of their employers; they have exhibited in emergencies not only rare judgment and knowledge, but conspicuous courage and devotion. They have greatly enlarged their experience as their employers have taken them further afield. They have visited Teneriffe, Alaska, and New Zealand, the Rocky Mountains and the Caucasus, the Andes and the Himalaya.

The raw material

THE RECOLLECTIONS OF A MOUNTAINEER 365

One of the veterans was AUGUSTE BALMAT, of Chamonix. Nearly forty years ago he was employed a good deal. He was an excellent guide then, and had some knowledge of foreign districts. He must have had considerable culture and many high qualities, or 'Wanderings in the High Alps' would hardly have been dedicated to him 'with feelings of hearty respect and affectionate regard.' Balmat was Forbes' attendant in his survey of the Mer de Glace, and made the first ascent of the Wetterhorn with Mr. Justice Wills. Upon the whole he was perhaps one of the best men that Chamonix ever produced. He was '*guide chef*' when the Emperor Napoleon visited that district after the annexation of Savoy and Nice; and he once described to me the awe with which he was inspired by the then master of many legions, and the charming manners of the Empress Eugénie, whom he had endeavoured to instruct in the right use of the alpenstock.

Forbes wrote that he was 'invaluable to any man with ever so slight a tincture of science,' and that his death 'was like a family bereavement.' Few men have 'better understood the real dignity of their calling, or occupied a more unique position in Alpine story.'

MICHEL CROZ was another of the pioneers. He came from the village of Tour, and affected to look down upon the Chamoniards. For six years, until his untimely death on the Matterhorn, he was in the very front rank. He made the first ascent of the Pic des Écrins with Mr. Whymper and Mr. Moore, and numerous expeditions of the highest order with the former and Mr. Reilly. Croz seemed never happy below the snow-line, but above it, and particularly in all cases of difficulty and danger, he was unsurpassed, for he had extraordinary physical powers and great geographical knowledge. Probably few guides have ever done so much good work in so short a time. He was known only to a few climbers, but all who knew him trusted and respected him.

Chamonix has never been fruitful in great guides. The absurd system by which a traveller is obliged to take the man whose name comes next on the roll, fetters the guide of genuine merit. How could it be otherwise? It can only be with feelings of irritation and disgust that a guide who is fit for the Dru or the Blaitière, is condemned to take his turn in conducting parties of young ladies and gentlemen to the Flegère or the Brévent. Such rules do not exist in other districts of the great playground. Mr. Cunningham, in his 'Pioneers of the Alps,' records that a certain Englishman in patent leather boots meeting Emile Rey at the Montanvert, and being in want of a guide, pointed to the Mer de Glace, uttering only the word 'Combiang?' 'Voilà les guides pour la Mer de Glace,' replied Rey, pointing to certain bad specimens of the Société des Guides; 'moi, je suis pour la grande montagne.' And yet there are some Chamonix guides whose names are household words amongst climbers, whose services are always appreciated, and whose memory in the time to come will always be held in honour; such are Edouard Cupelin, Michel and Alphonse Payot, and François Devouassoud.

One of the Old Guard

From Zermatt no really great guide has ever sprung. The Taugwalders were fair men for Monte Rosa and the various passes of the Pennine Alps; but neither they, nor 'funny Peter Perren,' nor 'Weisshorn Biener,' could ever be placed in the first rank. Good men, however, were reared in the immediate neighbourhood, at St. Niklaus, for instance, and at Saas; and Pollinger, Knubel, Franz Andermatten, Joseph Imboden, and Ferdinand Imseng have deserved well of modern mountaineers. If, however, we cross the range and pay a visit to Val Tournanche, there are to be found real masters of the art. MAQUIGNAZ, one of the greatest authorities on the south side of the Matterhorn, was cool, brave and resolute, and made many of the early ascents from Breuil. He was often in the service of Professor Tyndall, and was devoted to the eminent family of Sella, with some of whom he made the first conquest of the Aiguille du Géant. He, however, is no longer amongst us. Starting in August 1890 for an expedition on the western side of Mont Blanc with an Italian and another guide, named Antonio Castagneri, none of the party have ever been seen or heard of since. He was a typical Italian guide, and has left no survivor approaching him in merit in all his native valley. Within a week of his death another notable guide, who lived within a stone's-throw of him, died in the faithful discharge of his duty. There is no one who has frequented the Alps for the last thirty years but has seen or known JEAN-ANTOINE CARREL. With less charm of manner than Maquignaz, he greatly surpassed him in dogged and determined resolution. He was one of the most expert cragsmen in the Alps, and made many of the early attempts on the Matterhorn with Mr. Whymper, and might, but for some strange perversity, have shared in Mr. Whymper's triumph. He did in fact make the second ascent of that mountain on his own account, and the first from the Italian side. He can scarcely perhaps be called a pioneer, as most of the principal peaks had been conquered before he became known to fame. He was not only remarkable for his skill but for his tenacity and courage. Carrel led Mr. Morshead and myself to the summit of the great mountain twenty-one years ago, notwithstanding a severe snowstorm, and although Melchior Anderegg justly protested against the folly of the expedition. He accompanied Mr. Whymper to the Andes, and with him twice ascended Chimborazo, slept on the top of Cotopaxi, and vanquished many of the giant mountains of Ecuador. If he lacked the grace or the politeness of some of his colleagues, he was trustworthy beyond them all. He started on his last expedition in August 1890, and reached the hut on the south side of the Matterhorn on a cloudless day. A violent hurricane arose in the night, and continued without intermission for more than thirty-six hours; provisions ran short, and retreat was inevitable. The hut was left at nine in the morning, and, suffering greatly from starvation and frostbite, the party were engaged in a fierce battle with the elements for more than fourteen hours. Carrel, on whom the lion's share of the work fell, had only just succeeded in bringing his party to a place of comparative safety when he sat down and expired. Our annals contain no more conspicuous example of genuine fidelity and faithful devotion. English mountaineers never forget such services, and this was one of the many cases where they made ample provision for the widow and children of the man who had fallen, not without glory, in the front line of the battle.

The village of Courmayeur has also been the home of guides of great eminence. Of these LAURENT LANIER, who died a few years ago, was one of the most famous. A chamois hunter by profession, he was excellent in icemanship and brilliant on rocks, and many leading mountaineers had the benefit of his services. He was passionately fond of the chase, and the sight of a chamois on a mountain expedition filled him with a wild delight. I remember being on the Grandes Jorasses with him more than twenty years ago, and can still recall his intense excitement on finding four of these animals in a very perilous position, unable as he thought either to get up or down, within a hundred yards of our camping place. But the best of all the Courmayeur men is the still young and active EMILE REY. If the mountaineer, like the poet, is 'born not made,' then Rey is a born mountaineer, for few men ever had so natural an aptitude for a difficult profession. He is courageous without a trace of foolhardiness, temperate, honourable and refined. Those who have been fortunate enough to secure his co-operation may count upon success, and when at most two or three of his seniors take their inevitable place on the retired list, there is little doubt that he will be the first guide in the Alps.

The greatest guides of all have, however, been bred in the Oberland. The air of Grindelwald, Meiringen, or Lauterbrunnen, seems to favour the growth of the highest standard in mountaineering. The gallant BENNEN was a splendid example. He first ascended the Weisshorn with Professor Tyndall, and in his short career saved several lives by his capacity and his courage. He perished in an avalanche on the Haut de Cry at the early age of 40 years.

ANDREAS MAURER too, who was killed on the Wetterhorn, and JAKOB ANDEREGG, who died peacefully in his bed, were men whose memory will not easily fade. Of living Oberland guides fortunate is the traveller who can command the services of Kaufmann, Von Bergen, the two Baumanns, Ulrich Almer, or Johann Jaun. The oldest of the living Oberlanders who took to guiding as a profession is ULRICH LAUENER. He was one of three celebrated brothers, of whom one—Johann—was killed while chamois hunting on the rocks of the Roththal Sattel. The surviving brothers, ULRICH and CHRISTIAN, were guides of the very old times, and although both still climb a little, they are practically on the retired list; both are men of splendid physique, and were superb guides in their day; handsome, picturesque, manly, and brave, but not wholly devoid of swagger, they rendered conspicuous service to many an early mountaineer. But both, Ulrich especially, were lacking in the 'suaviter in modo.' Old Ulrich is over seventy now, and his brother but little younger, so probably in both cases their mountaineering days are over.

Of all the great Oberlanders whose names are held in honour amongst us, none have ever quite attained the reputation achieved by CHRISTIAN ALMER and MELCHIOR ANDEREGG. Nearly of the same age, equal in courage, judgment, and knowledge of their craft, they have been not only rivals without a tinge of jealousy, but have always entertained the highest regard for each other. The very thought of these two Bayards of mountaineering refreshes one like the cool air from a glacier, and recalls countless reminiscences of happy times. Almer began life as a cheesemaker, and afterwards became a soldier and a chamois hunter. He was with the party on the first ascent of the Wetterhorn in 1854, and for more than thirty years was engaged with earlier or later explorers in the various conquests of the great Alps. Possessing an equable temper and an inflexible resolution, his failures were few and far between, whilst his successes were probably greater than those which fell to the lot of any one of his colleagues. With him an almost unequalled Alpine experience was joined to the very soul of honour. Quite five and twenty years ago I had the good fortune to enjoy the benefit of his services. We crossed the ridge of the Lyskamm in a storm, made the first passage over the top of the Jägerhorn from Macugnaga to the Riffel, and the first ascent of the Lyskamm from the side of Gressoney. Would any of the younger guides seek to secure the reputation, or to rival the deeds of their fathers, let them take Christian Almer for an exemplar ! Unhappily he was severely frostbitten in a winter ascent of the Jungfrau in 1885, and suffered the loss of all the toes of one foot, but his courage and determination have enabled him to triumph over even this misfortune, and last year he was still climbing, though more than sixty-five years of age, with all his early skill and much of his former power.

What can I say of MELCHIOR ANDEREGG of Meiringen except that he is Melchior Anderegg? Unequalled as a pathfinder, he has no superior either upon rocks or snow. He is prudence itself, but it is the prudence which is born of knowledge. He has always had the courage to turn back when he was convinced that discretion was the better part of valour, and no amateur, however good, ever had the temerity to question his decision. Everyone knows the memorable answer

THE RECOLLECTIONS OF A MOUNTAINEER 371

he once gave to a first-rate climber who wanted him to take a course which the great guide deemed to be imprudent. 'Es geht, Melchior,' said the climber. 'Ja,' was the immediate answer, 'es geht; aber ich gehe nicht.' He has always been emphatically a safe man. No accidents worth recording, or which human foresight could prevent, ever happened under his leadership. Brilliant in all times of difficulty, morally unconquerable in all times of danger, incapable of carelessness, and insensible to fatigue, he is not only admired but almost worshipped by travellers and by guides. Even the Chamonix men, who last year distinguished themselves by the endeavour to levy a tax upon foreign guides who were mountaineering in their district, dare not say a word to Melchior. How difficult the task is to set down plain facts about such a man and yet to avoid the charge of exaggeration, can be known only to those who, like myself, have had the priceless advantage of his friendship and his companionship for more than thirty years. I could express no better wish for the dearest friend of mine than that when his turn comes to climb the steep hills of life, he may have such a companion to share the perils of the journey, or such a leader to point out the way.

'Rude donatus'

But it is not as a great mountaineering leader only that the memory of Melchior will be held in honour. No guide was ever more courteous, more gentlemanly, more sweet-tempered, or more considerate for the feelings of others. If he has a true man's courage, he has a woman's tenderness. It is impossible for me to forget how on one occasion, with a guide attached to our party severely wounded by a falling stone on the Dent d'Erin, we arrived after many hours of arduous labour on the top of the Col de Valpelline at ten o'clock at night, and had no alternative but to sit out in the open and wait for dawn. With what anxious care he ministered to the sufferings of the wounded man, and how he took off his own coat that his friend might be the better protected, whilst he himself braved the night cold in his shirt-sleeves, will be ever fresh in my memory. Fortunately he is still amongst us, though over sixty-four years of age. It is inexpressibly sad even to think of the time when the eye will become dim and the natural force abated, but living or dead he will have an abiding place in the hearts of many men, who owe him a debt of gratitude to repay which would be impossible.

Such are some of the great professionals who have assisted in creating the noble sport which these pages describe, but the list might be indefinitely extended.

As mountaineering became more and more popular, it became obvious that it was a sport which must necessarily be attended by certain possibilities of danger. Men differed greatly as to what the danger really was, but not as to the various means by which it could be avoided. The older climbers beyond all doubt over-estimated the real dangers of Alpine climbing. Glaciers and séracs, crevasses and cornices were imperfectly understood, and the precautions taken to avoid accidents were excessive. Onlookers regarded the pursuit as a certain proof of lunacy; and climbers themselves exaggerated the dangers which they had possibly encountered, and over-coloured the difficulties through which they had successfully passed. The error was a venial one. They had the dread as well as the charm of the unknown before them. Sir Charles Fellowes ascended Mont Blanc so far back as 1827; there was no special difficulty in that particular expedition, and yet he stated his opinion 'that if it ever fall to my lot to dissuade a friend from attempting what we have gone through, I shall consider that I have saved his life.' Untrained men, youths fresh from school, even ladies now climb, or are hauled up, the Alps. They under-estimate the difficulties they meet with just as Sir Charles Fellowes overestimated them; and their mistake is unquestionably the greater of the two.

In the ten years closing with 1860 no serious accident occurred on the Alps, with one unimportant exception. The founders of the craft did not attempt the gymnastic feats which are common in these days, but their work was sufficiently arduous and they climbed under exceptional disadvantages. The first lesson they learnt was to take every conceivable precaution against danger. Some say that they were over-cautious, but the fact remains that the early history of mountaineering is not stained by a single fatality worthy of record. Then the pendulum swung back. It had been thought that knowledge, experience, and training were indispensable preliminaries to a severe Alpine expedition. But men rushed out from England desirous of repeating in a hurry the feats that others had undertaken with a serious sense of responsibility. They refused to serve any apprenticeship. They believed 'that by a little happy audacity, and the expenditure of enough money, they could leap over all preparatory stages.' The result was that the mountains made an emphatic protest against the ignorance and incompetence of the tyro, and enforced a needful warning by a series of terrible disasters. Then men began to understand the meaning of 'the mountain gloom'; and were taught to treat a great peak with the respect which it deserves. As a matter of fact, since the season of 1860 about one hundred and fifty persons have lost their lives in strictly Alpine accidents, that is, in accidents that have happened to climbers at work. Familiarity has bred contempt, even ordinary precautions have not been observed. I fear that these accidents are on the increase, and am sure that if any considerable portion of the increased number of tourists who now visit the Alps takes to mountaineering, many further calamities will occur. It is not surprising that the outside world should have often inquired whether the dangers of mountaineering were not too great to justify a person of reasonable prudence in undertaking it. All really noble pursuits contain some element of danger, or they would lose some part of their charm; but apparently no one proposes to discontinue them on that account. Almost every fatal accident which has yet occurred in the mountains can be traced distinctly to ignorance, rashness, carelessness, or the culpable neglect of some well-known precaution.

The main rules for the guidance of mountaineers have been thus admirably summarised by Mr. Leslie Stephen:—' Any mountain may be climbed by trained and practised mountaineers with perfect safety, provided that fine weather, good guides, and favourable conditions of rocks or snow have been first secured. There is no mountain in the Alps which may not be excessively dangerous if the weather is bad, the guides incompetent, the climbers inexperienced, the conditions of rock or snow unfavourable.' Grave difficulties may arise from sudden storms or falling stones, but all other accidents are preventible. If disasters occur from neglect of the rope or its improper use, from the folly of climbing in bad weather, from the incompetence or inexperience of climbers or of guides, or from the neglect of obvious precautions, then the sport of mountaineering must not be discredited; but the foolish persons who refuse to be bound by the ordinary canons of prudence and good sense must take the blame upon their own shoulders. The greatest dangers are always incurred by those who ignorantly or wilfully engage in expeditions which are beyond their strength, and for which they have not deemed it necessary to make adequate preparation: the evils which result from overtaxing the physical powers are quite as great as those which arise from mountaineering accidents. Novices must learn that, apart altogether from physical gifts, serious training and considerable experience are absolutely necessary to enable them to climb for sixteen or eighteen hours without overtaxing their powers. Confidence is excellent when not misplaced, want of care is the root of all evil. The wise and prudent climber will scarcely deem any precaution excessive :

> The little more and how much it is,
> And the little less, and what worlds away.

What is it that makes mountaineering the noblest pastime in the world? Of course it is too much to expect that everyone should share this view, but each one of us can ride his own hobby-horse without wanting anybody else to get up behind. I should be the last to undervalue any of the great sports of which Englishmen are justly proud. I can understand the delight of a severely contested game of tennis or racquets, or the fascination of a hard-fought cricket match under fair summer skies. Football justly claims many votaries, and yachting has been extolled on the ground (amongst others) that it gives the maximum of appetite with the minimum of exertion. I can appreciate the charm of a straight ride across country on a good horse, and I know how the pulse beats when the University boats shoot under Barnes Bridge with their bows dead level, to the music of a roaring crowd ; and yet there is no sport like mountaineering. The reasons are not far to seek. In the first place, we are out for a long holiday and are escaping for a time from the strife and turmoil of life. We go out wearied with ten months of toil and labour, and begin to understand what freedom means when we catch our first glimpses of the Alps. After all, we are the products of an over-civilisation, and it is a protest to broil our own trout in some mountain hut, and consume them with a pocket knife on a sheet of the 'Journal de Genève.' Again, the friends we meet seem, in more than one sense, to be always on a higher level than when at home. The woman is in a holiday humour, and the man is at his unpremeditated best. The troubles of life seem to fall away in the presence of the everlasting hills, like the burden of sin which fell from

Christian's back in the famous allegory. We get far from the madding crowd

> And the cares that infest the day
> Fold their tents like the Arabs,
> And as silently steal away.

Everything seems to be in our favour. We make our daily expeditions in the midst of the most glorious natural beauty. No cruelty is connected with our craft. We breathe a diviner air. We watch the clear streams bounding out of the mountain sides and racing laughingly down to the green fields as if, as Mr. Ruskin says, 'the day were all too short for them to get down the hill.' We see the great forests wave and the great rivers roll. We are hushed to silence in the early morning by the awful beauty of the rose of dawn. We see the rising sun strike the snow peaks with a crimson flush ; we see the western horizon in the evening, one vast sea of fire. We hear the crash of the avalanche and the roar of the torrent, and yet ' Beyond these voices there is peace.' Of all the gifts in the purse of Fortunatus there is none comparable to the maximum capacity for enjoyment, and we start with this invaluable possession. We spend long days in the open air, and nights in that kind of sleep which 'knits up the ravelled sleeve of care.' The rarer air of the mountain quickens the blood and fills us with a strange vitality. Nothing is ordinary or commonplace. We walk in the early summer through a very paradise of bloom, with a vague regret we crush beneath our feet the lily and the crocus, or we greet with delight the splendid blue of the gentian, fairest of Alpine flowers, as it peeps coyly through the melting snows. We find acres of lovely ranunculus by the mountain tarns, we are knee-deep in harebells, auriculas, and forget-me-nots. We rest by the side of some moraine, and watch the rarest ferns holding their revels at our feet as in some Midsummer Night's dream. Masses of pink rhododendron form a foreground of glorious colour, and above them rise the solemn pines, and then the fairy fastnesses of the great séracs slashed with such green and blue as no one can understand who has not seen, and above them the soaring Aiguille or the stately dôme, and, over all, the infinite blue.

Bad weather sometimes troubles us, but even this is not without its charms. There are few sights more wonderfully impressive than the lifting of the mists after rain, under the influence of the cool north wind, when the sunlight strikes through the rifted clouds and reveals some majestic pinnacle that has been long concealed, and which looks so lofty and inaccessible that even the idea of climbing it would seem to the uninitiated to be an empty dream.

There is scarcely an expedition of any magnitude which is not associated in our minds with some happy recollection. The night bivouac by the glacier under the stars, the moonlight amongst the snows, the jest and laughter by the camp fire, when old friends long parted recall the experience of bygone years. Some great peak is to-morrow's goal : shall we be successful or not? None of us can tell; and the uncertainty adds a zest to the expedition. We can rely upon ourselves it is true, for we are in thorough training. We are a strong party of four, let us say, two men and two guides : the right strength for any Alpine expedition. We can thoroughly trust our guides, for we have often tried them in sunshine and in storm : but shall we have fine weather? Is the snow yet melted from the higher rocks, or shall we find our mountain glazed with ice? Shall we have much stepcutting? Can we do it in the time? Such are the questions we ask each other, whilst the guides express themselves after the manner of the ancient oracles, but as a rule decline to prophesy until they know; and so we start, sometimes in confidence, sometimes in doubt, always in hope. We pit ourselves with our training and experience, and, above all, with our absolute reliance upon one another, against the unknown difficulties of the mountain. We carefully husband our resources, for who can tell whether we shall be out for twelve hours, or fifteen, or even twenty? Until the sun gets up we march almost in silence. The man to back is the man who is uncommunicative, if not actually sulky, between two and five in the morning. We watch with intense and never-failing interest the actions of the leading guide. We evade crevasses, we thread our way through séracs, we wrestle with the rocks, we pull each other up or let each other down, we clamber up great staircases hewn in the blue ice, we wade through the snows, we get nearer and nearer our summit, we reach the final ridge, we hear a shout of joy from the leader, and we are all standing upon our lofty peak—the happier if it has been hitherto unconquered. We sweep the horizon on every side, and greet the familiar faces of scores of mountain friends. We have reaped the natural reward of our labours, and have the satisfaction that success, after well-planned and arduous labour, must necessarily bring to every well-regulated mind. Another day in our calendar must ever be marked with a white stone.

And if we fail; if the cold is too severe for us, or ice difficulties cannot be surmounted, or rocks for the time are too dangerous, or the snow is too soft, or the weather too doubtful, we are after all only beaten for a time; and our failure has not been without profit, or even without pleasure. If success brings rapture, at least defeat is unattended by despondency. And with what keen pleasure do we return from the successful expedition. The very bells that hang round the necks of the cattle seem to discourse more excellent music as we reach the confines of the habitable world; the patient cows, which for generations have experienced no fear of man, seem to give a silent greeting, and the goats are anxious to bear us company.

We can never over-estimate what we owe to the Alps. We are indebted to them and to all their charming associations for the greatest of all blessings, friendship and health. Let anyone who has been in the great playground for twenty or thirty years try for a moment to erase from his memory the many friends whom but for the love of mountaineering he would never have known. What terrible gaps would be left in his history, what happy recollections would be effaced, what warm sunshine would be taken out of life! Such friendships, too, are always perpetual; sharing in common difficulties and common dangers must of necessity bind men closer to each other. The brightest recollections of youth or manhood form the most abiding solace of old age. But years form no barrier to the enjoyment of mountaineering. It has been conclusively proved that of all sports it is the one which can be protracted to the greatest age. It is in the mountains that our youth is renewed. Young, middle-aged, or old, we go out too often jaded and worn in mind and body, and we return invigorated, renewed, restored; fitted for the fresh labours and duties of life. On the outward journey we are not unlike the troubled and turbid river which flows into the great Geneva lake to find transfiguration by its steadfast blue; but when homeward bound we resemble what Byron calls

The blue rushing of the arrowy Rhone.

To know the great mountains wholly is impossible for any of us, but reverently to learn the lessons they can teach, and heartily to enjoy the happiness they can bring, is possible to us all. May succeeding generations find in them, as we have found, the great restorers of the waste of life, and may their lives be enriched, as ours have been, by the sunny memories which no mere money could ever have bought, and which no possible misfortune can ever take away.

Zermatterinne

CHAPTER XIII

SKETCHING FOR CLIMBERS

BY H. G. WILLINK

Stonecrop

SOMEBODY was once asked whether he could play on the violin, and replied that 'He didn't know, for he had never tried.' This man has always been regarded as a wag, or as a presumptuous ignoramus. But something could be said on behalf of anyone who should return a similar answer if asked whether he could make a sketch from nature which would be of practical use for mountaineering purposes.

The object of such a sketch is not to create a picture, nor even to preserve a note from which a picture may be made, but simply to record bare facts of topography: skilful execution is immaterial; colour is not required; composition is worse than useless. The end is achieved if an intelligent person, notably the sketcher himself, can, with or without the aid of a map, recognise the main or the essential features of the depicted view, when he has changed his position ; and especially when he has walked right into that view, and is surrounded by those features. In other words, it sometimes happens to climbers, and even to mere walkers, that a right line would be taken, and a wrong one avoided, if the party had a clear recollection of how the ground looked from another point of view. The merit of a sketch is in proportion to its success in supplying such a recollection. It is not every scene that admits of being drawn in this rough-and-ready way ; but the essential portions of many can be treated separately ; and the attempt is always worth making, if only as being well calculated to train the faculty of observation.

Broadly speaking, the use of a sketch begins where that of a map leaves off. A map may guide you to a selected ridge or a face of rock ; it will very seldom tell you, when you get there, which particular spot on the ridge is to be your goal, or which of several gullies in the face is the one you chose in yesterday's reconnoissance as the best line of ascent, or which fork of that gully to follow, or where you should quit it, or whereabouts a traverse can probably be made.

Again, a map may tell a man standing on a peak or pass which valley to go down in order to reach a certain place. It will rarely tell him so much about the difficulties in doing so as he will himself see from his elevated standpoint. As he gets lower down, the subordinate features rise up and hide each other, and he will often wish he could be back on the top for a moment just to see that view again, in order to know which line to take. The points upon which he will then desire information are those which he should have sketched when he had the chance ; and this is doubly true when there is no map at all.

A sketch may, however, supply just this kind of information without possessing one well-drawn line or any artistic merit whatsoever, except a certain degree of intelligibility, if that indeed be an artistic merit. The sketcher must bear in mind two great principles, or rather he must shun two great evils— ambiguity and superfluity. His sketch must show just what is wanted and no more, and there must be no doubt, at any rate in his own mind, what it does show.

Any means are admissible, the simpler the better; a word or two jotted down (such as 'ware stones') may often take the place of actual delineation and save time and trouble, and complicated, not to say undecipherable marks. Everyone can devise for himself some form of draughtsman's shorthand by which a few symbols will denote certain often recurring features or objects.

Perhaps the best way of making the present section useful will be to mention, one by one, the chief points to which attention should be paid; and conclude with an example. Several of such points really concern the sketcher rather as a mountaineer than as a draughtsman, and may be found elsewhere in this book; but it is not possible to make distinctions in this respect. Most of this chapter must be read side by side with the chapter on Reconnoitring.

The only materials wanted are a tolerably hard pencil and a book of smooth hard paper, with elastic bands to keep the leaves from flapping. An elastic band will, on a pinch, serve to rub out pencil, but it is better to carry india-rubber, a small slice of which can be tied (by a long enough string) to the book or pencil. 'Sectional paper,' sold ruled into squares, may perhaps be useful as supplying the eye with a standard of verticality and horizontality, and also with a ready means of comparing the relative size of objects as drawn.

1. Lest the circumstances under which the sketch is made should be forgotten, the following *memoranda* should always be written down, viz. the date; the sketcher's standpoint— the description being such that anyone could readily identify it on a good map; the approximate compass-bearing of the centre of the view; and if any indication of sunshine and

SKETCHING FOR CLIMBERS 383

shadow is to be given (as below suggested), the hour of the day also.

2. *The extent*, from top to bottom, and side to side, of ground to be included, should be selected. This may conveniently be done by holding up the book or paper, and noticing how much of the view it covers. If only some little bit of the view is to be sketched, this does not matter so much.

It is always more easy to draw objects their apparent size. The apparent size upon a piece of paper varies with the distance at which the paper is held from the eye. This is clear if instead of paper a piece of glass, or a mere frame, be held up. Seen through it objects will occupy a larger or smaller portion of its area, according as it is held farther from or nearer to the eye. The distance at which the paper will be held from the eye while the sketch is being made, is for the present purpose the distance which will regulate the apparent size of objects.

The field of view covered by the paper, when held up at this distance, gives the field which can be drawn of its actual apparent size. If the paper, when held at a convenient distance for drawing upon, will not cover enough to include all that is required, it must be held closer to the eye during work, *or* the sketcher must consciously reduce as he works—quite a possible operation, but not easy.

3. *The dimensions of the principal masses*, and their proportions and positions with regard to each other, should be plotted. It is best to begin with the biggest, and then to work by careful comparisons and measurements. Objects can be measured by holding up the pencil against them, taking great care that it is always held at right angles to a line passing from the eye to the object, and that it is held at one and the same distance from the eye for all such measurements. This can most easily be done by always holding it at arm's length. The root of eye-training is the comparison of objects and parts of objects, as regards their relative sizes and positions; but if the

pencil is held up in this way at the same distance from the eye that the paper was held up when the extent of the sketch was determined, these pencil measurements may for practical purposes be taken as absolute and not merely relative. The process is then the same, really, as that of making a 'same size copy' of another picture by measurement. This is an unholy and inartistic suggestion, but these lines will not be read by artists ; they scorn short cuts, as is well known.

4. *General outline.*—Having thus fixed the crucial points, the outline may without much difficulty be roughly drawn in, with written notes as to probable heights and all features which are not readily drawn.

(*a*) *Skyline.*—In the case of range behind range it is better to draw each separately. Care must be taken not to make the angles too steep, a fault which real live artistic artists have actually been known to commit. Here it is useful often to hold the paper up alongside the view, and thus to check the slopes, as drawn, against the reality—sometimes a humiliating process, and obviously but seldom practised, as the poor Matterhorn can testify any day.

(*b*) *Main masses of rock.*—In this the points to notice are crests and ridges, their direction and character; stratification and quality of rock ; shape and position of gullies, especially of their tops and bottoms ; landmarks at crucial points, e.g. 'big red rock here' (written notes will be found especially useful in these connections) ; lastly, the general planes of rock faces, as to which some words will be found below in dealing with perspective.

(*c*) *Borders of glaciers.*—Showing or noting their general course, the position and run of bergschrunds and crevasses, situation and nature of ice-falls, accessibility of glacier from banks, &c.

(*d*) *Approximate snow-line*, if required, or shape of important snow patches. Here it should be observed that in outlining a patch of snow (or any definite space) much trouble will be saved if the limiting line be invariably made to return to the point whence it started, so as to completely enclose the space. Such spaces can be very rapidly lettered ' R ' for 'rock,' or as the case may be. Rough shading or monochrome can be added at leisure. If these limiting lines are not thus made continuous there is no end to the confusion which may arise, as the writer knows only too well. It may sometimes be useful to note the extent of an area of shade at a certain hour. A dotted line or some such symbol may be used for this purpose. A small round circle in the sky at the proper place is a convenient way of denoting the position of the sun if required.

(e) *Special features.*—These may be either actually drawn, or merely explained in writing or by a letter referring to a marginal note. When once the outline is fairly right, the rest is easy. In choosing special features for record, it is well to select those which can be readily identified by map, or otherwise, from another standpoint, and especially from their own immediate neighbourhood. In the main the choice will probably be regulated by mountaineering requirements. The map names may be written on the sketch.

5. One of the principal difficulties in making a good sketch is one which is a matter as much of eye as of hand. It is generally no less difficult to appreciate the foreshortening or *perspective* of a bit of mountain than to draw it when appreciated. The key, when there is a key, to one difficulty, however, gives the key to the other—very often the length of base will give it. Sometimes it is to be found in the shapes of shadows, or in the course of moraines, or glaciers, or streams. Whatever it is that tells the truth, to that must the attention of the sketcher be directed.

386   MOUNTAINEERING

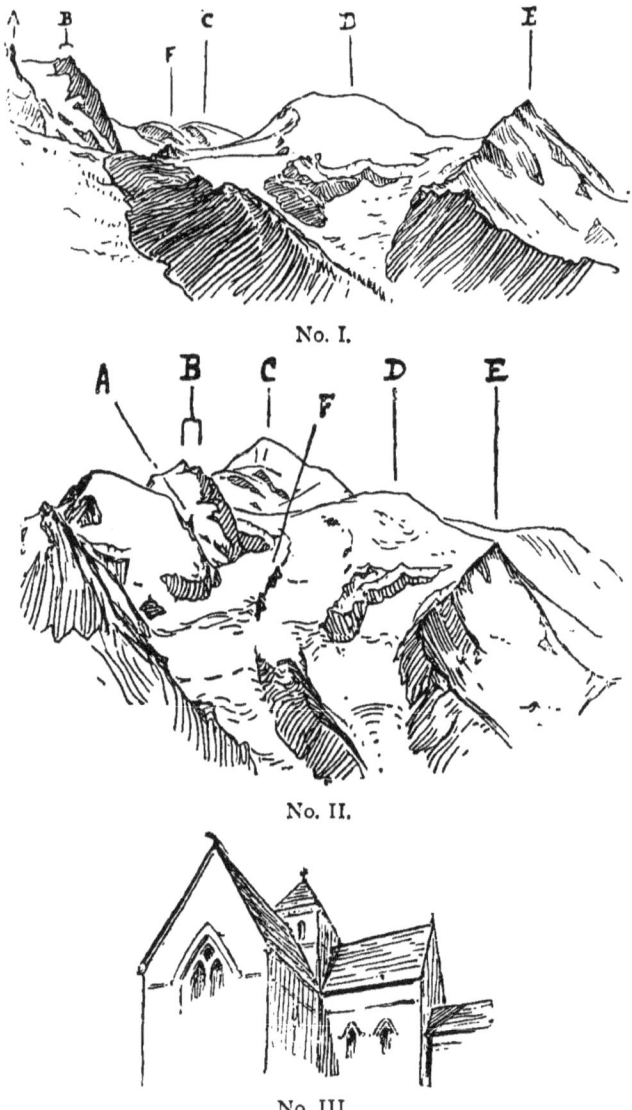

No. I.

No. II.

No. III.

A very familiar instance of perspective illusion is shown in figs. I., II., and III.   No. I. is an outline of the principal part of the Mont Blanc mass as seen from Chamonix, about 3,445 feet. No. II. gives the same features as seen from the Buet, 10,197 feet, the compass bearing being about the same, but the distance and elevation of the standpoint being much greater. The reader must not feel insulted by being offered the assistance of No. III., a simple rough piece of architectural perspective, to explain how it is that under some circumstances even valleys may appear to be exalted.

Lastly, a small camera obscura, by which the actual view (reduced) might be projected upon a piece of tracing-paper and then and there outlined, would be a useful addition to a mountaineering outfit. There could be little difficulty in constructing such an instrument, on the lines of the little photographic view-finders, to fold into very small space.[1] The suggestion may be taken for what it is worth. If feasible, anyone would be able to make a fairly accurate rough sketch in a very few moments.

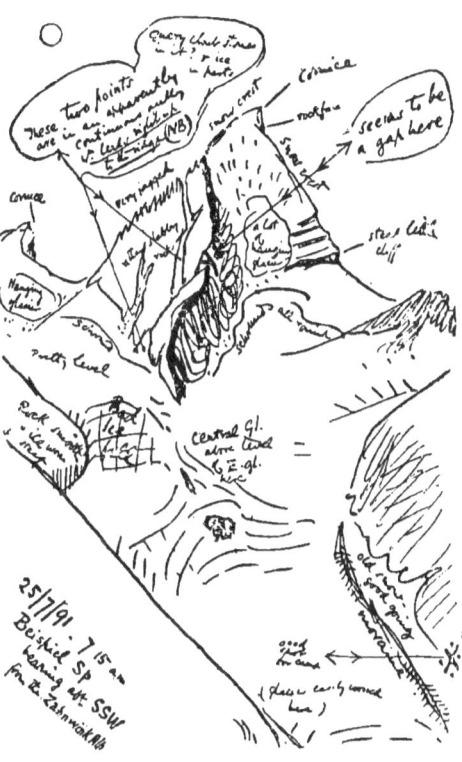

The foregoing remarks may now be illustrated by an example.

The above illustration represents such a rough sketch as

[1] Messrs. Ross & Co. make an instrument well adapted to the purpose.

might be supposed to have been made of the Beispielspitz (see p. 135) by a reconnoitrer unskilled in draughtsmanship. It sufficiently suggests some of the methods which have been advocated in this chapter, and shows how a very poor work of art may nevertheless record valuable information in a graphic form.

The writer cannot conclude without expressing his strong opinion that serious landscape painting is not nearly so much practised at high levels as it should be. Sportsmen may object that such considerations are altogether outside the scope of the Badminton Series. Mountaineering, however, as has been urged by other contributors to this volume, is to a greater extent than most sports bound up inextricably with pursuits and studies of a graver kind. If the gymnastic and even the holiday element were altogether eliminated, there would yet remain—there would perhaps even more clearly appear—in mountaineering the essential features which most entitle it to the respect and devotion of sensible men. And among these features must ever be found the opportunities for observation of some of the grandest of nature's scenery.

Besides, if the camera may speak, why not the paint-box? It cannot indeed be denied that photography is more within the reach of most persons as a successful operation than painting can be said to be. But both alike aim at preserving some form of representation of the mountain world, and only bear incidentally upon climbing pure and simple. And apart from this matter of utilitarian sketch-making, which has been above discussed and in which the draughtsman must always have certain advantages over the photographer, inasmuch as his work is immediately available as soon as the sketch is finished, and as he need never be in the ignominious position of a man who sees just what he wants to perpetuate but who cannot do so for lack of a plate, or a film, or a screw, or a cap, or some fiddling little mechanical contrivance—not to speak of the vexations of leaks in cameras, defective plates, 'doubled' exposures, misadventures to exposed plates by meddlers or otherwise, *et hoc genus omne*—apart from all this, and above it, is the matter of the paintability of high mountain scenery.[1]

Artists, except a few like M. Loppé (all honour to him), are wont to allege that such scenery is not adapted to picture-making. Possibly they fear to tread the untried region, because Angelico and the old masters did not rush in to the painting of glaciers—though Angelico soared high enough in his own way. If these cavillers had really tried, it is scarcely possible to believe that they would say such things. No doubt there is a typical 'fine-weather day,' on which snow mountains do almost bruise the eye by their hardness, when distance, 'middle distance' (save the mark !), and foreground are all as flat and dull as bad stage scenery, and when all sense of the

[1] Photography can, however, give us much. The process, if it has done nothing else, has made non-climbers familiar with many phenomena which have never been painted ; and can and will do more as photographers become better versed in the art of picture-making, though it is not likely that anyone will do better work than Mr. Donkin's. Certain points may be mentioned which are perhaps worth attention; the disregard of some of them has spoiled many an otherwise good photograph.

1. The principal feature in a view should not be in the plumb centre of the picture ; and its importance may often be enhanced by such a choice of position that the leading lines conduct the eye to it.

2. The values both of light and shade can be heightened by contrast. The device of bringing the highest light into proximity with the deepest shade may often be observed in effective compositions.

3. Cloud shadows may sometimes be utilised with great success to increase or reduce the prominence of given objects.

4. Cloud shadows are useful, moreover, for the purpose of indicating surface ; as also in foregrounds where there are any marks or things which suggest perspective ; e.g. footmarks, a path, a length of rope, cast shadows, a line of cornice, moraine, receding ridges, &c.

5. Mountain views often need something to give an idea of scale and distance. A succession of ranges or spurs will do a good deal in this way, or trees, chalets, cattle, or figures.

6. Although it is permissible and even right that dominant forms should be echoed by others for the sake of emphasis, yet actual parallelism of outline is always to be avoided; and, still more, awkward coincidences between edges and masses.

7. There is no end to the variety of interesting figure and cattle subjects for photography, as well as detail of all kinds in the High Alps. And the art as applied to the actual climber is still in its infancy. The sky too can never be hackneyed.

little colour that exists is scorched out of the retina by two seconds' gaze at snow or white paper. But such days are not unknown in even the lower parts of the world, though perhaps in a less acute degree. Mountain landscape is not to be judged by them alone. Men like Mantegna, Burne Jones, or Leighton (to take a curious group) have lavished all the resources of their genius upon composition of drapery. But who has bestowed one-tenth of the labour upon that subtle harmony of contour and line which abounds where snow, rock, and ice combine in the expression of the great forces of nature? Too many a woolly glacier and fluffy rock answer, 'Who, indeed?' Nor is this all. The charm of mountain scenery lies not only in statuesquely modelled surfaces, inexpressibly graceful though they be. Strength and tenderness of colour are there in inexhaustible profusion. Lowland foregrounds need, perhaps, fear no rival; but the limitless expanses of mountain distance are beyond compare. The mountain cloud-world has found no painter prophet. The mountain fairy-land is yet unlimned.

No man who has not taken the mountains for his friends and gone among them in all weathers and in all seasons, can tell, nor truly can he, what infinite variety and what exquisite beauty are to be found in these neglected regions of the sky and hills. It is not suggested that painters should, with the object of finding subjects, undertake difficult expeditions, the fascination of which consists, for many persons, chiefly in the difficulties to be overcome. Certainly they would find it impossible to climb seriously and to paint seriously on the same day. But they may reasonably be recommended to pay some attention to the craft of mountaineering as a means to an end. The experience thus gained would be of value in more ways than one. Not only would a man increase his knowledge of mountain anatomy and become familiar with a new class of noble scenery, but he would, by healthful and pleasurable exercise, acquire the capability of subsequently visiting at his leisure, without risk of undue fatigue, districts and points of view affording scope for his best talents.

SKETCHING FOR CLIMBERS 391

Mountain wandering is not to be confounded with mountain climbing pure and simple. The former is enough for an artist in search of work, and indeed amply satisfies many people who are skilled mountaineers; but it has dangers of its own, and is not a safe pursuit for any man who has not been well grounded in the practice of those principles which this book is designed to set forth.

On the Zermatt wall

CHAPTER XIV

CAMPING

By C. T. Dent

Travellers' Resting-places

EVEN in the little frequented portions of the Alps it is seldom necessary nowadays to camp out in the proper sense of the term. The huts that have been built by the foreign Alpine clubs, or provided by the enterprise of innkeepers, are so numerous and so well placed that some one of them can be utilised for almost any expedition. Hotels are now found at conveniently high levels. Moreover, the details of the various recognised expeditions are so minutely known that much time is saved, and a vast number of ascents can now be made in a single day for which it was formerly the custom to sleep out. Mont Blanc has been crossed within twenty-four hours, from Courmayeur to Chamonix; and the Matterhorn has been ascended and crossed in a single continuous walk. These, however, are rather feats of athleticism. Not many will care to undertake such fatiguing expeditions, during a great part of which the travellers are more like somnambulists than mountaineers. Nor is there any real advantage to be gained. It is undoubtedly better in every way, when possible, to sleep out in the huts provided. These huts are so well appointed that the traveller has nothing to take up beyond provisions and firewood; and he can make sure, unless indeed the place is crowded, of passing a comfortable night. It is well during the climbing season to ascertain beforehand the number of tenants a hut is likely to have. English mountaineers have been reproached with using these huts largely while neglecting to contribute to their maintenance, and to some extent the reproach is deserved. The up-keep of these places, added to the initial cost of building, which is often very considerable, falls rather heavily on the funds of the clubs. If English travellers were to make a practice of sending a contribution of, say, five francs, every time they use these shelters to the central committee of the club responsible for them, the money would be well laid out and the stigma would cease to exist. In some huts a caretaker resides during the summer, and a small payment is exacted. In Austria this is the usual custom.

In the meantime the majority of the club huts are free to all, and it is incumbent on every traveller to respect and observe rigidly the rules laid down with regard to them. These regulations are practically the same everywhere, and it will suffice here to quote one set drafted by the Swiss Alpine Club for the hut on the Dom :—

This hut is placed under the supervision of * *. The hut is always open, and is placed gratuitously at the disposal of tourists and guides, and entrusted to their care and protection.

Tourists and guides are particularly requested :—

1. Not to damage the hut in any way, and not to mess or dirty either the interior or the immediate vicinity.

2. To be careful in using the furniture and utensils, to wash the kitchen utensils after using, dry them, and replace them dry in their proper places.

3. To put the beds in order before leaving, to shake up the straw, hang up the blankets, put out the fire, and to shut up doors, windows and shutters.

4. It is absolutely forbidden to damage the hut or its furniture,

or to use any part as firewood. If tourists are compelled in emergency to take away such articles as spare rope or lantern, they are held bound to restore them to the hut with as little delay as possible.

5. Guides and tourists are held responsible for any damage done to the hut or its furniture.

6. Travellers must provide themselves with firewood.

7. Travellers are requested to write their names in the book provided for the purpose.

It is as well to enter also the names of the guides, and especially of porters, who may be taken only up to the hut and sent back the next day.

Travellers should make it their duty to see that the hut is left at least in as good and tidy a condition as it is found. An imperfectly closed door or shutter, or a broken window, may lead to the hut being filled with snow if bad weather should come on, and the next travellers may find the place uninhabitable. Before now, parties caught in these shelters and kept for two or three days by stress of weather, have been forced to use shutters or other parts of the hut for fuel. Such an occurrence should at the earliest opportunity be reported to the principal hotelkeeper in the valley below, and also to the central authority, giving an undertaking to defray all the expenses incidental to repairs. For those who are not afraid of solitude there is a great charm to be found in a two or three days' stay at many of these huts, most of which command magnificent views ; the photographer who has an eye for passing effects can hardly employ his time better, and can often obtain in twenty-four hours a series of pictures which a man who feels bound to descend to the valley every night may fail to collect in a whole season. But the time will hang heavy on the traveller who has not some definite scientific or artistic aim, and finds no delights in short wanderings at high levels.

A traveller, however, who uses only huts has never tasted the real pleasure of a camp, and will find himself at a great loss when mountaineering in districts in which every want of the tourist is less amply cared for than in Switzerland. A party

knowing little of the ground which they propose to traverse, and intending to camp out, may reckon, if they start from the valley, that a walk of some four or five hours will be necessary, and that in the months of July or August they should have selected their place for a bivouac not later than four o'clock in the afternoon. On slopes having a south aspect the tree level in Switzerland may be estimated at 6,000 to 8,000 feet, and on those with a north aspect about 1,000 feet lower. Supposing that the valley lies at a height of about 3,000 feet, and that the summit of the mountain is 14,000 feet, an endeavour should be made to find a suitable place at from 8,000 to 9,000 feet. Water is of course a necessity, but in the Alps springs are so abundant that there is never the slightest difficulty on this score, even if there are no snow patches near. Water when trickling down a flat steep surface of rock is not easy to collect. If a piece of stout paper or card is folded like a half-opened sheet of writing paper and the edges of one end held against the rock, the water will flow out down the trough.

If a traveller proposes to collect firewood on his way, he should allow at least an hour for doing so, and make sure that plenty is taken. Guides are parsimonious in collecting wood which they have to carry up, but spendthrift in its use. On a fine summer night, a party provided with sleeping bags or even a rug or blanket apiece, can sleep out in the open comfortably enough if not much above the tree level. But it is easy and better to have some sort of shelter. An overhanging rock will provide this well enough. In the Caucasus the natives cut down large branches and stick them upright in the ground closely together in a row; a fairly efficient shelter from the wind is thus provided. The branches of the cone-bearing trees however, such as are alone found at high levels, are not so well adapted for the purpose.

The earlier the party arrive at the camping ground the better; they have leisure to search for the best place, and time to make the camp fairly comfortable. Men will work astonishingly hard when they are taking measures to provide for their own comfort. When a tent is taken, everything is of course much simpler. A space of eight or ten feet square has to be thoroughly cleared, all imbedded stones being turned out. A very little projection under the canvas floor will destroy a night's rest, and such annoyances are not easily removed when the tent is once pitched and the baggage stored inside. A floor that is level save for a few slight concavities will answer, but a convex surface will provide exquisite discomfort. Artful people, before spreading the tent, scoop out in the ground small hollows to accommodate the hips when sleeping. Some nice calculation is necessary. The door should be directed away from the prevailing wind, and the opposite end should if possible be at a slightly higher level. If there is any intention of passing more than one night in the same place, it is wise to cut a trench six inches deep round three sides of the tent to allow the water to run off. However fine and promising the weather, the tent should be pitched as if a deluge were expected. See that the pegs are securely driven home; nothing is more annoying in the event of rain coming on than to find that shrinkage of the guy-ropes has dragged the pegs loose, and that the whole structure is in danger of falling in. The ropes are best slackened a little the last thing before turning in. Dew may cause shrinking. The 'dáterâm' principle of securing the ropes is seldom applicable in the mountains.

It should be one man's duty at once to unpack and arrange the sleeping things. All the articles that may be required for the next day's climbing, such as ropes, axes, gloves, gaiters, spectacles, and the like, should be arranged in a prominent place, and the travellers should see that the guides do not adopt a curious habit, so common to them, of hiding away these small articles in sequestered places, the situation of which they at once entirely forget. No one should be allowed to walk on the floor of the tent with his mountain boots on; the canvas bag in which the tent is packed will accommodate the boots, and the bag can be laid on the threshold and will serve as a door mat.

CAMPING

Not less important than the bedroom is the kitchen. A three-sided fireplace, constructed in a suitable spot, where the smoke will not blow into the tent, may be made of flat stones built up to a height of say twenty-four inches, the open space being to windward; a large upright flat stone or side of a rock forms the back of the stove. A stick can be laid across the top, and the cooking kettle slung on this. Stones retain heat for a great length of time, and are very comfortable companions inside the tent on a cold night, but the fireplace of course must not be destroyed in order that the travellers may be kept warm; a few extra stones can be placed near the fire. If there is any hot water to spare, a luxurious 'boule' may be made by filling a wine bottle. Take care that the firewood is at once put under shelter in a dry place.

Mr. Galton, in 'The Art of Travel,' describes an excellent fireplace which the natives make in India, and called a Chulha:—

They excavate a shallow, saucerlike hole in the ground, a foot or eighteen inches in diameter, and knead the soil so excavated into a circular wall with a doorway in the windward side; the upper surface is curved so as to leave three pointed turrets upon which the cooking vessel rests. Thus the wind enters at the doorway, and the flames issue through the curved depressions at the top and lick round the cooking vessel placed above.

The wall can best be built of stones. It is no easy matter to light a fire, or to feed it when once alight with the proper economy of fuel, and the traveller will do well to practise himself a little in the art. Mr. Galton (*op. cit.*) gives some admirable hints on this subject.

A good deal of mutual forbearance is demanded in tent life. Bad weather under such conditions in the mountains is doubly trying. The closer friends are packed together the more distant do they sometimes become. The most equable of companions will resent the man who is perpetually turning everything upside down in order to discover his pocket-knife or match-box. If you wish to test the possibility of travelling harmoniously with chosen companions, spend two or three nights (wet, for choice) with them under canvas. If you agree, then you will probably get on well together throughout. Each man should have his own place, and share of space, definitely assigned to him. The fellow-traveller who is selfish in the tent is quite as endurable as the man who is always ostentatiously making small sacrifices for the benefit of others. Englishmen, happily, have a method of accepting the verdict of a coin spun in the air as final and irrevocable. 'Tossing up' saves a multitude of small disputes. But let the man who wins the toss stick systematically to the advantage his luck has gained for him, and not give hostages to good fortune. So will harmony prevail, and the pleasures of life under canvas be enjoyed to the full.

If an overhanging rock is selected for the shelter, its direction should be observed with respect to the wind. Remember that a draught will prevent sleep much more effectively than mere cold. A rough semicircular wall two feet high can be quickly built up so as to give good shelter, the party sleeping with their heads close to it and their feet towards the fire. The interstices of the wall can be filled up with turf or moss or crumpled-up newspaper, and it is often well, if there is a spare blanket, to hang it over the wall and so keep out the draught. At the worst the ice-axes can be stuck into the ground and a blanket fastened to the picks, stones being placed below on the blanket to keep it close to the ground. Mackintosh coats answer admirably for this purpose, and so does the ever-useful shepherd's plaid folded lengthwise. Sheets of coarse paper— even newspaper will answer—stop draughts most effectively, but will not of course resist wet. Newspaper wrapped over the feet will keep them warm.

It is not easy to find a soft spot, as a rule, under an overhanging rock. If time allows, an excellent spring bed can be made by cutting down some of the dwarf rhododendron bushes which abound on the mountain-sides. The branches are spread in a thick layer, and on the top of them layers of dry turf are laid, and on top of these again the blankets. The traveller may remember that, the earth being a good conductor, he will lose more heat on the side on which he lies than on that exposed to the air. If there is not sufficient covering for the party, it will be warmer to lie on the blankets than to use them as a covering and lie on the ground.

The evening meal being over, the cooking things should be cleaned, the cooking pan filled with water, and all placed in readiness for the next morning, the provisions arranged, and the store for the next day packed up. Disputes as to the relative weights of the burdens are apt to arise among the porters at the hour of starting. Weigh all the sacks overnight, and apportion them. A small spring steelyard saves much difference of opinion on the part of the carriers. Loose shavings for lighting the fire in the morning should be got ready and placed in a dry place under shelter, the lantern hung up at the door of the tent, the boots overhauled and greased, and every preparation made, as for an immediate start, before the party turn in for the night.

A prudent traveller will carry with him, to the camp at least, a spare flannel shirt and an extra pair of thick socks. If the walk up has been a hot one, it is wise to change immediately on arrival at the camp, and to spread out the shirt and stockings that have been worn to dry. The garments, if dry, can be put on over the other clothes on turning in. On a cold night, and it must be remembered that cold nights commonly precede fine days, immense comfort will be found from a flannel cholera belt. To turn in for the night dry and warm is half the battle in camp, and will often insure a good night's rest. If any long dry grass is available the boots may be stuffed with it. Waking the party in the morning at the proper hour is generally the duty of the guide, and the most reliable man, or the one most given to snoring, should be told off to this duty. If engaged on an expedition of importance, the traveller may often keep awake half the night for fear of oversleeping in the morning. Old stagers drop off to sleep at once, for they know how to make themselves comfortable. But many find insomnia their portion in a hut or tent. Sleeplessness is less trying when there seems a sort of valid excuse for it. But the body does not really rest while the brain is active. Sleep can often be induced by any monotonous profitless exercise, such as counting backwards, just sufficient to divert the mind from more exciting thoughts. This is but self-hypnotising.

Few Swiss guides are experts in the details of a camp. One of the best that the writer can recollect was Andreas Maurer. He was attentive to every detail. Punctual to the moment the 'Herren' would be roused and stirred into activity by the announcement that breakfast would be ready in five minutes. Each man found his clothes neatly folded and arranged by his side, his boots greased and warmed, the provisions packed and a good fire blazing. As a rule, however, the Italian and French guides are far more handy, and many of them are excellent cooks. Still, if the traveller wishes everything to be really well managed, he must take the trouble to look after things himself.

Few pleasures in mountaineering dwell more in the memory than the recollections of these bivouacs. Even the most reserved will unbend under the stimulating influence of a crackling wood fire out in the open. The absolute sense of freedom is delightful in itself. The prospect of a good climb, and a little uncertainty about its achievement, strings every member of the party up to just the right degree of excitement. The closing in of the day and the sunset seem but a preparation made by nature for the pleasure of the morrow. It is certain that the oldest joke will be received with enthusiasm, that the most ancient story will be welcomed as the brilliant offspring of the teller's wit, that the cynic will become genial, the critic tolerant, and that the best qualities of companions and guides alike will come to the front, and this alone will make the camp out worth the trouble.

If the travellers do not propose to return to the camp, they should take at least as much pains to leave the place tidy as if they had been passing the night in a club hut. Some respect is due to nature as well as to man, and it is worse than slovenly to leave the site of the bivouac strewn with bits of paper, broken glass, and empty tins. Any unused firewood should be collected and stored in a sheltered place in case others pass the same way.

'Sub scopulo'

CHAPTER XV

PHOTOGRAPHY

BY C. T. DENT

A glass I

INCIDENTAL expenses form a rather serious item in a traveller's account. Among these the purchase of a photographic apparatus must now be reckoned. Fashion, which lays its ruthless hand on everything, has decreed that photography is a matter that must be taken up by everybody. Visitors to the mountains accordingly, not exempt from the epidemic craze, pack up a camera to take with them, even as, too often, they pack up all their respectable raiment,—and leave it behind. Seduced by specious advertisements, they rush out the day before starting on their tour, and purchase 'The young photographer's complete outfit,' and a little handbook on the subject, imagining that they then possess all that is needed to produce excellent pictures. No doubt photography has of late years been greatly simplified. In bulk and weight the apparatus is now reduced almost to a minimum. The tediousness of the old wet-plate process is now entirely superseded, though it may be remarked, in passing, that some of the finest photographs of the Alps have been taken by this same process.

The majority of tourists who carry a camera, as they carry a map, for occasional use, aim at little more than obtaining some pictures which may serve to amuse for a moment, or which may form a convenient form of diary. They expose a multitude of plates or films on the chance of some of them turning out to be worth printing. They carry out almost literally the dictum of the proprietors of photographic stores, who advertise that you have only to 'press the button, we do the rest.' The condition that 'we do the rest' on the understanding that you pay for it, is not so distinctly specified. However much apparatus may be simplified, and however small the necessity for learning or practising any of the details of photography in the case of a traveller who wishes to show a number of views or subjects which he can boast of as his own work, the fact cannot be got over that in photography, as in everything else, the personal equation must count for something.

With those who carry a camera purely as a plaything we have now no concern. As regards mountain photography, and photography under conditions of mountain travel, a few hints may not be out of place.

Photographers divide themselves naturally into two main classes ; first, those who desire to secure topographical views, which may be of interest or value to other climbers, which may serve to explain routes taken, views seen, or illustrate points of physical geography ; secondly, those who aim at obtaining pictorial records of the scenery through which they have travelled. Either of these classes, again, may make it their object to produce good photographs or photographs of good subjects. The nighest excellence is attained when the two objects are combined.

The apparatus, again, falls into one of two classes : the fixed, or camera fixed on the tripod stand, and the hand cameras, whose variety is infinite and whose popularity at the present moment is unbounded. In mountain photography the best results will be obtained with the fixed camera. Hand cameras are more portable, less troublesome to use, and better adapted for taking the 'instantaneous' views which are the fashion of the moment. But the best photographs of mountain scenes are taken on slow plates, and for these hand cameras are not suited.

For mountain photography the camera body should be thoroughly well made in all parts. The great variations of temperature will soon find out the weak points in apparatus made of unseasoned wood. The bellows should be made of stout leather. However great the fame of the maker, the camera should be carefully tested before starting, especially if it has been laid by for twelve months. If this precaution be taken the traveller may dispense with carrying any developing apparatus on his journey.

The late Mr. Donkin, who used almost invariably to carry his own apparatus, considered that the largest camera that could conveniently be carried in the mountains was one capable of taking a $7\frac{1}{2} \times 5$ in. plate. Mr. Donkin's outfit in the Alps consisted of a camera of this size, three lenses, three or four 'double-backs,' the whole fitting into a canvas waterproof case. The complete pack, exclusive of the legs, weighed about 15 lbs. A padlock on the strap of the case will prevent the curious from examining the slides to see how photographs are taken. If a larger machine is carried, it must either be entrusted to a specially strong porter or guide, or the apparatus be divided into two parcels, an arrangement that is never so convenient and makes the traveller less independent, while adding considerably to the cost. It is essential, for the photographer's peace of mind, that his apparatus should be simplified as much as possible, and that everything he uses should pack into one case and always occupy the same place.

The most useful lens for landscape work is undoubtedly the kind known as the rapid rectilinear or rapid symmetrical, of about nine inches focal length. The iris diaphragm, an arrangement whereby the loose stops are done away with, is an immense convenience, and in addition prevents dust from getting in between the combinations of the lens. A lens of some fifteen or sixteen inch focus is the best to add if a second is carried. The rapid rectilinear can be converted into a long focus lens by unscrewing the back combination. This will of course give a much smaller field, but pictures taken with it will give a much better idea of the height of the mountains. Unless the camera is extremely well constructed and the slides fit in with great accuracy, many pictures will be spoilt by the reflected rays, when taken on snow. To obviate this the focussing cloth should be made in the form of a tube so as to completely encircle the body of the camera ; loops or strings are required to secure it in case of wind. An excellent and simple form of level is a plumb-line made of wire fixed to the side of the camera. Knapsack-straps are usually provided with the canvas case, and the legs of the camera can be strapped either upon or below this. The four-fold legs will be found to get much less in the way when climbing than the longer form usually sold. A hard camera case is an awkward pack to carry, and a convenient plan is to stow the whole apparatus in a 'Rücksack.'

The weight of glass plates is undoubtedly a very serious drawback to their use. A box of twelve $7\frac{1}{2}$ by 5 plates weighs 3 lbs. 5 oz. and a store of six or eight dozen plates constitutes a rather formidable package, and one that cannot be transported without some expense. Further, the necessity of constantly changing the plates costs time. The recent invention of films does away with most of this trouble. The films may be either purchased cut up to the required size and used in the ordinary double-backs, or purchased in spools, each giving twenty-four, forty-eight, or more exposures. The latter arrangement necessitates the use of a 'roll-holder,' but enables the photographer to dispense with slides. The convenience of these films has made them extremely popular. The difference in weight is very great. Taking the 7½ by 5 size again, material sufficient for 192 exposures on thin films can be carried with a roll-holder, and weighs about the same as a single box of twelve glass plates. The result, as might have been expected, has been an enormous over-production of bad photographs. Probably all photographers of experience would agree that there is no thin film at present in the market which yields results equal to those that can be obtained on glass plates, though the films on celluloid, thirteen of which weigh about as much as one glass plate, can be thoroughly relied upon. These films are slightly more expensive than plates. A roll of films is necessarily of uniform sensitiveness throughout.

For landscape views in the Alps, for rendering the tones and shadows of snow and glacier as well as for giving the modelling of peaks and valleys, a moderately slow plate will yield the best results ; the very slow plates are harsh in gradation and give chalky pictures. Errors of exposure are less likely to occur when the calculations have not to be made in fractions of a second. Most of those who use films for mountain photography take the quick variety, for which there is the chief demand. The result is that their mountain landscapes are, for the most part, thin and poor, while the exposure is very seldom correct unless in experienced hands. A man who works with a roll of quick films may take fifty or sixty views in a day, comforting himself generally with the assurance that some of them are sure to turn out all right. As a rule they do not. If he develops his own negatives, he must either have extraordinary patience or unlimited leisure to work through some fifty negatives in the hope of finding two or three that are good. To take six or eight photographs in a day, to take them with every precaution, giving thought to the lighting, to the choice of subject, to the exposure and to all other details, is quite sufficient for the man who aims at doing his best, and at quality rather than quantity. If he brings back some sixty views from a mountain tour with a fair certainty that 75 per cent. will turn out respectably, he will be amply rewarded for his trouble and will save materially in pocket, although his method may not commend itself to the professional who is ready to do 'all the rest' for him.

At the risk of appearing old-fashioned, the writer still recommends that, when the conditions of travel allow of it, some glass plates should be included in the store. Successful photography depends, like everything else, on the capacity for taking trouble, and a man who has only a limited number of plates or celluloid films, being precluded from using them lavishly, is more likely to use them to good purpose. A limited number of quick films can be taken, for subjects requiring short exposures.

It is no part of the writer's business here to enter into the various kinds of developers; full descriptions abound in various books. If the photographer desires to make the most of his negatives, let him learn to develop them himself. The interest that he has in his own productions, aided by the memoranda that ought to have been noted down with reference to every view taken, such as light, exposure, lens, and the rest of it, will enable him to produce a far more satisfactory result than he will secure if he places his plates in the hands of a professional developer. At the worst, he will only have himself to blame for any imperfections that may be revealed. A batch of Alpine photographs taken by some one who, though a climber, yet has eyes for the beauties of the sub-Alpine world and for valley scenery, requires immensely different treatment in development if the utmost is to be made out of the pictures. A developer should therefore be selected capable of being used in varying strengths, and giving very different degrees of density. Probably most experienced Alpine photographers employ pyrogallic acid in some form, with ammonia or caustic potash. Eikonogen, with an alkaline carbonate, is strongly recommended by good authorities. Eikonogen 'cartridges' are convenient and portable.

Some of the advantages of hand cameras have been already stated, and it must be admitted that, if good photographs cannot always be obtained with these instruments, pictorial impressions on many subjects can be secured which the man with a fixed camera has no time even to attempt. Bear in mind that the more complicated the apparatus is, the more is it liable to get out of order. It will be found best in selecting a hand camera to choose a machine which has yielded good results in the hands of others, and not always to purchase the very latest type. The working of every single part should be mastered, and the instrument should be practised with before starting from home. The wise man who includes in his stores a pair of small pliers, a file and a bit of soft wire, can generally repair any defect or damage due to accident. An arrangement that allows for time exposures as well as for instantaneous is often supplied. Pictures taken however with the camera held in the hand where an exposure of, say, a second is given, will never be satisfactory. It is much better to rest the camera on some fixed object when time exposures are given. On the snow-fields this is not often possible. A level is quite as necessary on a hand camera as on one that is fixed.

The term 'instantaneous' has been used already rather in the sense in which beginners understand it. They imagine that the rapid films or plates have far greater capacity than they really possess, and that precisely the same snap exposures will do for a view over a glacier or snow-fields in brilliant sunlight at twelve o'clock in June, as for a valley scene in shadow at four o'clock in the afternoon in September. The difference of exposure required is really enormous. A purely snow view, for instance, may require an exposure of a second and a half; with the same lens and stop, and at the same hour of the day, a landscape with much dark foliage in the foreground or middle distance would require over thirty seconds. The use of quick plates or films will not affect the proportion, and yet a photographer with his hand camera often gives precisely the same snap exposure to the one as to the other. There is no special virtue in the hand camera that renders it superior to the rules laid down for guidance in the matter of exposure. Shutters can be bought which allow the exposure to be accurately graduated to small fractions of a second.

Ingenious contrivances are made, folding up into a small compass, which enable the photographer practically to carry about a little dark room and render him independent of elaborate arrangements for excluding all light in his bedroom or waiting till the dead of night before he can change his plates. A large bag of Turkey red twill made so as to tie round the body just below the armpits answers well enough for changing slow plates if the process is got through expeditiously; quick plates are very apt to suffer. A makeshift red lamp can be made out of an ordinary hock bottle of which the bottom has been knocked out. Most photographers when on their travels are content to develop a plate or two from time to time to make sure that the camera is all in good order; if duplicates are taken now and again for this purpose the process of fixing and washing may be dispensed with, and the plate thrown away as soon as it is seen to be satisfactory. In modern Alpine hotels facilities for complete development can often be found.

Alpine photography is probably as difficult a branch to succeed in as can be found, and the man who aims at attaining really good pictures, or, at any rate, up to the standard which he is able to secure at home, must not be disappointed if he has to throw away the whole of two or three seasons' work before he obtains a negative that really satisfies him. There is no doubt, however, that photography in the mountains can yield results as good as any that the limited nature of the process is capable of producing. In the high regions the intensity of the direct and reflected light is so great that the eye is unable to distinguish gradations of tone. A certain monotony of effect is the result. The photographic film is not, so to speak, liable to be dazzled. The soft contours of the snow-fields yield infinitely delicate shadows which can be caught in a photograph and made visible to others in a manner that perhaps no other process is capable of rendering as well. The massiveness and solidity of rock mountains, the great curving sweeps of the glaciers with their perfection of line, can be portrayed by photography in such a manner as at least to suggest the marvels of the high snow world. And for those who find a source of delight in the ever-varying beauties of cloud-land, the mountains offer the finest imaginable field for photographic work. The subjects are difficult, but worth any amount of trouble.

It is quite possible, even with so imperfect, colourless and limited a process as photography, to obtain pictures which show an artistic appreciation of the scenery. Though an infinite number of views have been taken in Switzerland, extremely few pictures have been produced. It must not be supposed that every view which looks beautiful to the eye will make a good photograph, or conversely, that an admirable photograph may not often be obtained out of a view which looks quite commonplace. It is not sufficient in choosing a view merely to introduce a proper amount of foreground and to balance it with a certain amount of sky. When the light is strong and the sky clear, some of the picture will certainly be either under- or over-exposed. By uncapping slowly from above downwards this defect can be partly remedied. Special shutters are made which open in this way. When the sky is cloudy, and in winter, the whole of the picture, if judiciously selected, may be brought into almost perfect harmony of tone. But the greatest attention should be paid to line, and the image thrown on the focussing screen should be regarded chiefly as a composition in line. Beware of endeavouring to include too much. As a rule, upright pictures give better balance of composition than those that are horizontal. The panorama views, very useful topographically, are in the highest degree unsatisfactory as pictures.

As regards the lighting, it must be remembered that objects near to the lens require longer exposure than those that are more distant. Endeavour, therefore, to secure views in which the light is stronger in the foreground than in the middle distance (to use an accepted but ungrammatical expression) and distance. This is especially necessary if any trees come into the foreground. Constantly in mountain photography views can be chosen in which the foreground is so far away as to be practically 'distant'; and these give excellent results.

While it is desirable not to attempt to get too much into a single view, a mistake in the opposite direction is almost equally to be avoided. A series of mountain views representing the main peak precisely in the centre, and suggesting that all mountains are alike and should be looked at in the same way, is but a further limitation of a pictorial process of which the field is already contracted enough. Here is a bold rugged mountain, seamed with gullies, plastered with snow and ice patches, its ridges torn and splintered. Weather-beaten and water-worn, ruggedness is its chief characteristic. The whole mass suggests a stubborn resistance to forces that will yet be too strong for it. Next to it, perhaps, in nature's order is an equally high peak draped from crown to foot with soft undulating downs of *névé* snow, with delicate curves and sweeping lines. It cannot be imagined that two such different subjects will look their best under the same conditions. Even the photographer ought to show that he understands something more than the mere technique of his business. A rock mountain looks its best when the sun is still comparatively low, and there are strong shadows to give the modelling and to accentuate the lines. Sharp focus is required and bold development. The other is seen to the greatest advantage when the sun is somewhat higher and the sky is not perfectly clear, thin veils of mist alternating with patches of blue. Delicacy of shading is the point chiefly to be secured. Slow development with weak solutions, the process not being carried too far, will answer best. Although the ultimate pictorial result may be deficient in many respects, it can at least show that the camera did not end where the plate was put in, but that there was an intelligence behind that again.

Far too many photographs are taken merely, as it were, to exhibit the qualities of the lens. There is almost a craze for giving the sharpest possible definition to the greatest amount of detail that can be got in. If the detail is so over-accentuated it is made the most important feature of the picture, and anyone looking at the print will miss entirely the general effect and concentrate his attention merely on the definition. He may be able to count almost every stone in a mass of screes, and get no idea at all of the picture really presented in the photograph.

Enlargement of photographs is now no longer a very troublesome or costly process. There can be no doubt that an enlarged photograph very often looks much better than a small one, and gives a much better idea of the scale. This is frequently due to the fact that the observer looks at the enlargement from a so to speak unfair point of view. In looking at any photograph, the paper ought to be held at a distance equal to the focal length of the lens with which it was taken. If enlarged, that distance must be proportionately increased; but it is often found that an enlargement looks much better when viewed from a point within the proper range, which is tantamount to saying that the photograph looks its best when the definition is less acute and hard than represented in the original view. It must not be supposed that every view will stand enlargement. Photographs taken with a very small stop on a clear day with pretty hard contrasts and a fair amount of detail will enlarge well enough, and will gain by the process. Photographs of cloud subjects will not usually bear much enlargement, nor will photographs taken with hand cameras, save in exceptional instances. If the view has been taken on a quick plate and with a large aperture to the lens, the effect of enlargement is to give but a thin, blurred and weak picture. An idea that seems generally prevalent, that a small hand camera does not require a well-made lens just as much as a larger machine, is accountable for many of the failures seen in enlargements from these little pictures.

The winter time is an admirable season for the mountain photographer. The days are indeed short, but the light is excellent during the hours that the sun is above the horizon.

PHOTOGRAPHY

The harsh contrasts are softened down, shadows everywhere abound as the sun is so low in its zenith, and the uniformity of tone renders the taking of successful pictures in the winter far more easy than in the ordinary holiday months. Yet this is too often the time at which the camera is packed up and stowed away. One caution may be given to those who take up winter photography for the first time. The exposure is six to eight times as long as in the summer months, and unless this is borne in mind the beginner will probably find that the whole of his views will be hopelessly under-exposed.

Camaraderie

GLOSSARY

ALP.—A summer pasturage.—*Swiss*.
ARÊTE.—The crest of a final ridge.
BERGSCHRUND.—The great crevasse separating the commencement of the snow-field from the face of the mountain.
CHIMNEY.—A narrow rock gully.
COL = PASS = BALLOCH, *Highlands*.—A marked depression in the line of watershed.
CORRY = CORRIE.—A crescent-shaped hollow on the side of a mountain.
COULOIR.—A steep gully in the side of a peak or wall.
CREST.—The highest line of a ridge.
CWM = CORRY.—*Welsh*.
DIRT-BANDS.—Wide curvilinear stripes extending across dry glaciers, due to a superficial deposit of fine débris. They are best seen when looking down on a glacier from a height in a slanting or faint light. They afford valuable information as to the rate of movement of different portions of the ice-stream. The dirt-bands are very plain on the portion of the Mer de Glace that corresponds to the Glacier du Géant.
EDGE—*Cumberland*.—A ridge.
GENDARME—*Alpine slang*.—A tower on a ridge. The word is usually applied to such towers on summit ridges.
GLACIÈRE.—Cave containing ice. In old Swiss books the *névés* are called glacières, as distinguished from the glaciers which descend into the valleys. ('Murray's Handbook.')
GLACIER-TABLE.—A large block of stone balanced on a column of ice, on a dry glacier.
GLÄNZENDES EIS.—The thin layer of ice 'glazing' rocks. The term is also applied to slopes of clear ice.
GRAT = RIDGE = ARÊTE.
GULLY = COULOIR.
HANGING GLACIER.—Masses of ice clinging to the rock walls.

GLOSSARY

HOT-PLATE.—An exposed cliff of rocks occurring in a glacier owing to the breaking away of the ice.

ICE-FALL.—The crevassed part of a *névé* or glacier where the ice-stream, bending sharply over a cliff, is subjected to extreme tension.

KAMM.—An irregular rock ridge.

KOSH.—A rude shelter used by shepherds.—*Caucasus.*

MOEL.—A round-topped hill.—*Welsh.*

MORAINES.—The débris resulting from the weathering of the rocks bounding the glacier that has fallen on the surface of the snow, and been stretched out in long lines by the downward movement of the ice-stream. The LATERAL MORAINES flank the sides of the glacier. Owing to the shrinking of the glaciers, there may be more than one lateral moraine on each side. MEDIAL MORAINES are formed by the union of the adjacent lateral moraines of two tributary glaciers, or by the débris that falls from a hot-plate. The latter are very small. TERMINAL MORAINES are the deposits at the lower extremity of the glacier. Owing to the recession of the glaciers, it is often possible to recognise several concentric terminal moraines.

MOULINS.—Shafts, extending through the entire thickness of dry glaciers bored by the action of water.

PITCH.—A small cut-away cliff in a gully.

RANDKLUFT = BERGSCHRUND.

REGELATION.—When two pieces of ice, with moistened surfaces, are placed in contact, they become cemented together by the freezing of the film of water between them (Tyndall). This phenomenon is termed regelation.

RIB, OF ROCK.—A long buttress running up the side or face of a mountain, best seen on the peaked mountains.

ROCHES MOUTONNÉES.—Rocks scraped and polished by the action of ice : the result of the movement of a glacier.

SAND-CONE.—A small conical elevation of clear ice covered with adherent grit, found on dry glaciers.

SCHRUND = BERGSCHRUND.

SCREES.—Long lines or heaps of small stones or débris lying piled up at the bases or against the sides of rock faces, due to the weathering and disintegration of the more exposed ridges and surfaces.

SÉRAC.—An ice-tower formed by the intersection of transverse and longitudinal crevasses.

SNOUT.—The lower termination of a glacier.

SNOW-LINE.—A term used rather vaguely in mountain literature. Perhaps best defined as the necessarily irregular level at which the amount of snow that collects in the winter is equal to the amount that melts in the course of the year.

SPALT—*German.*—A large crevasse.

STRATH.—A broad expanse of low ground between bounding hills. *Highlands.*

SU.—A small river.—*Caucasus.*

SUB-ALPINE.—The region where the true 'alps' or summer pasturages are found.

TAU.—A peak.—*Caucasus.*

TRAVERSE.—Sometimes used substantively to denote a surface of rock or snow that has to be crossed more or less horizontally.

VERGLAS.—A thin layer of ice glazing rocks. Under exceptional conditions of weather rain freezes as it falls on the rocks, and is then sometimes said to fall as 'verglas.'

WALL.—A term used to denote a steep face leading up to a ridge.

WAND—*German.*—A wall of rock.

A lengthy observation

'Auf Wiedersehen'

INDEX

ABNEY'S clinometer, 60
'Abode of Snow,' Andrew Wilson's, 291
Accidents, 373
Adlerjoch, the, 308
Aeronauts, subject to 'mountain sickness,' 84-86
Africa, 287, 292
Aiguille des Charmoz, 221
Aiguille du Dru, 308, 366
Aiguille du Géant, 232, 367
Aiguille du Goûter, 360
Aiguille Grise, 359
Aiguilles, 220-222, 255
Air-cushions, 64
Alagna region of Monte Rosa, 9
Alaska, 288
Alcohol, 77, 79, 210
Almer, Christian, guide, 164 n.; qualities and career of, 369
Almer, Ulrich, guide, 164, 369
Alpenstock, use of the, 14, 15, 25, 27, 29, 30, 36, 66, 67, 98, 99, 110-112, 165
Alpine Club, the, rope sanctioned by, 72, 101, 102 n.; map of Switzerland, 264, 353
'Alpine Guide,' Ball's, 275-277, 353
Alpine horn, the, 14
'Alpine Journal,' 2 n., 7 n., 21 n., 33 n., 35, 38, 271, 287

'Alps in 1864,' Moore's, 358
Altai range, the, 287
Anderegg, Jakob, guide, 369
Anderegg, Melchior, guide, 115, 133, 169, 173, 241, 321, 367; character and career of, 369-372
Andermatten, Franz, guide, 367
Andes, the, 24, 74, 289, 292, 367, 368
Aneroids, 60, 61
Anklets, 50
Anthropology, 305
Aosta, valley of, 349
Apennines, the, 22
Arêtes, 136, 202
Aristotle, quoted, 13
Arran, 342
Arran Mowddwy, 340
'Art of Travel,' Galton's, 397
Aspargatas, 257
Assynt, 341, 345
Atkins, H. M,, 30
Atlases, 264
August in the Alps, 124-126
Auldjo, Mr., 31, 192
Austen, Godwin, 284, 287
Austrian mountain lanterns, 65
'Autocrat of the Breakfast Table,' quoted, 75
Avalanches, 17, 127, 197-202
Aviemore, 346

E E 2

AVI

Avienus (Vogel of Glarus), 10
Axe, the ice; *see* Ice-axe
Axe-fiends, 111
Axe-slings, 70

BACON, Lord, quoted, 75
Bags, provision, 55; for sleeping in, 47, 64; wine, 56
Balaclava (Templar) caps, 48
Balfour, Francis Maitland, 361
Ball, John, 29, 35, 36, 275, 352; first President of the Alpine Club, 353; his works on the Alps, 353; his 'Notes of a Naturalist,' 354; character and influence of, 354; 362
Balmat, Auguste, guide, of Chamonix, 365
Balmat, Jacques, guide, 25, 74
Barometers, mercurial, 61
Baumanns, the, 369
Beaupré, Julien de, his ascent of Mont Aiguille, 7
Beddgelert, 340
Beispielspitz, the, 134, 162, 272, 387
Belts, 72, 79
Ben Alder, 345
Ben Lui, 345
Ben Nevis, 339, 342, 345
Bennen, guide, 29; fate of, 369
Beresford mountain lantern, the, 65
Bergen, Johann von, 241
Bergschrunds, 139, 153, 186-190, 192, 289
Bernese Oberland, the, 34, 35
Bert, Paul, 84, 85
Bethesda, 340
Bicarbonate of soda lozenges, 80
Bidein Druim-nan-Ramh, 344
'Biener, Weisshorn,' guide, 366

BRI

Bietschorn, the, 223, 251
Birkbeck, Mr., 360
Black Hills of Torridon, 345
Black Sail Pass, 337
Blaitière, the Aiguille, 366
Blencathra, 338
Blindness, snow, 15, 81
Bonney, Professor, 81
Boot-laces, 43
Boot-nails, 44, 59, 160
Boots, 42-45, 83, 399
Boric acid and cocaine for snow ophthalmia, 82
Bosses du Dromadaire, Mont Blanc, 360
Botany, 305
Bottles, 55; used as lanterns, 66, 409
Bouguer, 24
Bourrit, 3, 24-26
Bowfell, 333
Bowline knot, 100, 101; on a bight, 184
Brandreth, 337
Brandy, use of, 79, 210
Breithorn, the, 308
Brentari, O., quoted, 7 *n*.
Brenva Glacier, the, 360
Breuil, 367
Brévent, Mont, 133, 366
British Isles, hill climbing in the, 325; qualifications for a mountaineer, 325; learning to climb, 326; choice of routes, 327; climbing in cloud and mist, 327; practice on rocks, 328; dangers in winter on the English hills, 329; unwisdom of trusting to tufts of grass, &c., 330; best months for the hills, 330; guides in this country, 331; kit required, 332; a lesson-

excursion, 333-336 ; fancy climbs, 336 ; districts where mountaineering may be profitably studied, 337 ; on the watersheds of the Lake-country, 337, 338 ; in Yorkshire, Derbyshire, &c., 338 ; on Snowdon, 339, 340 ; Cader Idris, 340 ; in Ireland, 340 ; the Highlands of Scotland, 340-346 ; Ben Nevis, 342 ; Skye, 342-344 ; Sutherlandshire, 344, 345 ; lessons to be learned, 346

Broad Stand of Mickledore, the, 336

Bruhns' Life of Humboldt, 24 n.

Bryce, Mr., on Ararat, 6

Buckingham's Alpine Club rope, 102 n.

Bündner Oberland, the, 33

Burgener, Alexander, 241

Burnet, Bishop, 21

Burnet, William, 21

Burning-glasses, 58

Burnmoor, 337

Butler, Montagu, 362

Buttermere, 338

CADER IDRIS, 340, 358

Cairn Gorms, the, 341, 345

Cairns, 318

Calotte, the, Mont Blanc, 359

Camera obscura, the, 387

Cameras, Hand, 303, 403

Camping, 392 ; club huts, 393 ; Swiss Alpine Club rules for the use of the hut on the Dom, 393 ; hour for the bivouac, 395 ; collection of firewood, 395 ; pitching the tent, 396 ; arrangement of baggage, 396 ;

construction of fireplace, 397 ; sleeping under overhanging rocks, 398 ; making the bed, 398 ; preparation over night for the morning start, 399 ; attention to clothing, 399 ; use of cholera-belts, 399 ; securing sleep, 399 ; duties of guides, 400

Camps, 147 ; see Camping

Canigou, Pyrenees, 6, 284

Canteens, 64

Cantering, 98

Capel Curig, 340

Caps, Balaclava, 41 ; Dundee whaler's, 48 ; woollen and leather, 48

Cardigan jackets, 47

Carnedd Davydd, 340

Carnedd Llewellyn, 331, 340

Carrel, Jean-Antoine, guide, 241 ; character and career of, 367 ; manner of death, 368

Carrel, Louis, guide, 296

Carrel's Ledge, 232

Cascade range, the, 288

Castagneri, Antonio, guide, 367

Caucasus, the, for mountaineering purposes, 48, 55, 62, 64, 117, 134, 159, 160, 176, 177, 213, 254, 262, 264, 279, 286, 292, 357, 325 ; maps of, 291

Central Asia, 287, 292

Chalk cliffs, 339

Chamois, 11, 158, 260, 368

Chamonix, as a mountaineering centre, 3, 24, 26, 27, 29, 35, 96, 113, 133, 315, 355, 357, 359, 363-365, 387, 392

Champagne, 79

Changlung Barma La Pass, 88

Charles Emmanuel II., 5

Charles VIII. of France, 7

CHI

Chili, 289
Chimborazo, 24, 84, 85, 158, 159, 296, 368
Chlorodyne, 80
Chocolate, 80
Cholera-belts, 47, 79, 399
Chulha, a, 397
Clark, Dr. Edmund, 29, 31
Clement, Count, 5
'Climbers' Guide to the Pennine Alps,' 277
Climbers' shelters, 282
Climbing-irons, 15, 27, 28, 73
Climbing outfit for ladies, 50–52
Clinometers, 59, 60
Clissold, F., 28 *n.*, 29, 31
Clothing, 30, 40–42, 332, 399
Clouds, 122
Club huts, 282, 393
Coats, 40, 41; waterproof, 47, 48
Cocaine, 82
Coignach Hills, the, 345
Col d'Erin, 352
Col de la Pilatte, 193
Col de Miage, 360
Col d'Ollen, 9
Col de Valpelline, 372
Col delle Loccie, 103
Col du Géant, 27, 29, 352
Col du Mont Rouge, 364
Cold Cream, 81
Comforters, woollen, 47
Compass, use of the, 50, 209, 267
Condamine, 24
Coniston, 337, 338
Connemara, 340
Conway, W. M., 34, 38 *n.*, 302
Cooking apparatus, Rob Roy, 65
Coolidge, Rev. W. A. B., his ascent of Mont Aiguille, 8;

DIA

quoted, 14 *n.*, 15–17, 19 *n.*, 38 *n.*; his 'Swiss Travel and Swiss Guide-books,' 273
Coolin Hills, the, 282, 342-344
Cornices, snow, 161–164, 188, 202, 319
Corrie nan Creiche, 344
Cotopaxi, 368
Couloirs; *see* Gullies
Courmayeur, 3, 357, 359, 364, 368, 392
Couttet, François, guide and hotel-keeper, 363
Coxwell, Mr., aëronaut, 84
Crampons, 15, 27, 28, 73, 74
Crevasses, 103, 104, 141, 142, 152–156, 179, 180–195, 208–210
Crib Goch ridge of Snowdon, the, 133, 336, 339, 340
Crocé-Spinelli, 85
Croz, Michel, guide, 164, 241, 365
Cuchullins, the, 262, 342–344
Cunningham, Mr., 366
Cupelin, Edouard, 366
Cuvier, 362
Cwm Tryfan, 340
Cynicht, 340

Dante, as a mountaineer, 7
Darjiling, 301
Darwin, 361
Daudet, Alphonse, quoted, 33
Dauphiné, 284, 351
De Saussure, 3, 24–27, 28 *n.*, 29, 31, 85, 351
Dent Blanche, 317
Dent d'Erin, 372
Derbyshire, 338
Dévouassoud, François, 296, 366
Diablerets, the, 284
Diagrams of bergschrunds, 189;

DIA

of glaciers, 154; of crevasses, 155
Dialogue between the Niesen and the Stockhorn, 18-20
Diarrhœa, 80
'Diary of a Traveller over Alps and Apennines,' quoted, 3;
Diet, 77
Dolgelly, 358
Dolomite rocks, 219, 254
Dom, the, 393
Donkin, W. F., photographer, 389 n., 404
Doone Valley, North Devon, Ordnance Survey of, 300
Dow Crags, the, 338
Dragons in the Alps, legends, 18
Dresses, ladies', 50, 51
Drinking cups, leathern and metal, 56
Dru, the Aiguille du, 308, 366
Dufourspitze, the, of Monte Rosa, 37
Duilier, Nicolas Fatio, 21
Dundee whalers' caps, 48
Dungeon Ghyll, 333, 338
Durham, 338
Durier, M. Charles, quoted, 3 n., 5 n., 25, 31

ECKENSTEIN, O., his experiments with ropes, 102 n.
Ecrins, the, 251, 365
Ecuador, Great Andes of, 24 n., 289, 368
Eczema, 80
Eigenthal, the, 11
Eikonogen cartridges, 407
Eisbeil, the, 29
Elbruz, 31, 74, 159, 357
Ellicott, Bishop, 362
Engadine snow-shoes, 74
Engel Pass, the, 88

EQU

Engelhardt, 362
English guides, 331
Equatorial Andes, the; see Andes
Equipment and outfit, 39; clothing, 40-42; the hat, 42; boots, 42; boot-laces, 43; boot-nails, 44, 59, 160; boot preservation, 45; stockings, 45; the garter, 46; shirts, 46; pyjama sleeping suits, 47; Cardigan jackets, 47; cholera-belts, 47, 79, 399; waterproof coats, 47, 48; woollen and leathern caps, 48; gloves, 48; gaiters, 49; outfit for ladies, 50-52; knapsacks, 52; the rücksack, 52-54; havresacs, 54; provision bags, 55; self-cooking tins of preserved foods, 55; bottles, 55; wine-bags, gourds, and flasks, 55, 56; leathern and metal drinking cups, 56; snow-spectacles, 56; knife, 57; whistle, 58; field-glass, 58; matches, 58; needles, thread, and buttons, 59; medical stores, 59; scientific apparatus, 59; compasses, 59; clinometers, 60; aneroids, 60; mercurial barometers, 61; pocket divider and protractor, 61; maps, 61; tents, 62; the Mummery tent, 63; sleeping bags, 64; the Tuckett bag, 64; inflating mattresses and air-cushions, 64; the canteen, 64; lanterns, 65; alpenstocks, 66; ice-axes, 67-71; walking-sticks, 71; the rope, 71; belts, 72; crampons or climbing-irons, 73; snow-shoes, 74

EQU

Eskdale, 335, 336
Esk Hause Pass, 335
Esquimaux snow-spectacles, 57
Etiquette on the mountains, 116, 316
Etive, 342
Eugénie, Empress, 365
Excelsior mountain lanterns, 65
Eyes, the, 81, 82

FACE masks, 52
Fairfield, 338
Falling stones, xix., 241, 257-262, 317
Fatigue, 77-79
February in the Alps, 126
Feet, to keep warm, 211
Felley, Benjamin, of Lourtier, chasseur, 363
Fellowes, Sir Charles, 372, 373
Festiniog, 340
Field-glasses, 58
Figure-of-eight rope tie, 102
Finsteraarhorn, the, 35, 361
Finsteraarjoch, the, 357
Firewood, 394, 395
Fisherman's bend, in tying rope, 101, 102
Fixed ropes, 255
Flag signalling, 119, 120
Flannel shirts, 47
Flasks, 56
Flegère, the, 366
Fluids, drinking, 78
Fog, 208
Föhnwind, the, 121, 127, 211, 212
Food, 78, 79, 210, 302; preserved, 55
Forbes, James David, his works and influence on mountaineering, 351, 352; friendship with Adams Reilly, 354

GLE

Forbes, Dr. John, 362
Forsyth, Sir Douglas, 88
Fortin's mercurial barometers, 61
Frankland, Dr., 84
'Fremden Industrie,' the, 282
French Alpine Club, the, section de l'Isère, 8
Freshfield, Mr. Douglas W., quoted, 7 *n*., 8, 9, 33-35, 38 *n*., 357
Frost-bite, 82, 83
Fuel, 394, 395

GABELHORN, the, 164, 321
Gaiters, 49, 50
Galton, Mr., quoted, 62, 64, 397
Galway, 340
Gardiner, Mr. F., 8
Garters, 46
Gautier, Théophile, quoted, 280
'Gendarmes,' mountain, 142 *n*.
Geological maps, 263
Geology, 150, 218, 304
George, Rev. H. B., 17
Gerard, quoted, 29
Gesner, Conrad, of Zürich, quoted, 9, 10; his ascent of Pilatus, 11, 12-14, 22, 23, 37
Glacier des Bossons, the, 25, 28
Glacier-nails, 28
Glacier streams, 155; tables, 154
Glaciers, serpent-formed, 18, 19; map diagrams of, 154; 20, 21, 103, 142-146, 150-153, 155, 179-195, 210, 211, 283, 317, 384
'Glaciers of the Alps,' Tyndall's, 29
Glaisher, Mr., 84
Glas Lyn, 339
Glazed rocks, 249
Glen Coe, 342, 345

INDEX

GLE
Gletscher, the, canton of Berne, 21
Glissading, 192, 194-197, 320
Gloves, 48, 49, 52
Glycerine, 81
Glyder Fawr, 340
Glyders, the, 357
Gourds, 56
Graham, Mr. W. W., 85, 158
Grande Chartreuse, the, 22
Grandes Jorasses, the, 368
Grands Mulets, the, 30, 133, 359
Granite mountains, 222
Grapnels, 232
Grass-covered rocks, 246
Grass-soled shoes, 256
Gray (the poet), quoted, 22
Greadaidh, 344
Great Andes of Ecuador, the, 289; *see* Andes
Great Britain; see British Isles
Great Gable, 337, 338
Great Napes, 338
Great St. Bernard, the, 277
Gredig, Herr, of the Krone, Pontresina, 350
Green Gable, 337
Gressoney, 370
Grey Knots, 337
Grimsel, the, 35
Grindelwald, as a mountaineering centre, 34-36, 121, 296, 315, 349, 364, 369; glaciers, 19 *n.*, 20, 21
Grombchevsky, Captain, 291
Gruner, 22
Guide-books, usual character of their compilation, 271; really valuable information, 271; account of an imaginary ascent of the Beispielspitz, 272; Coolidge's 'Swiss Travel and Swiss Guide-books,' 273;

GUI
changing condition of information concerning mountaineering, 273; assistance that can be afforded to editors of guides by readers, 271, 274; best system of structure and arrangement for a climbers' guide, 275; Ball's 'Alpine Guide,' 275; catering for Alpine travellers as well as climbers, 276; sketch of an ideal guide-book, 277; in what does a difference of route consist, 278; necessity of a work for Caucasian travellers, 279
Guideless climbing, 307; considerations of weather, 308; the capability of the members of the party, 309; necessity for topographical insight, 310; mental notes of landmarks on route, 311; individual preparation for difficult ascents, 311; one of the party to be a good iceman, 312; all must have had long experience under good guides, 312; things to be learnt from guides and held in remembrance, 313; question of a three or four member party, 314; the great centres to be at first avoided, 315; with another party on the same mountain, 316; advantages of out of the way districts for beginners, 316; general precautions to be observed, 317; on glaciers, 317; on the upper snowfields, 318; cairns as guides on return, 318; in gullies, 318; on snow and ice slopes, 319; use of the axe,

GUI

319; dangers of cornices, 319; halting on the summit, 320; glissading, 320; preparations for a start, 321; comparative skill of amateurs and guides, 321; summary of advice, 322-324

Guides, 77, 112, 113, 123, 212, 294-297, 322, 331, 367-372, 400

Gullies, 141, 145, 175, 176, 318

Güssfeldt, Dr., 289, 297

HALLER, Albert von, his poem, 22, 23

Hamel, Dr., 29

Hanging Knots, 335

Hardy, John Frederick, the 'King of the Riffel,' 361

Harta Corrie, 344

Harter Fell, 337

Haslithal, 121

Hats, 42, 51

Haut de Cry, 369

Havresacs, 54

Health, considered in connection with mountaineering, 75; physical capacity for climbing, 75; will, muscles, and intellect simultaneously in action, 75; the age for mountaineering, 76; training in preparation, 76; appropriate diet, 77; the question of alcoholic stimulants, 77; water for drinking, 77; avoidance and counteraction of fatigue, 77, 78; a tiring man to be fed early and fed often, 79; obviating intestinal derangements, 79; use of the cholera-belt, 79; light eating after heavy exertion,

HIS

79; remedies for stomachic ailments, 80; food before starting on an expedition, 80; remedies for sunburn, 80; snow-blindness and its treatment, 81, 82; treatment of frost-bite, 82, 83; mountain sickness and other prejudicial effects due to rarefied air, 84-88; hints to those essaying lofty summits, 86-88

Heartburn, 80

Helvellyn, 338

Highlands of Scotland, 340

Himalaya, the, 74, 85, 88, 134, 159, 284, 287, 291, 292

'Himalayan Journal,' Sir Joseph Hooker's, 291

Hinchliff, Thomas Woodbine, character of, 356; influence of his ' Summer Months among the Alps' on mountain exploration, 356

Hindu Kush, the, 287

'Hints to Travellers,' 302, 305

History, early, of mountaineering, 2; classical climbers, 4; an eleventh-century chronicle of an attempt on the Roche Melon, 5; Rotario d'Asti's ascent of same in fourteenth century, 5; Peter III. of Aragon's account of his marvellous experiences on Canigou, Pyrenees, 6; Petrarch on Mont Ventoux, 7; Dante as a mountaineer, 7; Julien de Beaupré's ascent of the Mont Aiguille in the fifteenth century, 7; Mr. Coolidge's notes of ascent of same in 1881, 8; Leonardo da Vinci as a mountaineer and a delineator of

HIS

mountain forms, 8; mountaineering zeal of the humanists of Switzerland in the sixteenth century, 9; Count Gesner's monograph, 9-14; ascents of Pilatus in 1518, 11; Gesner's account of Pilatus, 11; Rhellicanus's Latin hexameters on the Stockhorn in 1536, 14; Simler's treatise on the Alps, 14-17; arrest of the mountaineering impulse in the seventeenth century, 18; Wagner the naturalist's account (1680) of dragons in the Alps, 18; Scheucher's enhancement of the dragon legend, 18; Rebman's verse-dialogue between the Niesen and the Stockhorn, 18; Merian's 'Topographia Helvetiæ,' 20; the scientific world's interest excited in glaciers, 20; Justel's account of the Gletscher, 21; Bishop Burnet's estimate of mountain height, 21; the poet Gray's description of the Grande Chartreuse and Mont Cenis, 22; the romantic and sentimental movement of interest in wild nature, 22; Von Haller's poem on the Swiss Alps and the virtues of the Switzers, 23; J.-J. Rousseau's place in the history of Alpine exploration, 23; first ascent of the Titlis (1739), 23; Bouguer and Condamine in the Cordilleras, 24; Humboldt on Chimborazo (1802), 24; Windham and Pococke's visit (1741) to the glaciers of

ICE

Chamonix, 24; early attempts on Mont Blanc, 25; Bourrit and De Saussure, 25, 26; apparatus used in early mountain climbing, 27-29; inventory of clothing worn on an ascent of Mont Blanc, 30; record of distressing sensations in mountaineering, 30; physical conditions of past and modern pedestrians, 31-33; Albert Smith's ascent of Mont Blanc, 32; Father Placidus in the Bündner Oberland, 33; a quotation (1824) on pedestrian excursions in the Alps, 34; the Meyers and Studers in the Oberland, 34; first ascents of the Jungfrau and Finsteraarhorn, 35; Gottlieb Studer on the Strahleck (1840), 35; John Ball on the same and on Monte Rosa (1859), 35, 36; Mr. Justice Wills's ascent of the Wetterhorn, 36; influence of his writings on Englishmen, 37; Hudson and Kennedy's experiences and observations, 37

Home sickness, 296
Hooker, Sir Joseph, 283, 287, 291, 292
Horn, the Alpine, 14
Hudson, Charles, 37, 359; his character and mountaineering career, 360
Humboldt, A. von, 24, 283
Huts, club, 282; rules concerning use of, 393

ICE-AXE, the, use of, 28-30, 34, 36, 67-71, 98, 99, 110.

ICE

Ice, 112, 164, 166-178, 236, 319
Ice-bridges, 187, 192
Ice-falls, crevasses in, 183-185
Iceland, 285, 293
Illimani, 290
Imboden, Joseph, guide, 367
Imseng, Ferdinand, guide, 241, 367
India-rubber wine-bags, 56
Instruments, mountaineering, 305
Interlaken, 349
Interpreters, 305
Intestinal derangements, 79
Ireland, 340
Irons, climbing; *see* Climbing-irons
Italian Alpine Club, 65, 358
Italian guides, 297, 367

JACKETS, 40; Cardigan, 47
Jaegerhorn, the, 370
Jaeger's wool, 64
January in the Alps, 126
Jaun, Johann, guide, 369
Jerseys, 47
Johnson, Mr., 85 *n.*, 287
Johnston, Mr., 292
Jökulls, the, 285
July in the Alps, 124-126
June in the Alps, 125, 126
Jungfrau, the, 35, 126, 352, 361, 370
Jungfraujoch, the, 28, 103, 192, 308
Justel, Monsieur, 21

KABRU, 85
Kandersteg, 121
Kanderthal, 121

LAS

Karakorum Pass, the, 88, 287, 291, 302
Kasbek, 357
Kashmir, 287, 292
Kaufmann, guide, 369
Kenia, 288
Kennedy, Mr. E. S., 37
Khardong Pass, the, 88
Kibo, 287
Kilimanjaro, 287, 290
Killarney, 340
Kimawenzi, 287
Kinchinjanga group, the, 301
Kirkfell, 337
Knapsacks, 52-54
Knickerbockers, 40, 41, 51
Knitted woollen caps, 48
Knives, 57
Knots, to tie, 101; to untie, 110
Knubel, guide, 367
Kola chocolate, 80

LA GRAVE, 357
Ladakh Tartars' snow-spectacles, 57
Ladders, 28
Ladies' mountaineering outfit, 50-52
Lake District, the, 333, 336-338
Lancashire, 338
Landmarks, identification of, 132
Landscape-painting, 388
Langdale Pikes, 338
Lanier, Laurent, of Courmayeur, guide and chamois-hunter, 368
Lanoline, 81
Lanterns, 65
Lassitude, 30

INDEX

LAT

Lateral moraines, 156
Lauber, Herr, 362
Lauener, Christian, guide, 369
Lauener, Johann, guide, 369
Lauener, Ulrich, guide, 37, 369
Lauterbrunnen, 349, 369
Leader, the, of a mountaineering expedition, duties of, 123, 293
Leaf, Mr. W., 362
Leathern drinking cups, 56
Legros, Prof. L. A., his experiments with ropes, 102 n.
Leila Peaks, the, 294
Lightfoot, Mr., 362
Limestone mountains, 218, 338
Little Loch Broom, 345
Livy, quoted, 4
Llanberis, 330
Lliwedd, 339
Lloyd, quoted, 29 n.
Local traditions, 395
Loch Duich, 342,
Loch Hourn, 342
Loch Inver, 342, 345
Loch Long, 342
Loch Maree, 341
Longitudinal crevasses, 153
Longman, William, his 'Modern Mountaineering and the History of the Alpine Club,' 34, 35 n.; literary and other services of, to mountaineering, 360
Loppé, M., on winter in the Alps, 127; on mountain landscape-painting, 389
Lough Derg, 356
Luchon, 284
Luzern, 10, 11
Lys Glacier, the, 268
Lyskamm, the, 272, 361, 370

MER

McGillicuddy Reeks, 340
Macugnaga, 370
Main, Mrs., 50 n.
Malkin, A. T., 362
Malkin's Diaries, 271
Manilla hemp rope, 72, 100
Mannering, Mr. G. E., 294
Map-diagrams of glaciers, 154
Maps, 61; geological 263; the Siegfried Atlas the model of a map, 264; Alpine Club map of Switzerland, 264; Mr. Adams Reilly's maps of the Mont Blanc range, &c., 264; Russian maps of the Caucasus, 264; main uses of maps, 264–267; the prismatic compass, in conjunction with, 267, 268; map-making, 268; topographical descriptions of routes, 269, 270; Mr. Malkin's and Mr. A. W. Moore's accounts of mountain expeditions, 271; 300, 381
Maquignaz, Italian guide, his career and fate, 367
Marginal crevasses, 152
Marius, in Numidia, 4
Masks for the face, 52
Matches, 58
Mathews, Mr. W., 362, 363
Matterhorn, the, 132, 223, 232, 251, 280, 281, 317, 321, 360, 365, 367, 368, 392
Mattresses, inflating, 64
Maurer, Andreas, guide, fate of, 369; 400
Maurers, the, 296
Medial moraines, 156
Medical stores, 59
Meiringen, 121, 369
Mer de Glace, origin of the name, 25; 352, 365, 366

MER

Merian's 'Topographia Helvetiæ,' 20
Metal drinking cups, 56
Metcalf, Dr., 360
Mexico, 289
Meyer, Dr., 292
Meyers, the, 34
Mhadaidh, 344
Mickledore, 336-338
Middleman noose, in rope-tying, 101, 102 n.
Milk, 80
Mist, 208, 327
Mocassins, 257
'Modern Mountaineering,' Mr. William Longman's, 34, 35, 360
Moel Wyn, 340
Moming Pass, the, 164, 308
Monboso, 8
Mönchjoch, the, 195
Mons Romuleus (Roche Melon), 5
Mont Aiguille, near Grenoble, 7, 8
Mont Blanc, 21, 24-26, 29, 31, 32, 36, 37, 74, 84-86, 96, 113, 126, 133, 198, 220, 255, 355-357, 359, 367, 372, 387, 392
Mont Brévent, 133, 366
Mont Cenis, 22 ; tunnel, 349
Mont Cervin Hotel, 362
Mont Perdu, 284
Mont Ventoux, near Vaucluse, 7
Montanvert, 25, 26, 366
Monte Rosa, 8, 9, 36, 37, 289, 304, 355, 356, 366
Monte Viso, 223
Moore, A. W., 271, 301 ; official career of, 356 ; in the Alps and in the Caucasus, 357 ; his extensive knowledge of

NOR

the Alps, 357 ; anecdotes concerning, 357, 358 ; his literary abilities, 358, 365
Moraines, 156, 253
Morse code of signalling, 119, 120
Morshead, Mr. F., 359, 367
Morteratsch, Piz, 308
Mount Ararat, 6
Mount Brown, 288 n.
Mount Etna, 361
Mount Everest, 85, 86, 89
Mount Hercules, 290
Mount Hooker, 288 n.
Mount St. Elias, 288
Mount Shasta, 289
Mountain sickness, 30, 31, 37, 84-86, 202, 295
Mountaineer, definition of a, 1
Mountains, geological formation of, 150
Mourne Mountains, 340
Müller, J., on the Stockhorn, 14
Mummery, A. F., his portable tent, 63, 64
Mürren, 349

NAILS, boot, 28, 44, 45, 59
Nansen, Dr., 74, 288
Napoleon III., Emperor, 365
Natural history, 305
Needles, thread, and buttons, 59
Névé, 156, 159, 292
New Guinea, 290
New Zealand, 213, 290, 292, 294, 301
Niesen, the, and the Stockhorn, dialogue between, 18-20
Nomenclature of peaks, 303
Norfolk jackets, 40

NOR

North America, 288, 293
Northumberland, 338
Norway, 285, 291
Note-books, 303
'Notes of a Naturalist,' Ball's, 354

OBAN, 342
Oberaarjoch, the, 360
Oberland, the, 127, 369
Oehlmann, E., on Alpine Passes, 2 *n.*
Ogwen Valley, 340
Oiled silk bags, 55
Ophthalmia, snow, 57, 81, 82
Ordnance Survey of Doone Valley, North Devon, 300
Osenbrüggen, 20 *n.*
Outfit ; *see* Equipment
Overhand knots, 101, 102 *n.*
Owen, Harry, 357
Oxygen, 78 ; inhalation of, as a remedy for mountain-sickness, 85

PACCARD, Michel, 25
Pace, in mountaineering, 94, 104, 114
Packe, Mr. C., 284
Pamirs, the, 291
Parson's Nose, the, Snowdon, 340
Payot, Alphonse, guide, 366
Payot, Michel, guide, 366
Peaked mountains, 222
'Peaks in Pen and Pencil for Students of Alpine Scenery,' Elijah Walton's, 218 *n.*
Peaks, nomenclature of, 303
'Peaks, Passes, and Glaciers,' 353, 360

PHO

Pennine Alps, the, 277, 284
Pen y-gwryd, 357
Permanganate of potash, solution of, 85 *n.*
Perren, Peter, guide, 366
Perrin, A., his History of Chamonix, 3 *n.*
Perspective illusion, 386
Peru, 289
Peter III. of Aragon, his ascent of Canigou, in the Pyrenees, 6
Petersgrat Pass, the, 246, 307
Petrarch, his ascent of Mont Ventoux, 7
Peyer, quoted, 11 *n.*, 20 *n.*
Philip of Macedon, in the Balkans, 4
Philips, Mr., his account of the ascent of Mont Blanc, 32, 33
Photography, 132, 179, 201, 301, 305, 388 ; points to be borne in mind in mountain work, 389 *n.* ; the wet-plate process, 402 ; applied to mountain scenery, 403 ; topographical views and pictorial records of scenery, 403 ; suitable apparatus, 403, 404 ; fixed and hand cameras, 403, 404 ; testing the camera, 404 ; size and weight of apparatus suited to the mountains, 404 ; the rapid rectilinear or rapid symmetrical lens, 404 ; the iris diaphragm, 404 ; glass plates, 405, 407 ; films, 405 ; slow plates, 406 ; developing, 407 ; eikonogen cartridges, 407 ; hand cameras, 408 ; 'instantaneous,' 408 ; a portable dark room, 409 ; difficulties of the art in Alpine scenery, 409 ; choice of view,

PHO

410; length of exposure, 410; variations in character of picture, 411; enlargement of photographs, 412; winter practice, 412
Pic des Ecrins, 365
Pic du Midi, 284
Pierre à l'Echelle, 25, 28
Pilatus, Mont, 10-12, 14
Pillar Rock, the, Lake-country, 336-338
Piolets, 29
'Pioneers of the Alps,' Cunningham and Abney's, 164 *n.*; 366
Piz Bernina, the, 304
Placidus a Spescha of Disentis, 33
Pocket dividers, 61
Pococke, 3, 22, 24
Pollinger, guide, 367
Pomponius Mela, 11 *n.*
Ponchos, waterproof, 48, 62
Pontius Pilate, legend concerning, 10, 14
Pontresina, 350, 364
Porters, 113, 114, 413
Pratt, Mr., 361
Preserved foods, 55
Principles of mountaineering, 90; qualities of the true mountaineer, 90; fallacy of the climbing test, 90; how to walk up hill, 91; preservation of silence while ascending, 93; regulation of pace, 94; adoption of the zigzag in ascents, 94; dealing with slips, 95; proper rate of speed, 95, 96; short cuts, 96; how to descend, 96; cantering, 98; most rapid mode of descent, 99; down grass slopes, 99; use of the

REC

rope, 100-110; use of the axe and alpenstock, 98, 99, 110-112; choice of guides, 112-114; proper order of a party when roped, 114-116; etiquette towards another party, 116; preparations for starting on an expedition, 116-118; emergencies on the mountains, 119; flag signalling, 119, 120; judging the weather, 121-123; selection of, and duties of, leader, 123; most suitable months for mountaineering, 124-128
Prismatic compasses, 267
Protractors, horn, 61, 267
Provision bags, 55
Ptarmigan, 14
Putties (list bandages), 50
Pyjama sleeping suits, 47
Pyrenees, the, 257, 284, 285

RANDKLUFT, 186
Rannoch, Bogs of, 331
Rarefied air, symptoms due to, 31, 37, 84-86, 202
Ray, quoted, 21 *n.*
Rebman, his verses on the Alps, 18-20
Recipe for snow-ophthalmia, 82
Recollections of a mountaineer, 348; changes effected by railways in Switzerland, 348, 349; old inns and new hotels, 350; some of the makers of mountaineering, 351-361; some of the most excellent guides, 362-372; over and under estimation of mountain risks, 372, 373; Alpine accidents, 373; main rules for the

INDEX

REC

guidance of mountaineers, 374; the charms of mountaineering, 375–379

Reconnoitring, 129; observation of peak before ascent, 129; a mountain to be considered first as a whole and subsequently in detail, 130; best time for starting, 131; identification of landmarks, 132; usefulness of photographs of mountains and passes, 132; appreciation of scale of surrounding scenery, 133; estimation of heights and distances, 134; selection of spot for comprehensive view of a mountain, 134; reconnoissance of a typical mountain, 135–148

Regelation, 151, 160

Regions beyond the Alps, mountaineering in, 280; true view of mountaineering, 280; loss by the Alps of their primitive charms, 282; the vulgarisation of summits, 282; character of the Pyrenees, 284; Norway, 285; Iceland, 285, 293; grandeur and extent of the Caucasus, 286, 292; Central Asia, 287, 292; Africa, 287, 290, 292; North America, 288, 293; Mount St. Elias, 288; the United States, 289; Mexico, 289; South America, 289; the Great Andes of Ecuador, 289, 292; Peru and Chili, 289; New Zealand, 290, 294; season to select for exploration, 291; the Himalaya, 283, 287, 292; constitution of an ex-

ROC

ploring party, 293; selection of guides, 294–296; physical and moral effects of needless hardships, 295; effects of home sickness on expatriated Swiss Guides, 296; interpreters, 297; hiring porters and dealing with natives, 298; making plans of routes, 299; idiosyncrasies of maps, 300; reconnoitring, 301; photographs, 301; variations in the conditions of snow and ice in different countries, 301; best time for starting on an ascent, 302; food, 302; note-books, 303; nomenclature of peaks, 303; geology, 304; botany and natural history, 305; photography, 305; anthropology, 305; local traditions, 305; instruments, 305

Reilly, Adams, his maps of the Mont Blanc range, 264, 352, 354; takes up the mantle of J. D. Forbes, 355; his character and mountaineering work, 355; burial-place of, 355; 365

Reuss Valley, the, 121

Rey, Emile, guide, 241, 366, 368

Rhellicanus (J. Müller of Rhellikon), 14

Richardson, Miss, 50 n.

Richter, E., on glacier passes, 2 n.

Riffel, the, 132, 356, 362, 370

Riffelberg, the, 362

Riffelhorn, the, 59

Rob Roy cooking apparatus, 65

Roche Melon, the, 5

F F

ROC

Rock climbing, 215 ; nature of its attraction, 215 ; its dangers contrasted with those on snow, 216 ; reasons for its popularity, 217 ; on limestone mountains, 218 ; characteristics of dolomite, 219 ; the aiguilles, 220-222 ; granitic mountains, 222 ; the peaked mountains, 222 ; advisability of a climber reverting to the quadrumanous type, 223 ; handhold and foothold, 223 ; starting an ascent correctly, 224 ; on rock slopes, 224 ; straight up a rock face, 226 ; use of the rope and axe, 228 ; progress in zigzags, 229 ; length and position of steps, 229 ; action of limbs, 230 ; in a cramped position, 230 ; climbing with bent knee, 230 ; ascending narrow rock gullies, 231 ; steep walls of rock with no foothold, 231 ; Carrel's Ledge on the Matterhorn, 232 ; use of lassos and grapnels, 232 ; descending rocks, 233 ; difficulties of descent contrasted with those of ascent, 233 ; different aspect of route in ascent and descent, 234 ; true method of observation of line of ascent, 234 ; the first man in ascent and last man in descent, 234 ; balance, 235 ; two methods of descent, 235 use of axe in descent, 236 ; utilisation of friction as brake power, 236 ; placing the feet, 236 ; the danger of wedged-in feet, 236-238 ; right position in use of the hands, 238 ;

RUS

variations of method of descent, 239 ; brilliant gymnastic feats versus safe climbing, 239 ; the essence of good climbing, 240 ; dislodgment of stones, 241 ; dealing with insecure foothold and slips, 242-246 ; use of the rope, 244 ; on grass-covered rocks, 246-249 ; on glazed rocks, 249, 250 ; on rotten rocks, 250-252 ; crossing loose scree, 252 ; walking over terminal moraines, 253 ; on dolomite rocks, 254 ; use of fixed ropes, 255, 256 ; special shoes, 256 ; guarding against falling stones, 257-262

Rocky Mountains, the, 289

Rope, use of the, 15, 16, 27, 71, 72, 100 ; length, 100 ; knots, 101 ; Committee of the Alpine Club on, 101 ; fixed, 102 ; on snow-fields, 103 ; putting on the, 103 ; pace with the, 104, 105 ; on rocks, 106 ; misuse of, in slips, 108 ; untying knots, 110 ; 114, 115, 116, 184, 216, 244, 255, 343

Ross & Co., 387 *n*.

Ross-shire, Western, 342

Rotario d'Asti, his ascent of Roche Melon, 5, 6 *n*.

Roththal Sattel, the, 369

Rousseau, Jean-Jacques, quoted, 10, 22, 23, 26

Rowing, as a preparatory exercise for mountaineering, 76

Royal Geographical Society, 302, 306

Ruapehu, 290

Rücksack, the, 52-54

Ruskin, John, 362, 376

INDEX 435

RUS

Russell, Count Henry, 284
Russian hemp rope, 72
Russian maps of the Caucasus, 244, 291
Ruwenzori, 288, 290

SAAS, 367
Saccharine tabloids, 55
St. Gotthard, 349
St. Niklaus, 367
Sal volatile, 79
Salimbene, Fra, quoted, 6
Sallust, quoted, 4
San Giacomo d'Ayas, 353
Sand-cones, 155
Sasser Pass, the, 88
Scafell district, the, 333-338
Scheuchzer, quoted, 11 *n.*, 18, 19 *n.*, 21 *n.*
Schild-Lawine, the, 127
Schlagintweit, H. and A., 287
Schreckhorn, the, 126, 317
Schwarz, quoted, 21 *n.*
Schwarz Thor, the, 67, 353
Scientific mountaineering apparatus, 59
Scotland, 331
Scottish Mountaineering Club, the, 341
'Scrambles amongst the Alps,' Whymper's, 232, 357
Scree, 252
Seatoller, 338
Seiler, Alexander, proprietor of the hotels at Zermatt and the Riffel, 362
Self-cooking tins of preserved foods, 55
Selkirks, the, 62, 288
Sella, Signor Quintino, President and Founder of the Italian Alpine Club, 358 ; on Mont Blanc ; 359 ; 367

SKE

September in the Alps, 125, 126
Séracs, 153
Seven Wonders of Dauphiné, the, 7
Sgurr Alisdair, 344
Sgurr Dearg, 344
Sgurr Dubh, 254
Sgurr nan Eag, 254
Sgurr nan Gillean, 336, 343
Shelters, climbers', 282
Shepherds' plaids, 64
Sherwill, 29
Shetland wool, 47
Shirts, 46, 47
Shoes, 256
Short cuts, 96
Sickness, mountain ; *see* Mountain sickness
Siegfried Atlas, the, 264
Sierra Nevada, Spain, 284
Signalling, 119, 120
Sikkim, 292
Silk shirts, 47 ; underclothing, 47, 52 ; waterproofs, 48
Simler, Josias, quoted, 9, 14-17, 27
Simplon, the, 277
Sion, 9
Sisal hemp rope, 72
Sivell, M., 85
Sketch-books, 303
Sketching, for mountaineering purposes, 380 ; utilitarian limits, 380 ; use of a sketch in conjunction with a map, 381 ; points to which attention should be paid, 382 ; suitable paper, 382; memoranda to be written down, 382; extent of ground to be included, 383 ; dimensions and positions of the principal masses, 383; general outline, 384 ; skyline, 384 ;

F F 2

SKE

main masses of rock, 384;
borders of glaciers, 384;
approximate snow-line, 384;
special features, 385; foreshortening, 385; perspective illusion, 386; the camera obscura, 387; rough sketch of the Beispielspitz, 387; landscape painting, 388; photography in aid, 389 n.

Ski, 74
Skiddaw, 338
Skirts, ladies', 50
Skye, 59, 254, 262, 336, 342
Sleat, 342
Sleep, 399, 400
Sleepiness, 203
Sleeping-bags, 47, 64, 395
Sleeplessness, 400
Sligachan, Skye, 342
Slings, axe, 70
Slippers, 45
Sloane, Hans, 21
Smith, Albert, quoted, 4, 24, 32, 33, 181
Smock frocks, 30
Smyth, on Monte Rosa, 37
Snow-blindness, 15, 81
Snow-bridges, 185, 188-190
Snowcraft, 149; phenomena connected with mountain upheaval and glacier formation, 149, 150; movement of the glaciers, 150; the snow-line in the Alps, 151; average rate of glacier movement, 151; marginal crevasses, 152; transverse crevasses, 152, 179, 180; longitudinal crevasses, 153; séracs, 153; bergschrunds, 153; map-diagram of a glacier, 154, 155; glacier streams, 155; sand-cones, 155; lateral and

SNO

medial moraines, 156; névé, 156; soft snow, 156; crusted snow, 157, 197; hummocky snow, 158; chamois tracks, 158; granular or powdered snow, 158; snow slopes and gullies, 159, 160; deep snow, 160; ridges and cornices, 161-164; use of the ice-axe, 164; the alpenstock, 165; scraping out steps, 165; step-cutting, 166-178; the pick used as an anchor, 178; hand-holds, 178; snow-probing, 178; anticipating the presence of crevasses, 179; photography as an aid in detecting crevasses, 179; rules to be observed in crossing crevassed regions, 180; extricating those who have fallen into crevasses, 182; extricating from crevasses in ice-falls, 183-185; doubtful snow-bridges, 185, 188-190; accidents on dry glaciers, 186; character of bergschrunds or Randklufts, 186; crossing bergschrunds, 188-192; shooting a crevasse, 192; jumping a crevasse, 193; art of glissading, 192, 194-197; action in case of avalanches, 197-202; use of photography in showing the track of avalanches, 201; arêtes, 202; mountain sickness, 202; guarding against sleepiness, 203; on the summit ridges, 203, 204; 'variation' routes, 204; on the mountains in bad weather, 206; tracing out old tracks, 206; how to act when the

SNO

way is lost, 207 ; in mist and fog, 208 ; use of the compass, 209 ; when benighted, 210 ; sheltering in crevasses, 210 ; dispensing food, 210 ; best application of spirits, 211 ; keeping the feet dry and warm, 211 ; in driving snowstorms, 211 ; the Föhnwind, 211, 212 ; characteristics of the Swiss guides, 212, 213
Snowdon, 133, 330, 336, 339, 357
Snow-line, the, in the Alps, 151
Snow-shoes, 15, 16, 74
Snow-spectacles, 15, 16, 56, 57, 82
Snowstorms, 211
Socks, 46
Solar topees, 42
Soup, 79
South America, 289
Spadrilles, 257
Spectacles, snow, 15, 16, 56, 57, 82
Spey, valley of the, 346
Stack Polly, 345
Stanley, H. M., 288
Starch-powder, 81
Starting on expeditions, 116-118, 131, 302
Steinbock, 11
Step-cutting, 166-178, 308, 319
Stephen, Mr. Leslie, 17, 219, 362, 374
Stimulants, 77-79, 210
Stockhorn, ascent of the, in 1536, 14 ; verse dialogue with the Niesen, 18-20
Stockings, 45, 46, 51
Stones, falling, xix., 241, 257-262, 317
Strachey, Sir Richard, 284, 287

TIT

Strahleck Pass, the 35, 299, 357, 361
Straits of Magellan, 290
Strathaird, 342
Studer, Gottlieb, 34, 35, 36 *n.*, 38, 362
Styhead Pass, 336
Suanetia, 289, 294
Sub-Alpine regions in the winter, 127
Suilven, 345
'Summer Months among the Alps,' Hinchliff's, 356
Sunburn, 80, 81
Susa, 5, 6 *n.*
Sutherlandshire, 344, 345
Swiss Alpine Club, the, 393
Swiss guides, 77, 212, 213 ; railways, 349
' Swiss Travel and Swiss Guidebooks,' Coolidge's, 2 *n.*, 273

TAIRRAZ, 181
Taugwalders, the, 366
Tay, river, 345
Taylor, Mr., 362
Tea, 79
Teallach Hills, 345
Temper, 105
Tents, 62-64
Terminal moraines, 253
Tetnuld, 296
Theodul Pass, the, 103, 106, 246, 322, 360
Thirst, 78
Thunderstorms, 212
Tibet, 287
Tinder paper, 58
Tinner, John, his exploits with dragons in the Alps, 18
Tissandier, M., 85
Titlis, the, 23, 219

TOB

Toboganning, 127
Toilet Lanoline, for sunburns, 81
Torridon, 341, 345
Towers, 142
Traditions, local, 305
Transaltai range, the, 287
Transverse crevasses, 152, 179, 180
'Travels in the Alps of Savoy,' Forbes's, 351
Tree level in Switzerland, 395
Trolliet, Bernard, of Bagnes, chasseur, 364
Tryfan, 340, 357
Tschingel Pass, the, 106, 322
Tucker, 357
Tuckett, Mr. F. F., 362
Tuckett's sleeping bag, 64
Tweeds as material for clothing, 40
Twelve Pins of Connemara, 340
Tyndall, Prof., 17, 29, 37, 281, 367, 369
Typical mountain, reconnoissance of a, 135-148
Tyrol, the, 73, 74
Tyrolese guides, 297

ULLAPOOL, 341
Ulrich, Duke of Wurtemberg, his ascent of Pilatus, 11
Ushba, 289
Uzielli, Signor, quoted, 6 n., 8 n.

VADIANUS, 11 n.
Val Anzasca, 289
Val de Bagnes, 363
Val Sesia, 8
Val Tournanche, 268, 367
Valais, the, 9
Valleys, origin of, 150
Vallot, M., 86

WIL

Verglas, 249
Vinci, Leonardo da, as a mountaineer, 8-10
Visp, 349
Vispthal, the, 282
Vogel of Glarus, 10
Von Bergen, J., guide, 369
Vulcanite gourds, 56

WAGNER (the naturalist) on dragons in the Alps, 18
Waistcoats, 40, 41
Wales, 339
Walker, Frank, mountaineering career of, 360
Walking-sticks, 71
Walking uphill, 91-96
'Wanderings in the High Alps,' Wills's, 356, 365
Wastdale Head, 333, 334, 336-338
Water, the drinking of, 12, 13, 77
Waterproof capes, 47; coats, 47, 48; gloves, 48; ponchos, 48
Weather signs, 121-124
Weisshorn, the, 133, 223, 308, 321, 369
Wengern Alp, the, 349
Wetterhorn, the, 36, 365, 369, 370
Whistles, 58
White wine, 77
Whymper, Mr. Edward, 24 n., 31, 60-62, 84-86, 158, 159, 193, 232, 256, 281, 284, 289, 292, 302, 357, 362, 365, 367
Wilderwurm Gletscher, the, 19
Wildstrubel, the, 219
Willesden canvas tents, 62, 63
Wills, Mr. Justice, 36, 356, 362, 365

INDEX 439

WIL

Wilson, Andrew, 291
Windham, 3, 22, 24, 25
Wine, 77 ; bags for, 56
Winter mountaineering, 126, 127
Woollen comforters, 47
'Writing-desk' peaks, 219

YÂKS, efficiency of, in mountain travel, 88

ZSI

Yorkshire, North, 338
Younghusband, Captain, 291, 295

ZERMATT, as a mountaineering centre, 132, 282, 296, 315, 349, 353, 356, 362, 363, 366
Zigzag, the, in ascents, 94
Zinc ointment for sunburns, 81
Zsigmondy, Dr. Emil, 255

'Finis coronat opus.'

PRINTED BY
SPOTTISWOODE AND CO., NEW-STREET SQUARE
LONDON

THE BADMINTON LIBRARY.
Edited by the DUKE OF BEAUFORT, K.G. and A. E. T. WATSON

HUNTING. By the DUKE OF BEAUFORT, K.G. and MOWBRAY MORRIS. With Contributions by the EARL OF SUFFOLK AND BERKSHIRE, Rev. E. W. L. DAVIES, DIGBY COLLINS, and ALFRED E. T. WATSON. With 53 Illustrations. Crown 8vo. 10s. 6d.

FISHING. By H. CHOLMONDELEY-PENNELL. With Contributions by the MARQUIS OF EXETER, HENRY R. FRANCIS, Major JOHN P. TRAHERNE, FREDERIC M. HALFORD, G. CHRISTOPHER DAVIES, R. B. MARSTON, &c.
Vol. I. Salmon and Trout. With 8 Plates and 150 Woodcuts &c. Crown 8vo. 10s. 6d.
Vol. II. Pike and other Coarse Fish. With 7 Plates and 126 Woodcuts &c. Crown 8vo. 10s. 6d.

RACING AND STEEPLE-CHASING. *Racing*: By the EARL OF SUFFOLK AND BERKSHIRE and W. G. CRAVEN. With a Contribution by the Hon. F. LAWLEY. *Steeple-chasing*: By ARTHUR COVENTRY and ALFRED E. T. WATSON. With 56 Illustrations. Crown 8vo. 10s. 6d.

SHOOTING. By LORD WALSINGHAM and Sir RALPH PAYNE GALLWEY, Bart. With Contributions by LORD LOVAT, LORD CHARLES LENNOX KERR, the Hon. G. LASCELLES, and A. J. STUART-WORTLEY.
Vol. I. Field and Covert. With 12 Plates and 93 Woodcuts &c. Crown 8vo. 10s. 6d.
Vol. II. Moor and Marsh. With 8 Plates and 57 Woodcuts &c. Crown 8vo. 10s. 6d.

CYCLING. By VISCOUNT BURY, K.C.M.G. (the Earl of Albemarle), and G. LACY HILLIER. With 19 Plates and 70 Woodcuts &c. Crown 8vo. 10s. 6d.

ATHLETICS AND FOOTBALL. By MONTAGUE SHEARMAN. With an Introduction by Sir RICHARD WEBSTER, Q.C. M.P. With 6 Plates and 45 Woodcuts &c. Crown 8vo. 10s. 6d.

BOATING. By W. B. WOODGATE. With an Introduction by the Rev. EDMOND WARRE, D.D. and a Chapter on 'Rowing at Eton' by R. HARVEY MASON. With 10 Plates and 39 Woodcuts &c. Crown 8vo. 10s. 6d.

CRICKET. By A. G. STEEL and the Hon. R. H LYTTELTON. With Contributions by ANDREW LANG, R. A. H. MITCHELL, W. G. GRACE, and F. GALE. With 11 Plates and 52 Woodcuts &c. Crown 8vo. 10s. 6d. [*Continued.*

London : LONGMANS, GREEN, & CO.

THE BADMINTON LIBRARY.
Edited by the DUKE OF BEAUFORT, K.G. and A. E. T. WATSON.

DRIVING. By His Grace the DUKE OF BEAUFORT, K.G. With 11 Plates and 54 Woodcuts &c. Crown 8vo. 10s. 6d.

FENCING, BOXING, and WRESTLING. By WALTER H. POLLOCK, F. C. GROVE, C. PREVOST, E. B. MICHELL, and WALTER ARMSTRONG. With 18 Intaglio Plates and 24 Woodcuts &c. Crown 8vo. 10s. 6d.

GOLF. By HORACE G. HUTCHINSON, the Right Hon. A. J. BALFOUR, M.P. Sir WALTER G. SIMPSON, Bart. LORD WELLWOOD, H. S. C. EVERARD, ANDREW LANG, and other Writers. With 22 Plates and 69 Woodcuts &c. Crown 8vo. 10s. 6d.

TENNIS, LAWN TENNIS, RACKETS, and FIVES. By J. M. and C. G. HEATHCOTE, E. O. PLEYDELL-BOUVERIE, and A. C. AINGER. With Contributions by the Hon. A. LYTTELTON, W. C. MARSHALL, Miss L. DOD, H. W. W. WILBERFORCE, H. F. LAWFORD, &c. With 12 Plates and 67 Woodcuts &c. Crown 8vo. 10s. 6d.

RIDING AND POLO. By Captain ROBERT WEIR, Riding Master, R.H.G. and J. MORAY BROWN. With Contributions by the DUKE OF BEAUFORT, the EARL OF SUFFOLK AND BERKSHIRE, the EARL OF ONSLOW, E. L. ANDERSON, and ALFRED E. T. WATSON. With 18 Plates and 41 Woodcuts &c. Crown 8vo. 10s. 6d.

SKATING, CURLING, TOBOGGANING, and other ICE SPORTS. By J. M. HEATHCOTE, C. G. TEBBUTT, T. MAXWELL WITHAM, the Rev. JOHN KERR, ORMOND HAKE, and HENRY A. BUCK. With 12 Plates and 272 Woodcuts &c. by C. WHYMPER and Captain ALEXANDER. Crown 8vo. 10s. 6d.

MOUNTAINEERING. By C. T. DENT, with Contributions by W. M. CONWAY, D. W. FRESHFIELD, C. E. MATHEWS, C. PILKINGTON, Sir F. POLLOCK, H. G. WILLINK, and an Introduction by Mr. JUSTICE WILLS. With 13 Plates and 95 Woodcuts &c. by H. G. WILLINK and others. Crown 8vo. 10s. 6d.

BIG GAME SHOOTING. 2 vols. By C. PHILLIPPS-WOLLEY, W. G. LITTLEDALE, Major H. PERCY, Captain C. MARKHAM, R.N. and W. A. BAILLIE GROHMAN. With Contributions by other Writers. [*In preparation.*

COURSING AND FALCONRY. By the EARL OF KILMOREY, HARDING COX, the Hon. G. LASCELLES, and other Writers. [*In preparation.*

SWIMMING, &c. By ARCHIBALD SINCLAIR and WILLIAM HENRY, Hon. Secs. of the Life Saving Society. [*In preparation.*

BILLIARDS. By H. SAVILE CLARKE. [*In preparation.*

YACHTING.

London : LONGMANS, GREEN, & CO.

www.ingramcontent.com/pod-product-compliance
Lightning Source LLC
Chambersburg PA
CBHW021427300426
44114CB00010B/690